RECHARGING CHINA IN WAR AND
REVOLUTION, 1882–1955

Recharging China in War and Revolution, 1882–1955

Ying Jia Tan

CORNELL UNIVERSITY PRESS

ITHACA AND LONDON

First published 2021 by Cornell University Press

Library of Congress Cataloging-in-Publication Data
Names: Tan, Ying Jia, 1978– author.
Title: Recharging China in war and revolution, 1882–1955 / Ying Jia Tan.
Description: Ithaca [New York]: Cornell University Press, [2021] |
    Includes bibliographical references and index.
Identifiers: LCCN 2020058201 (print) | LCCN 2020058202 (ebook) |
    ISBN 9781501758959 (paperback) | ISBN 9781501758966 (pdf) |
    ISBN 9781501758973 (epub)
Subjects: LCSH: Electric industries—China—History—19th century. |
    Electric industries—China—History—20th century. | Electric industries—
    Economic aspects—China. | Electric industries—Political aspects—
    China. | Electric power production—China—History—19th century.
    | Electric power production—China—History—20th century.
Classification: LCC HD9697.A3 T36 2021 (print) | LCC HD9697.
    A3 (ebook) | DDC 333.793/2095109041—dc23
LC record available at https://lccn.loc.gov/2020058201
LC ebook record available at https://lccn.loc.gov/2020058202

Cover illustration: Sheng, Cijun. *Protect our factory: Guard against fire, burglars, and
agents' activities*. Poster. China: People's Arts Publishing House, 1951. From Poster
collection, CC 120, Hoover Institution Library & Archives, Stanford, California.

CONTENTS

List of Illustrations ix

Acknowledgments xi

List of Abbreviations xv

Note on Transliteration xvii

INTRODUCTION
Forging Resilience 1

CHAPTER 1
Spinning the Threads of Discontent 18

CHAPTER 2
Defending the Public Good 38

CHAPTER 3
Unleashing Fire and Fury 62

CHAPTER 4
Dawning of the Copper Age 86

CHAPTER 5
Turning the Tide 112

CHAPTER 6
Waging Electrical Warfare 138

CHAPTER 7
Manufacturing Technocracy 163

CONCLUSION
Hauntings from Past Energy Transitions 185

List of Chinese and Japanese Terms 197

Notes 205

Bibliography 239

Index 259

ILLUSTRATIONS

FIGURES

1.1. Map of China 2

1.1. Map of Shanghai's main power station and
major textile mills, circa 1929 19

3.1. The Final Days of the Anti-Japanese Resistance (1941) 63

4.1. *Products of the Central Electrical Manufacturing Works* (1948) 88

5.1. *Artist Impression of John Lucian Savage's Proposed Yangtze Gorges Dam* 113

5.2. Profile of the Tennessee River (1944) 121

5.3. Map of the Yangtze Gorges site depicting the extent of Savage's survey 125

6.1. Fushun Power Plant after the Soviet takeover, circa 1945 139

TABLES

3.1. Electrical industries under Japanese and
Guomindang control, circa 1939 79

7.1. Power output figures for Shanghai, 1949–1956 174

ACKNOWLEDGMENTS

It has been a decade since the triple disaster of an earthquake, tsunami, and nu-clear meltdown struck Fukushima in March 2011. This book has grown out of a series of questions arising from that massive catastrophe: Why are high-energy societies so reliant on fossil fuels and nuclear power, despite their risk to human health and political stability? Have East Asian societies experienced large-scale energy crises that can show us how governments and people cope with a cata-clysmic collapse of the energy infrastructure? I am indebted to the people and institutions that helped me on my quest to address these questions.

My research into the history of electricity began at Yale in March 2011. After aborting research plans that would have taken me to Japan, I had a long and serious conversation with Valerie Hansen, who suggested that I make a foray into energy history, since it was closely related to my interest in the history of cartography. Valerie guided me through the writing process by setting clear and achievable targets and reminding me to express my ideas clearly and succinctly. Peter Perdue, Frank Snowden, and Bill Rankin were always on hand to provide much-needed intervention in the early stages of this work and helped me develop a sharper thematic focus for this book. Many people at Yale provided encour-agement for this project, especially Paola Bertucci, Mary Augusta Brazelton, Ivano dal Prete, Fabian Drixler, Daniel Kevles, Leon Rocha, Naomi Rogers, Richard Sosa, Brian Turner, Courtney Thompson, Jinping Wang, and John Harley Warner.

My colleagues and students at Wesleyan University encouraged me to re-imagine and reshape this project. My mentor Bill Johnston repeatedly returned to Fukushima to photograph choreographer-dancer Eiko Otake performing in the area near the damaged nuclear reactor. Besides offering feedback on every draft chapter of the manuscript, Bill inspired me to think of disasters on multiple geographic and time scales, as he tirelessly read through every word of the manuscript. Mary Alice Haddad not only introduced my work to social scientists, but also offered pointed criticism that helped me reorga-nize the introduction and conclusion. My research apprentice Riyu Ling, who painstakingly assembled the data for this book, reminded me to keep in mind a fundamental question: What was electricity used for? During my time as a faculty fellow at the Center for the Humanities, my colleagues Andrew Curran, Axelle Karera, Ethan Kleinberg, Stephanie Koscak, Vijay Pinch, and Gabrielle Ponce-Hegenauer pushed me to think about how the history of electrification straddled the divide between "grand narrative" and "modest proposals." I also thank Steven Angle, Joan Cho, Paul Erickson, Demetrius Eudell, Courtney Fullilove, Erik Grimmer-Solem, Valeria Lopez-Fadul, Bruce Masters, Jim McGuire, Don Moon, Peter Rutland, Ron Schatz, Gary Shaw, Victoria Smolkin, Jennifer Tucker, Laura Ann Twagira, and Duffy White for their valuable input on this project.

I am grateful for the financial support that has helped me bring this book to fruition. Funding sources include Yale University Graduate School of Arts and Sciences, Charles Kao Fund and Summer Travel and Research grants administered by the Council on East Asian Studies at Yale, and the Taiwan Fellowship. A generous postdoctoral fellowship with the D. Kim Foundation in the 2018–2019 academic year afforded me the time to complete the manuscript. The Office of Academic Affairs at Wesleyan University offered financial support through the Grants in Support of Scholarship. I am also grateful to my colleagues in the History Department for approving my applications for two supplemental grants and a publication subvention from the "Colonel Return Jonathan Meigs (1740–1829) Fund."

This book draws on sources from archives in China, Taiwan, Japan, and the United States. The research would not have been possible without the generosity of many host institutions and archivists. I thank Zhang Baichun and Sun Lie for giving me a home at the Institute for the History of Natural Sciences in Beijing, Han Zhaoqing at Fudan, and Cao Shuji at Shanghai Jiaotong University. I am appreciative of the flexibility granted by archivists at Beijing Municipal Archives and Shanghai Municipal Archives. My term as Taiwan fellow in the summer of 2017 was instrumental to the completion of this book. Wen-de Huang and Melodie Wu at the National Central Library made sure that I was well settled in Taiwan. At Academia Sinica, Mike Liu arranged for me to be hosted at the Institute of Taiwan History and gave me unfettered access to a special collection of sources in his research office. I also had the pleasure of discussing my work with Chang Che-chia, Chen Chien-shou, Chiah Sing Ho, Paul Katz, Lin Lanfang, Sean Lei, and Shu-li Wang. At Academia Historica and later Academia Sinica in Taiwan, Ya-hung Hsiao not only helped me secure access to the documents but also directed me to leading scholars in the field. Outside of Academia Sinica, Chiao-lun Cheng, Ming-sho Ho, Wen-hua Kuo, and Sao Yang Hong, Jen-shen

Wu, spent many hours discussing my project with me and pushed me to articulate my underlying arguments more clearly.

Through the years of research and writing, members of the scholarly community have challenged me to connect electricity to political and economic issues. Members of the Civil War workshop at Yale in May 2016, Cheow Thia Chan, David Cheng Chang, Shiu-on Chu, Rob Culp, Fa-ti Fan, Joshua Freeman, James Gao, Judd Kinzley, Diana Lary, Joe Lawson, Rebecca Nedostup, Philip Thai, Xian Wang, Dominic Yang, and Shirley Ye inspired me to situate electricity within multiple modes of warfare. The working group on electricity based at the University of Leeds, namely Animesh Chatterjee, Graeme Gooday, Phil Judkins, Michael Kay, and Daniel Zapico, kept me abreast of developments in the history of technology. At the Anthropocene workshop in August 2018, Arjun Guneratne, Prasannan Parthasarathi, Kenneth Pomeranz, and Julia Adeney Thomas encouraged me to engage with the scholarly literature about the Anthropocene. I would also like to thank Seung-joo Yoon for inviting me to present the contents of this book at the Distinguished Ott Family Lecture series at Carleton College.

I would also like to extend a heartfelt thanks to all those who offered important advice after reading parts or all of this work: Mark Baker, Nicole Barnes, Nick Bartlett, Peter Braden, Corey Byrnes, Chen Zhongchun, Howard Chiang, Angelina Chin, William Chou, Marwa Elshakry, Victoria Frede-Montemayor, Gao Ming, Li Hou, Kathryn Ibata-Arens, Lijing Jiang, Abishek Kaicker, Stefan Krebs, Chien-wen Kung, Shigehisa Kuriyama, Ulug Kuzuoglu, Eugenia Lean, Thomas Lean, Kan Li, James Lin, Helena Lopes, David Luesink, Christine Luk, Timothy Mitchell, the late Aaron Moore, Seung-Youn Oh, Chang Woei Ong, Elizabeth Perry, Jiwei Qian, Kerry Ratigan, Christopher Reed, Daniel Sargent, Corinna Schlombs, Phil Scranton, Victor Seow, Qin Shao, Yubin Shen, Misato Shimizu, Tuan-hwee Sng, Wayne Soon, Fubing Su, Nicolas Tackett, Christopher Tong, Yijun Wang, Heike Weber, R. Bin Wong, Shellen Wu, Xie Shi, Nobuhiro Yamane, Wen-hsin Yeh, Yuan Yi, Zhaojin Zeng, Ling Zhang, and Dongxin Zou.

I am grateful to have worked with Emily Andrew, Allegra Martschenko, and Alexis Siemon at Cornell University Press. The constructive feedback from the two reviewers pushed me to articulate the main questions motivating the research. Micah Muscolino identified himself as one of the reviewers. Wen-Qing Ngoei, who published with Cornell University Press, offered much needed advice in responding to reviewer comments. I also thank Katherine Thompson for her editorial assistance in the early stages of revision, and Mike Bechthold for professionally made maps. Wendi Field-Murray, Christine Ho, Hsiao-ting Lin, and Covell Meyskens offered invaluable advice to help me locate the cover image. I am grateful to Jessica Ryan for her meticulous copyediting, and Ihsan Taylor at Longleaf Services for guiding this first-time author through the production process.

Finally, the research and writing would not have been possible without my supportive family. My wife, Teoh Song Keng, took care of Ke Qing, Ke Ren, and Keda while I was away on research trips or presenting my work at some academic conference. I hope that this book will raise our awareness of the precarity of carbon-fueled economies and inspire conversations about how to leave behind a sustainable world for our future generations.

AH	Academia Historica
AHC	American Heritage Center
AS	Archives, Institute of Modern History, Academia Sinica
BMA	Beijing Municipal Archives
CAS	Chinese Academy of Sciences
CCP	Chinese Communist Party
FMSXSL	Cheng Yu-feng and Cheng Yu-huang, ed., *Ziyuan weiyuanhui jishu renyuan fu mei shixi shiliao* [Archives on the National Resources Commission Technicians' Training in the United States]. Taipei: Academia Historica, 1988.
hp	horsepower—a unit of power equivalent to 550 fount pounds per second
Hz	hertz—number of cycles per second
kW	kilowatt—one thousand watts, with a watt being one unit of electrical power equal to one ampere under the pressure of one volt
kWh	kilowatt-hour—unit of electrical energy equivalent to one thousand watts of power supplied to or taken from an electrical circuit steadily for one hour
NARA	National Archives and Record Administration
NCC	National Construction Commission
NDL	National Diet Library
NRC	National Resources Commission
SGML	Seeley G. Mudd Library
SMA	Shanghai Municipal Archives
SMC	Shanghai Municipal Council
V	volt, the SI unit for potential difference. One volt is the potential difference that appears across a resistance of one ohm when a current of one ampere flows through that resistance.

Note: Definitions of scientific units based on the glossary of the US Energy Information Administration.

NOTE ON TRANSLITERATION

This book mostly uses *hanyu pinyin*, the official romanization system used in mainland China for Chinese names, terms, and expressions. The transliteration is also used for the titles of Chinese publications. Names of individuals are written in the Chinese way, surname first.

Some names and places have traditional Wade-Giles spellings that are in common usage. I have decided to use the original spelling and included the *pinyin* in parenthesis at first usage and also in the list of Chinese characters. Examples include Sun Yat-sen (Sun Zhongshan), Chiang Kai-shek (Jiang Jieshi), T.V. Soong (Song Ziwen),

I have also decided not to impose *pinyin* on place names and political figures in Taiwan. Taipei is spelled as such, rather than converted into Taibei. The name of the electrical engineer who became the Premier in Taiwan is rendered as Sun Yun-suan rather than Sun Yun-xuan.

RECHARGING CHINA IN WAR AND
REVOLUTION, 1882–1955

Introduction

Forging Resilience

I N 2010, CHINA OVERTOOK the United States as the world's largest consumer of power.[1] This was a remarkable achievement considering that the People's Republic had struggled to keep the lights on during its early years. The Communist regime greeted this accomplishment not with celebratory cheer but with trepidation. Seemingly aware of the destructive and precarious nature of its carbon-intensive development, Zhu Yongpeng, the general manager of the state-owned power company Guodian Corporation, had in 2009 called for the transition from fossil fuels to clean energy to "promote sustainable economic development" in an editorial for *Qiushi*, the bi-monthly political theory periodical published by the Chinese Communist Party. He noted that the fossil fuels that had fueled China's exponential growth "have an end-use energy conversion rate of only 30 percent and account for over 40 percent of global greenhouse gas emission." Zhu thus called on Chinese state power companies to fully harness China's clean energy potential to "better balance economic and social development with environmental concerns."[2] Like his predecessors who had built China's electrical infrastructure through decades of perpetual warfare, Zhu not only was facing a trade-off between short-term needs and long-term sustainability but also had to build a power system resilient enough to withstand the vagaries of war and economic turmoil.

Three separate modes of warfare shaped the electrification of China through an age of revolutionary upheaval and armed conflict between 1882 and 1955. The organizational resilience of China's electrical power sector, and by extension the national economy, emerged from the lessons learned from resolving electrical crises. First, electricity came onto the scene during an age of economic warfare between Chinese and foreign textile firms from the late Qing to the early republic. Chinese mill owners learned to adapt to the unequal distribution of energy resources and leveraged their political connections to survive the onslaught of global economic competition. Second, following the catastrophic loss of 97 percent of China's generating capacity to the Japanese after the outbreak of the

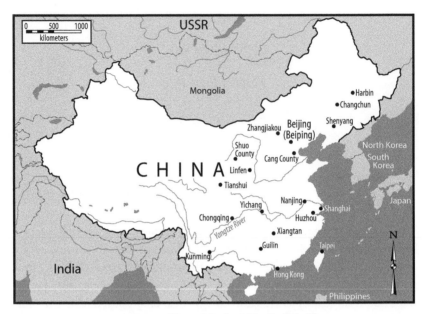

FIGURE I.I. China. Map by Mike Bechthold.

second Sino-Japanese war in 1937, competing regimes that vied for supremacy in China capitalized on the exigencies of military warfare to consolidate and nationalize a highly fragmented electrical power sector. Nationalist forces that retreated to Southwest China were thus able to sustain the resistance against the Japanese by powering its defense industries with makeshift power stations. Finally, by using electricity as an instrument to wage the people's war, the Communist regime not only was able to inherit the electrical infrastructure largely intact after a bloody military conflict but also established the mobilization structures for production campaigns in the early years of the People's Republic.[3] The image on the cover of this book, showing a worker defending a factory from sabotage, suggests that political power did not simply emerge from the barrel of the gun but arose from the state's ability to defend its power infrastructure and recharge itself in the face of adversity.

The ability of China's electrical industries to recover from catastrophe raises a deeper question: Why did China's industries achieve such rapid growth during these turbulent years despite material scarcity? To answer this question, this book focuses on the formative years of China's electrification and explores how the interplay of strong and weak organizational forces within a compressed time frame strengthened the capacity of the electrical infrastructure to adapt and recover under adversity, which in turn strengthened the organizational resilience of China's political regimes. In the opening sentence of his study on administered mass organization, Gregory Kasza proclaimed, "technology made it possible to organize people on an unprecedented scale."[4]

Electricity is a tool for the state to strengthen its control over the populace. And electrical industries served as a site of policy innovations for weak institutions, which, according to Yuen Yuen Ang, played a crucial role in helping China escape from the poverty trap. Its troubled beginnings offer empirical evidence for the "mixture of top-down direction and bottom-up improvisation that lays the foundation for coevolutionary processes of radical change."[5] Unlike Andrew Nathan's work that pointed to the political processes within the Chinese Communist Party as the main reasons for its survival, this study sees the electrical infrastructure as the source of China's organizational resilience.[6]

The compressed development under conditions of material scarcity also shaped the economic growth models of China and its East Asian neighbors. These institutional constraints shaped China's electrical industry in these turbulent decades of war and formed the foundations of what Huang Xiaoming described as the East Asian growth model, which began in the second half of the twentieth century as "a desperate quest for survival and catch up." These developments also pushed the East Asian economies into "a form of growth that derived its power for rapid and prolonged expansion from institutional manipulation and national reorganization for international competitiveness."[7] Material scarcity forced power companies and state regulators to operate with limited capital and labor. Understaffed political institutions responded to energy crises by introducing adaptations to improve the efficiency of electrical power production, distribution, and consumption. The successful resolution of energy crises lent greater credibility to state agencies that then exercised their newfound authority to dictate the pace of China's industrialization.

Dark forces, however, lurk beneath growth through adversity. Far from being a triumphant narrative about system builders overcoming great difficulties to set the stage for China's transformation into an industrial powerhouse, this book highlights the immense cost of China's electrical infrastructure development at the national, international, and global scale. At the local and national level, the compressed development of China's electrical industry resulted in residual inefficiencies, as system builders resorted to low-cost fixes to meet immediate demands for electricity during moments of crises. Power companies saved time and cost by installing new equipment without dismantling the outdated versions, and defects embedded in the system lowered operational efficiency in the long run and even caused catastrophic failure further down the road. These practices led to the persistence of obsolete technology.

On the international level, resistance against foreign control over China's electrical power sector became a source of diplomatic tensions. China's industrialists and policy makers saw Chinese control over the electrical industries as the key to defending the nation's economic sovereignty. Prior to the Japanese invasion of China in July 1937, a handful of foreign-owned power stations accounted for most of China's electric power supply, while multinational corporations cornered the market on electrical supplies. The economic blockade during the War of Anti-Japanese Resistance revealed the dangers of overreliance on foreign capital in electrical power infrastructure. The nationalistic objectives of China's state-sponsored electrical sector were at odds with foreign partners that came to its aid. Even as the state-owned power sector relied heavily on technology transfer from British and American corporations, their system builders rigorously defended its autonomy by fending off attempts to impose foreign technical standards on China. Technical cooperation was fraught with disagreements over the value of intellectual property and access to proprietary knowledge, which soured the relations between China and its foreign partners. These conflicts persisted over time and are at the root of conflicts that reemerged in 2018 during the trade war between China and the United States under the Trump administration.

Finally, at the planetary level, the electrification of wartime China laid the groundwork for the phenomenal industrial growth that transformed China into a major contributor to the "Great Acceleration" characterizing the Anthropocene. The *Anthropocene*, a term coined by Paul Crutzen and Eugene Stoermer in 2000, suggests that the Earth has entered a new geological epoch through the force of human activity. Unprecedented spikes in fossil fuel consumption and the appearance of vast quantities of manufactured materials in sediments suggest that the Earth is stratigraphically and functionally different than in earlier epochs.[8] Clive Hamilton went further to argue that the Anthropocene is "a proposition that human activities have ruptured the old Earth system and produced a new one."[9] The War of Anti-Japanese Resistance marks a point of rupture, during which the frenzied struggle for survival catalyzed the growth of China's electrical power sector. This book delves deeper into questions raised by the contributors to the 2014 special issue of the *Journal of Asian Studies*, in particular the impact of the Anthropocene on the vulnerability and resilience of Asian populations.[10]

The formative years of China's electrical industries offer a window into China's chaotic transition from, to borrow E. A. Wrigley's terms, an "organic economy" to a "carbon economy."[11] As China's economy became integrated with the industrialized nations in the West, energy shortage limited the scope of its economic growth. For most of the period covered in this book, China was what Micah Muscolino termed an "advanced organic economy."[12] The introduction of the steam engine did not lead to exponential growth in manufacturing output, as the quantity of raw materials remained constrained by the pace of the carbon and nitrogen cycles of this "organic economy." The creation of a military-industrial complex during the relentless state of war between 1937 and 1954 sped up the extraction of carbon resources—a trend mirrored in Europe and the United States. Persistent energy crises forced system builders to focus on short-term gains at the expense of long-term sustainability. As Vaclav Smil has shown, energy intensity, the ratio of total primary energy supply to the gross domestic product (GDP), followed an overall upward trend to the end of Mao's rule.[13] The final chapter of this book identifies 1955 as the starting point of the Great Acceleration of China's electrical industries. In this year, the Ministry of Fuel Resources abandoned a course of development based on optimal use of existing equipment and embarked on a massive expansion of different sectors within the electrical industries. It ordered the installation of large numbers of generators and the building of hundreds of small dams, paving the way for the energy-intensive period of the Great Leap Forward.

Thrust onto the front line of war and ideological struggle, government engineers, whom I refer to as engineer-bureaucrats, acted as the brokers of change as they built China's electrical infrastructure under the stress of material scarcity. This history of China's electrification and state-building follows these engineers and examines how they confronted the cost of compressed development at the national, international, and planetary level. It adopts Bruno Latour's approach of "following the scientist" to open the black box and better understand science in the making. The settings of the individual chapters are what Latour would consider to be "centers of calculation" where these engineers devised solutions by drawing together facts and mechanisms gathered from earlier observations.[14] These engineer-bureaucrats were not technocrats in the strictest sense of the word; nor should China be regarded as a "carbon technocracy."[15] Primarily concerned with the state's proper functioning through the implementation, administration, and regulation of policies, they were bureaucrats first and foremost. Engineering was the means used to address the problems at hand.

These engineer-bureaucrats resembled the scholar-officials who had managed the Yellow River and Grand Canal hydraulic systems. Just as the rivers provided the lifeblood of agrarian economies, the electric grid supplied the driving force for new industrial and commercial activities that transformed urban life in modern China. The first generation of electrical engineers saw Confucian scholar-generals as their role models. Yun Zhen (1901–1994), an engineer who appears in every chapter in this book, studied at the Government Institute of Technology (predecessor of Shanghai Jiaotong University) when Neo-Confucian scholar Tang Wenzhi served as president. Yun fondly remembered how Tang revived the ideals of "unity of thought and mind" advocated by Ming Neo-Confucian scholar Wang Yangming (1472–1529) and taught his students to "seek truth from facts."[16] Engineer-bureaucrats were "specialists in a system that admired generalists, technologists in a system that prized knowledge of Confucian ethics," a situation similar to Confucian scholar-officials working on hydraulic projects during late-imperial China in Randall Dodgen's study.[17] Unlike their predecessors, the engineer-bureaucrats of the early twentieth century entered the electrical industry after completing formal engineering education and therefore did not have to learn the technical details on the job.

The "carbon lock-in" arising from the state-centered approach to electrical infrastructure development parallels the "hydraulic lock-in" from three thousand years of unsustainable growth fueled by state-sponsored irrigation projects. As Mark Elvin argued, the "paradoxical combination of increased stability and increased instability" arose because Chinese authorities expended considerable resources to prop up existing hydraulic systems, which stabilized short-term tax revenue but destabilized the hydrological regime further down the road.[18] Engineer-bureaucrats, who administered the state-owned power industries, shared the state-centered outlook of their predecessors. Just as the state's ability to control the Yellow River–Grand Canal hydraulic systems came to be seen as an indicator of dynastic vitality, the engineer-bureaucrats considered the scale of the state-owned power industry as a proxy for state authority. Seeking to expand the state-controlled regional power grids under conditions of material scarcity, engineer-bureaucrats built on top of the ruins of defunct power systems, thereby embedding the legacy systems' flaws into the new infrastructure.

Engineer-bureaucrats looked toward the examples of scholar-generals from the Ming and Qing Dynasties as they came to concurrently assume civilian and military responsibilities. Yun Zhen was familiar with the writings of Wang Yangming, a Ming Dynasty scholar not only well known for his writings about the unity of knowledge and action but also for his military exploits. Zeng Guofan, the self-strengthening reformer who organized the provincial militia that helped the Qing defeat the Taiping rebels in 1864, was another role model for these engineer-bureaucrats. Early in their careers, engineer-bureaucrats began assuming a dual civil-military identity. They worked for the National Resources Commission (NRC), a government agency established in 1932 for industrial development in anticipation of an all-out war against Japan.[19] The NRC's predecessor was the National Defense Planning Commission—the name a clear indication of its military function. When Sino-Japanese hostilities broke out in 1937, the electrical power industry became a key aspect of military logistics. Conferred military titles, the engineer-bureaucrats dismantled electrical equipment in coastal cities besieged by Japanese forces and transported them hundreds of miles to mountainous inland cities. They painstakingly assembled electrical equipment from disparate sources to build from the ground up the electrical infrastructure that powered the defense industries. These battlefield experiences morphed into strategies for accelerated industrial growth under conditions of material scarcity.

Reinscribing Patterns of Historical Change

In addition to identifying the electrical infrastructure as the source of institutional resilience, placing electrification at the center of modern Chinese history unsettles the linear conception of historical progress focused on the revolutionary movements led by China's political leaders. This book is not a survey of the history of China's electrical industries.[20] It focuses on the alternating current (AC) as a revolutionary force in two senses of the word. In the literal sense, the AC is an electric current that periodically reverses direction, as the flow of the electric current changes direction with the wire loop rotating in a magnetic field. In the figurative sense, the AC was revolutionary: it fueled the rise and fall of regimes vying for supremacy in modern China and served as the driving force of historical change. This overview of modern Chinese history offers a sneak preview not only of how the strong dominates the weak through the control of electricity but also of how weak insurgents capitalized on vulnerabilities within the power network to overthrow strong incumbents.

Networked electricity came to China just as foreign powers carved out spheres of influence along coastal cities and border regions of the Qing Empire. China's first power station was built in 1882 in Shanghai—a treaty port established after the Qing Empire's defeat by the British in the Opium War (1839–1842). British, American, French, and Japanese capitalists invested in power stations in treaty ports all over China, using them as instruments to assert their economic hegemony.[21]

China's reformers and revolutionaries imagined electricity as a force for accelerated development that would propel China into the ranks of industrially advanced nations. Sun Yat-sen, the founding father of the Chinese Republic, saw electricity as a magical force. In his June 1894 letter to Viceroy Li Hongzhang, a key advocate of the self-strengthening movement launched to preserve the Qing Dynasty by maintaining traditional values while embracing Western technology, Sun wrote, "Electricity, for example, has neither form nor substance; it resembles but is not water. Its energy pervades all things and circulates throughout the universe. Its uses are more extensive than anything else, and it is most versatile."[22] Viceroy Li had long been aware of the vast potential of electrical power. Back in December 1888, Li approved the installation of a 15 kW generator to power the electric lights in Empress Dowager Cixi's palace at the West Garden (Xiyuan).[23] This former imperial garden is known today as Zhongnanhai and is the headquarters of the Chinese Communist Party and State Council. Coincidentally, the site of China's first power station outside of the treaty ports is now the focal point of central political authority.

With the collapse of the Qing Empire in 1911 following the armed uprising in the treaty port of Wuchang, Sun and his successors saw the nationalization of the power sector as a necessary step toward national unity. Between 1916 and 1928, control of China was divided among military cliques and regional factions. Chiang Kai-shek, the commandant of the Whampoa Military Academy who emerged as Sun Yat-sen's successor in the Guomindang, declared victory over the warlords and established the national capital in Nanjing in 1927. Chiang's government tried to fulfill Sun Yat-sen's vision of bringing the electrical industry under national ownership. As we will see, a lack of state capital and fierce opposition from the gentry not only derailed these plans for nationalization but also forced the Nationalist government to safeguard the monopolies of small power plants littered across the Lower Yangtze. The fragmented nature of China's electrical industries was an expression of the limited capabilities of the Guomindang regime.

Electricity embodied the space-time compression experienced during an age of political upheaval. Science fiction writers of the late Qing had already envisioned an age in which people communicated with brain electricity, or used electricity as an instrument of mind control.[24] By 1928, Hu Shi, a Columbia-trained intellectual of the May Fourth Movement, was expressing amazement that "my people have traveled with me from the vegetable oil lamp to electricity, from the wheel-barrow to the Ford car, if not the airplane, and this in less than forty years' time!"[25]

With the outbreak of the War of Anti-Japanese Resistance in 1937, romantic visions of linear progress toward modernity evaporated, as China grappled with the catastrophic loss of its electrical power sector. Back in 1933, Japan's electrical industry was 1,756 times that of China.[26] This translated into a higher industrial capacity for the Japanese, which in turn strengthened their military capability. The Japanese were thus able to quickly capture the coastal cities, causing the Guomindang regime to lose 95 percent of its sources for industrial production.[27] The war created sudden shocks to China's electrical industries. As discussed later in this book, the most heavily electrified regions in the Lower Yangtze suffered a decline, as Japanese occupation forces shifted their industrial production northward to take advantage of rich coal deposits. The generating capacity of Southwest China increased dramatically as the Guomindang's engineer-bureaucrats brought electricity to these far-flung regions to power the defense industries. Electricity thereby completed its transformation from an exotic commodity into a weapon for state power.

Riding the waves of massive upturns and downturns, the Communists, who had played a peripheral role in the War of Anti-Japanese Resistance, came to defeat the technologically superior and well-armed armies of Chiang Kai-shek in 1949. The Communists, who had entered into an uneasy anti-Japanese alliance with Chiang in December 1936, ruled a desperately poor Shaanxi-Gansu-Ningxia area with a population of only 1.4 million people.[28] But by October 1949, they were firmly in control of the cities. This is remarkable considering that the Communists had limited experience in administering large cities. As we will see, the struggle to control the electrical industries paved the way for their control over the urban economy. They gained the support of Western-educated engineer-bureaucrats serving the Guomindang regime, which allowed them to inherit the existing power infrastructure without significant damage. Contrary to the conventional narrative that the People's Liberation Army simply took over the cities by surrounding them from the countryside, a closer look at the role of electricity in urban warfare brings to attention the Communists' economic management strategies that helped them gain popular support among urban residents.

The perpetual state of war after the founding of the People's Republic in October 1949 also gave rise to a siege mentality within the electric power sector. The "recovery period for the national economy" (Ch. *guomin jingji huifu qi*) between 1949 and 1953 overlapped with China's entry into the Korean War. Well aware of the strategic importance of electrical power, military representatives took control of the electrical industries and, as we will learn, military power on its own was insufficient to command authority. This was also the period when technical experts relinquished decision-making powers to nonexperts, as any increases in electrical output was achieved through the reorganization of industrial activity rather than through installing additional generating capacity. In line with the ideal of workers seizing the means of production, electrical workers without formal engineering training replaced engineers as power plant superintendents.

The electrical industry embodied the Leninist and Maoist idea of "development as a unity of opposites." Despite the presence of Soviet advisers, the PRC's power sector operated independently of foreign influence. In a December 1952 essay celebrating Sino-Soviet friendship month, translators in the Ministry of Fuel Resources thanked their Soviet compatriots for unleashing the full potential of China's existing generating capacity, but added that the Soviets had "aroused Chinese national pride, encouraged us to be self-sufficient, . . . and praised the Chinese for relying on domestically made electrical products."[29] Not only were the Chinese determined to chart their own path of development, they went further than their "Soviet Big Brother" by entrusting the means of production not to highly educated technocrats but to rank-and-file workers. Economic plans served as a double-edged sword for the electrical industry. On one hand, performance metrics provided benchmarks to measure efficiency and reliability. On the other, the pressure to meet numerical targets forced system operators to risk catastrophic failure by pushing existing equipment to breaking point. All the while, the electrical industry not only was told to curb the expansion of capacity but was also ordered to increase output faster than other industries.

Bridging the Divide

Besides remapping the contours of historical change in modern China, this book bridges the divide between Chinese political economy and history of technology by highlighting the interdependence between energy infrastructure and economic development. Dealing with China's transition into the carbon economy, an issue tangentially addressed in existing literature about the causes of China's rapid economic growth, it also answers a call from the social sciences for greater awareness of "how energy transitions had unfolded in other places and times," in light of emerging concerns about environmental degradation, resource scarcity, and sustainability.[30]

The earliest histories of China's electrical industries emerged out of Cold War research aimed at evaluating the economic potential of Communist China. In his groundbreaking study *Foreign Trade and Industrial Development of China: An Historical and Integrated Analysis Through 1948* published in 1956, Chinese economist Zheng Youkui (Y. K. Cheng), who was then a visiting fellow at the Brookings Institution, argued for the rapid industrialization of China under government guidance to prevent industrialized nations from profiting off its raw material exports. Zheng had previously joined the NRC's Hong Kong office in 1940, then moved to the Chinese embassy in Washington, DC, where he was appointed assistant trade counselor in 1948. He pointed to China's hydropower potential, of which 99 percent was untapped, as the key to unleashing China's full industrial potential.[31] At the same time, another Chinese economist working for the US Air Force Project RAND, Kung-chia Yeh (Ye Kongjia), published a working paper titled "Electric Power Development in mainland China: Prewar and Postwar." Using statistical digests published by the Communist regime and Xinhua News Agency press releases, the paper pointed out that the Communist government exaggerated the extent of its rehabilitation efforts but had largely been able to increase power output to meet the demands of its heavy industries.[32] Building on Yeh's earlier work, Robert Carin at the Union Research Institute at Hong Kong continued the surveillance on China's electrical power sector.[33] With no direct access to Chinese government data, Carin's 1969 report drew extensively from *Past and Present Industries* published in 1958 by the Industrial Statistical Department in Beijing.

China's entry into the reform and opening-up era after 1978 led to a reexamination of China's pre-1949 industrialization. Zheng Youkui, who had been branded an American spy during the Cultural Revolution, returned to Shanghai Academy of Social Sciences in 1979 after political rehabilitation. He and Cheng Linsun corrected the misperception that the NRC was an instrument of "bureaucratic capitalism of the reactionary Guomindang regime" and detailed how it laid the groundwork for the government to coordinate industrial development through long-term planning.[34] In 1984, the deputy minister of Water Resources and Electric Power Li Daigeng authored two books on the early history of China's electrification with the explicit goal of "cautioning the younger generation against the nihilistic view that everything foreign is good, everything Chinese is backward." Li, who helped the Communists secure the power network in the Lower Yangtze in the 1950s, saw the history of China's electrical industries as a reminder of "how imperialists monopolized capital and exploited the Chinese people."[35] His techno-nationalist outlook reflected the views of the Chinese leadership in the 1980s.

China scholars in the West recognized the importance of China's electrical industries but focused on two key aspects: their impact on Chinese economic growth and their role as a training ground for China's future leaders. Czech-Canadian scientist and policy analyst Vaclav Smil, who published dozens of books about China's energy system, entered the field after the Organization of the Petroleum Exporting Countries (OPEC) crisis and published his first survey piece on China in 1976 after two years of "economic archaeology." Smil's research appeared just as the American government sought information about the Chinese economy ahead of the normalization of diplomatic ties with China. In 1978, the Joint Economic Committee of the US Congress published a compendium of papers on the post-Mao economy. The papers addressed one key question: "Has the economy of the People's Republic of China settled down to a stable, continuous process of economic growth?"[36] Smil and William Clarke offered the most updated appraisal of China's electrical industry in their contributions. Clarke characterized the electrical power sector as a "vanguard industry" that had to grow at a rate of 1.4 times that of general industrial growth and concluded that if all the problems of China's electrical industry were solved, it would be possible for China to "achieve [an] industrial growth rate of 10 percent for a sustained period commencing about 1981."[37]

Continuities between the Nationalist and Communist period became the focus for China scholars of the late 1980s and 1990s, as they sought to identify common factors that contributed to rapid economic growth in mainland China and Taiwan. Men trained as scientists and engineers began assuming top leadership positions in the Communist Party in mainland China and the Guomindang in Taiwan by the 1980s. Cheng Li and Lynn White point to the "shared technocratic orientation" of mainland China and Taiwan, noting that Taiwan's premier between 1978 and 1984 Sun Yun-suan and China's premier between 1988 and 1998 Li Peng, who both presided over fast-growing economies, had started out as electrical engineers.[38] William Kirby pointed out that electrification was the first order of business after Chiang Kai-shek established the Nationalist government in Nanjing in 1928.[39] He also identified legacies in the Nationalist period that provided the "approaches, institutions and individual skills" for the PRC to carry out Sino-foreign technology transfer and industrial development. These continuities include the retention of "central-government control over industrial and technological development," the maintenance of "close but controlled trading and technological relationship with an advanced industrial power," and the "devotion to central economic planning and research."[40] This current study builds on the insights of this earlier generation of scholars to explore how electrification became intertwined with state-building.

Although electricity fueled China's rapid economic growth, it did not become an object of inquiry among historians of modern China.[41] This is not the case in the history of technology, where historical studies about electrical power infrastructure have constantly yielded new theoretical insights for more than thirty years. In *Networks of Power*, Thomas Hughes put forth a four-phase model of

technological systems, starting with invention, continuing to technology transfer and system growth, and culminating in momentum. He generalized this developmental pattern by studying the growth of small intercity lighting systems
of the 1880s into regional power networks by the 1930s in Europe and America.[42]
Hughes's model came under criticism with the emergence of the "Social Construction of Technology" (SCOT) in the 1990s. In *Electrifying America*, David
Nye objected to Hughes's overemphasis on the sequence of inventions and the
entrepreneurs who built the electrical industry and chose instead to describe
electrification as "the story of increasing technical potentialities embedded in
social processes."[43] As Brian Larkin pointed out, Hughes's system-centered analysis deemphasized the technology and brought into focus "non-technological
elements" that helped produce the technical system, which led subsequent generations of scholars to rely on actor-network theory to "trace associations across
heterogeneous networks."[44]

The history of electrification emerged as fertile ground for comparative history. Jonathan Coopersmith's *The Electrification of Russia*, which also aligns itself with SCOT, noted that Hughes's model was not representative of the world
outside the Western industrialized world.[45] The different emphasis of Nye's and
Coopersmith's narratives stemmed from economic and societal differences between the United States and Russia. The use-based history of technology that
emerged in the early 2000s not only moved further away from technological
aspects but also decried the innovation-centric accounts that privileged wealthy
industrialized nations and called for a truly global perspective that considered
the impact of technology on poor and nonwhite populations. Making the above
point in *The Shock of the Old*, David Edgerton called on historians of technology
to "shift attention from the new to the old, the big to the small, the spectacular
to the mundane, the masculine to the feminine, the rich to the poor."[46] Electrical
power systems, having been around for more than a century, serve as an ideal
object of historical inquiry to explore the comingling of new and old.

Histories of electrification, more broadly energy history, emphasize the inte-
gral role of energy delivery systems in shaping the modern political order. They
do so either by identifying universal trends in energy infrastructure development
or by exploring the impact of local politics on site-specific developments. David
Ekbladh, and more recently Christopher Sneddon, explored how the United
States exported integrative river-basin development of the Tennessee Valley
Authority (TVA) as a mode of liberal regional and social planning.[47] Timothy
Mitchell's *Carbon Democracy* showed how coal miners pushed their demands
for democracy as they were able to control the chokepoints of fuel distribution,

while the sociotechnical arrangements of oil production prevented oil workers from making similar demands through industrial action.[48] The examples of the TVA's hydropower development and coal and oil extraction by European and American energy corporations form the basis for broad universal claims about the impact of energy extraction on politics. Yet, the experiences of developing nations are barely discussed. Two recent studies on electrification in India and Palestine have corrected this imbalance and highlighted the tragic human cost of nation-building through electrification. Sunila Kale's 2014 work on electrification in India asks why India's national politicians failed to make electricity universally available to all Indians, creating severe inequalities across states. Fredrik Meiton's *Electrical Palestine* documented the way that electrification contributed to Jewish statehood and Palestinian statelessness.[49]

Developments in the history of technology have sparked a revival in East Asian energy history in the past five years. Coal, which is the largest fuel source for China's power sector, offers a natural starting point for Chinese energy history. Back in 1984, Tim Wright's study on coal mining between 1895 and 1937 challenged the idea that pre-1949 China suffered from economic stagnation. It took nearly two decades before Elspeth Thomson followed up with her study on coal mining in the People's Republic. Another decade would pass before the publication of Victor Seow's dissertation about Fushun, the coal capital of the Japanese empire in Manchuria, and Shellen Wu's *Empires of Coal*. Wu opened up a new path of inquiry into the legacy of imperialism on resource extraction by exploring how coal transformed into "an essential fuel of the Chinese drive to wealth and power" during the Qing-Republican transition.[50]

This book represents an effort to connect studies about energy history with broader discussions about the reconfiguration of the social and economic order during decades of revolution and war. Several contributors of the 1990 edited volume by Joseph Esherick and Mary Rankin on the Chinese elite have shown how the local elite seized on new opportunities presented through commercialization and industrialization.[51] Energy history offers insights into the power dynamics created by new modes of production during an age of political upheaval. Scholars led by Tajima Toshio at the University of Tokyo have offered a preliminary response by showing how an electrical industry that grew in fits and starts led to the creation of China's "economies of insufficiencies."[52] Micah Muscolino's concept of "military metabolism," through which the mobilization for war provided the impetus to exploit new energy sources, features prominently in this work.[53] Judd Kinzley's book on resource extraction in Xinjiang, which uncovered various layers of surveys, capital investments, and political institutions, offers another source of inspiration. This work traces the expansion of China's energy frontiers, similar to how Kinzley "revealed the long-term legacies of formal and informal empires all along China's long borders."[54]

Chapter Overview

Each of the following chapters examines how various regimes vying for dominance in modern China strengthened the resilience of economic and political institutions, as they grappled with chronic power shortage and recovered from catastrophic power loss. Unlike studies about resource extraction centered on a single site, this history of electrification of wartime China requires a more fluid conception of geographical space. The front lines in the battle for control over China's electrical industries constantly shifted during these decades of ceaseless warfare. The narrative begins with the founding of the first power station in Shanghai in 1882. Instead of staying put in Shanghai, the narrative moves across the Lower Yangtze; North China; Kunming; Knoxville, Tennessee; and Beiping (Beijing), before returning to Shanghai around 1950. Shifting between different geographic scales, it explores broader national, regional, and transnational implications of local electrification projects. The fragmented state of China's electrical industry in the early chapters is an expression of China's political state. With the central government playing an increasingly active role in the resolution of electricity crises, electricity became an instrument of state power, which resulted in the loss of local autonomy and transfer of political authority to the national government. Wartime mobilization catalyzed this transformation. As Kenneth Pomeranz argued in *The Great Divergence*, the vast distance between the rich coal deposits in North China and the fuel-hungry centers of textile production in the Lower Yangtze increased the cost of adopting fossil fuel technologies, such as the steam engine.[55] Additionally, Japanese occupation forces working under conditions of wartime austerity took steps to reduce fuel transportation costs by moving industrial activity closer to North China and away from the Lower Yangtze.

The first two chapters focus on the power shortages confronting the cotton and silk industries of the Lower Yangtze between 1882 and 1937. In both cases, the central governments surrendered control over the electrical industries, giving free rein to private enterprises to settle disputes on power distribution among themselves. Chapter 1 examines how the Shanghai Municipal Council (SMC), the governing body of Shanghai's International Settlement, expanded its electrical infrastructure to cater to the increased power demands from British and

Japanese cotton mill owners. The SMC was unable to make use of its control over electrical power to achieve its political objectives. During the general strike following the May Thirtieth Incident of 1925, it cut off electricity to force British and Japanese mill owners to negotiate a settlement to end the strike. The threat led to accusations that the SMC politicized the supply of electric power to bolster the ambitions of foreign imperial powers. A few years later, its ratepayers approved the sale of the power assets to a private company to prevent political interference in the distribution of electric power.

The Guomindang regime had even less control over the electrical industries when compared to the SMC. The electrification of Huzhou's silk industry and its entanglement with local and national politics is the focus of the second chapter. Facing competition from mechanically woven silk from abroad, the gentry in the silk production center of Huzhou promoted the electrification of silk weaving. Just like the foreign capitalists, the local elite used their control over the power supply as a means to secure social and economic dominance.[56] Having rescued the silk industry with no assistance from the national government, Huzhou's power plant owners campaigned vigorously against the nationalization efforts of the Guomindang regime. The weak institutions of the newly established Nanjing government backed down from nationalization, as they lacked the capital to acquire hundreds of power stations across the nation.[57] In the face of a dominant local elite, the engineer-bureaucrats refocused on their roles as industry regulators. They stepped in to mediate a dispute between silk weavers in Huzhou and the local power company. Besides systemizing pricing rules within the electrical power sector, however, the regulators did little to address the lack of standardization and poor performance of smallholding power companies.

The three subsequent chapters deal with the energy crises that broke out following the Japanese invasion in July 1937 and discuss how wartime infrastructure development led to the centralization of state authority. In chapter 3, both the Japanese and Guomindang regime turned to coal as the primary source of fuel for the electrical industries not only because of its relative abundance but also due to the ease of centralizing its production and distribution. The exigencies of war increased the urgency to rationalize the distribution of fuel resources. This led the Japanese to expand the power grid in North China at the expense of the highly electrified Lower Yangtze region. Similarly, the Guomindang regime in Southwest China vertically integrated the coal and electrical industry in order to secure fuel sources. Chapter 4 explores how the state agencies of the Guomindang regime overcame the shortage of electrical equipment through a combination of technology transfer and applied research. Adapting to material scarcity, the wartime electrical equipment manufacturers pushed for greater standardization to lower production costs and thus vigorously defended the national voltage standards. The war also presented an opportunity for China to work with its allies to unleash its unfulfilled hydropower potential. Chapter 5 turns to the technological diplomacy between the NRC and the Tennessee Valley Authority (TVA) between 1942 and 1945, which culminated in the Three Gorges Dam survey by American dam builder John Lucian Savage. The attempt to take on the construction of a megadam of national significance ultimately forced the Nationalist regime mired in war and political upheaval to spread its limited resources too thin.

The final chapters trace the transformation of electricity from a weapon of urban warfare into a tool for mass mobilization. Chapter 6 looks at the decisive role of electrical power in urban warfare during the Chinese Civil War. It begins by exploring how the Guomindang regime's failure to take over electrical assets in Northeast China led to an energy crisis that destabilized the national economy. The Communists were able to win over the hearts and minds of the Guomindang's engineering elite, which then allowed them to inherit the electrical infrastructure largely intact after a bloody civil war. In the final chapter, we see how the perpetual state of warfare shaped the electrical infrastructure development in the early years of the People's Republic. The fear of a Nationalist comeback and mobilization for the Korean War led to increased military presence in the electrical industries. The Communists were also reluctant to increase generating capacity. They instead redistributed power demand by micromanaging the production schedules of hundreds of factories to spread out industrial activity over a twenty-four-hour period, leading to the transformation of a workforce into an army of industrial conscripts.

The trauma of war lingers into the present. To this date, policy planners and engineers on both sides of the Taiwan Strait remain anxious about disruptions to energy flows that could potentially derail economic development. Lessons from the early stages of China's electrification serve as stark reminders of the vulnerabilities of centralized power networks hastily built on the ruins of war and revolution. They continue to walk a tightrope as they strive to balance short-term economic gain with long-term political and environmental sustainability. This is the predicament of the Anthropocene. The extent of massive carbon lock-in, along with a paranoia of sabotage, returns to haunt mainland China and Taiwan as they attempt to transition toward renewables.

CHAPTER 1

Spinning the Threads of Discontent

T RAVELERS TO 1920S SHANGHAI found themselves captivated by the sights and sounds of a modern city powered by electricity. Electric street lamps had been illuminating the Bund, Nanjing Road, and Broadway since 1883, and Shanghai's architectural internationalism was on display all day and night. City lights accentuated the beauty of the European-style buildings throughout the Bund and French Concession and electric tram lines zipped through the International Settlement. There was never a dull moment at night. Shops lit up their signboards with neon lights; restaurants and dance halls welcomed customers late into the night; theaters dazzled the audience with the astute use of lighting in their performances.

Where did the energy that fueled Shanghai's prosperity come from? Follow the map in figure 1.1 to locate Shanghai's engine room. Turn away from the bright lights on the Bund. Cross Suzhou Creek on Garden Bridge (now known as Waibaidu Bridge), walk along Broadway, then continue east onto Yangshupu Road. The "Paris of the Orient" fades into the background as one ventures into the "Manchester of the East." One would pass by numerous cotton mills on both sides of Yangshupu Road and see a few others along Ward Road, which runs parallel to Yangshupu Road. Steam and coal ash permeates the air. Some of that ash comes from the chimneys of cotton mills, as workers fed coal in the furnaces to keep the old steam engines and generators chugging along. Some comes from the Riverside Power Station on the eastern end of Yangshupu Road. In fact, Chinese, Japanese, and British cotton mills were the largest clients of this power station, which was completed in 1912 and controlled by Shanghai Municipal Council up until 1929.

This hypothetical journey from the Bund into the Eastern Textile District traces the developmental trajectory of electrical power in Shanghai. The lights in Shanghai, which captured the imagination of Chinese novelist Mao Dun and Japanese writers who visited Shanghai, such as Akutagawa Ryūnosuke and Tanizaki Jun'ichirō, served as the spark that led to the creation of the city's

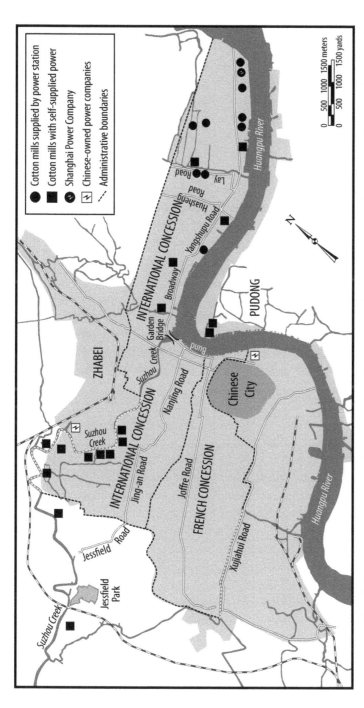

FIGURE 1.1. Shanghai's main power station and major textile mills, circa 1929. The power station on the eastern end of Yangshupu Road supplied electricity to cotton mills in the International Settlement. Many Chinese-owned power mills, the cluster of Japanese cotton mills along the Suzhou Creek established by Naigai, and Donghua mills at Lujiazui in Pudong generated their own electricity to power their operations. Source: 1932 Hochi News Map of Shanghai, 1929 Statistical Investigation of Electrical Power. Map by Mike Bechthold.

electrical power systems. Had Shanghai's electrical industry relied solely on electric lighting for its revenue, it would have died an early death. It was the huge power demand from cotton mills that catalyzed the expansion of Shanghai's electrical infrastructure. The struggle over access to electrical power became a key aspect of the commercial warfare raging between Chinese and foreign cotton mill owners.[1] Its outcome shaped the pattern of accelerated economic development under conditions of capital scarcity, inspired conceptions of economic sovereignty, and generated residual inefficiencies that had broad environmental implications.

A journey into the engine room of China's early industrialization not only allows us to track the development of Shanghai's electrical infrastructure alongside its key industries, but it also exposes problems created by the uneven pace of technological development. The mechanization of cotton production had begun before the introduction of electricity. Shanghai's first cotton mills were set up in 1878, four years before the first power station was built. The first generation of cotton mills installed their own power-generation equipment, as the generating capacity of power stations was too low to meet their demands. The vast expansion of Shanghai's electrical utilities after 1914 did not lead to the end of the "self-supplied power" operational model. Japanese cotton mills became the greatest beneficiaries of this newly available source of electricity, while Chinese plant owners maintained the practice of generating their own electricity, or even installed obsolete machinery hooked up to steam engines. The persistence of obsolescence contributed to residual inefficiencies within Shanghai's textile industries.

Self-supplied power sheltered Chinese cotton mills from the vagaries of the "transnational colonialism" within the Shanghai Municipal Council (SMC). I concur with Isabella Jackson that the colonial nonstate actors from different nations within the British-dominated SMC cooperated in a manner that transcended their own national interests.[2] That said, the SMC remained prone to accusations of restricting Chinese businesses from accessing public utilities. This multinational cooperation also broke down during the May Thirtieth Movement in 1925, as the British used electricity as a bargaining chip to force a resolution of the strike. Interestingly, Chinese cotton mills that had been denied access to the power grid and installed their own power equipment were able to resume operations faster than the cotton mills that depended on the Electricity Department for their power supply.

The mechanization of cotton milling in China has a profound environmental impact, triggering China's transition from an organic economy to a carbon-based one. While I agree with Sven Beckert's assertion that "cotton became the launching pad for the broader Industrial Revolution," I take issue with his claim that China and India became more "subservient to the Europe-centered empire of cotton."[3] The electrification of cotton milling in China set the stage for the Great Acceleration of the Anthropocene, during which the massive expansion of human economic activity impacted the functioning of the Earth system at the global level. This chapter addresses Amitav Ghosh's question: "Could it be the case that imperialism actually delayed the onset of the climate crisis by retarding the expansion of Asian and African economies?"[4] I contend that imperialist forces were responsible not only for introducing the carbon-intensive modes of industrial production by financing Shanghai's cotton industries but also for creating the inequalities that increased the energy burden of China's textile industries.

From Light to Power, 1882–1915

Shanghai's fragmented power industry mirrored its urban politics. Located in Shanghai's International Settlement, China's first power station began operations on July 26, 1882. In that very year, Thomas Edison lit up Wall Street with direct-current generators at the Pearl Street Power Station. The former chairman of the Shanghai Municipal Council Robert Little raised funds to convert a warehouse at the intersection of Nanjing East Road and Jiangxi Road (near present-day Nanjing East Road Subway Station) into a small power station outfitted with a 16 kW direct current (DC) generator.

New entrants swooped into Shanghai's electrical power market in 1897 and 1911 to capitalize on its huge growth potential. The French Concession established its power station at Yangjingbang in 1897, then in 1906 it merged the power company with the tramway company to establish the Compagnie Francaise de Tramways et d'Eclairage Electrique de Shanghai. The Qing government built the South City Power Company in 1898, which was then sold to Chinese businessman Li Pingshu and his partners in 1906. The South City Power Company later merged with Chinese Merchants Tramways Company in 1918 and was reincorporated as the Chinese Merchants' Electric Company. In October 1911, Li Pingshu completed the construction of Zhabei Electric and Water Works in the Chinese-controlled Zhabei district.[5] Having carved up Shanghai into separate franchise areas, these late entrants started out with electrical illumination as their main revenue source, but diversified their client base by supplying electricity for industrial use. In 1907, 92 percent of the electrical power sold in Shanghai went toward public and private lighting. By 1935 nationwide, electric light only accounted for 25 percent of power consumption (390,307 kWh), while electrical power amounted to 57 percent (892,046 kWh). The remaining 18 percent was lost via power transmission.[6]

Electric lighting faced stiff competition from gas and other alternatives, which limited its scope for growth. Yang Yan's research shows that the first generation of electric arc lamps were less cost competitive than gas lamps. In February 1883, Little placed a bid to replace 155 gas lamps along Nanjing Road, the Bund, and Broadway with thirty-five arc lamps at an annual cost of 9,100 taels of silver. The gas company counteracted by placing a competing bid for sixty-two new gas lights with an annual cost of 4,542 taels. While the number of arc lamps increased from thirty-five in 1883 to eighty-one in 1892, the number of gas lamps at this time also went up from 398 to 483. Up until 1892, the cost of gas lighting was about 40 percent lower than electric lighting. Little's Shanghai Electric Company suffered from manpower and capital shortage and was unable to promptly repair broken generators and arc lamps.[7] It was reincorporated in 1888 and was acquired by the Shanghai Municipal Council in August 1893.[8] The cost of electric lighting went below that of gas lighting that same year when the Shanghai Municipal Council assumed ownership of the power company.

The newly emergent alternating current (AC) system contributed to the expansion of the power generation network. In DC systems adopted by the early power companies, voltage fell significantly when the current traveled for more than one mile. Little could have tried to resolve the problem of transmission loss by moving his small power station half a mile north to another warehouse in Zhapu Road. As Ernest Freeberg pointed out, AC systems developed by Westinghouse and the Thomson-Houston companies began to displace Edison's DC system by 1889, as they were much more efficient and flexible.[9] Transmitting alternating current at a high voltage reduced line loss. The network of transformers could safely step down the voltage for safe use by consumers. Placed under municipal ownership, the Shanghai Power Company adopted the AC system when it established a new power station on Fearon Road (present-day Jiulong Road), which was completed in 1896. The electrical power network in Shanghai expanded rapidly, and the electrical engineer of Shanghai Municipal Council T. H. U. Aldridge reported that the mileage of electrical mains increased by 210 miles between 1903 and 1907, reaching a total of 400 miles by 1907.[10]

Revenue growth from electric lighting began slowing down during this period of rapid expansion. Between 1883 and 1907, the number of arc lamps lighting Shanghai's streets increased from 35 to 238. Average revenue growth from energy-intensive arc lamps fell a few years after 1886. Between 1882 and 1893, the number of arc lamps increased at an annualized rate of 19.2 percent.[11] Revenue per lamp peaked at 253 taels in 1886, remained stagnant for a few years, then steadily declined to 140 taels in 1893.[12] The annualized rate of increase decreased to 4.0 percent for the next thirteen-year period (1893–1906). In 1907, only eleven additional arc lamps were erected.[13] Residential lighting and smaller street lamps that used less electricity became increasingly popular but did not help the power company boost its revenue. Lighting the side streets and outlying roads of Shanghai were 736 incandescent lamps. Between 1900 and 1907, the number of lights installed grew at an annualized rate of 32.6 percent, but electrical power units sold only grew by 22 percent.[14] Electrical illumination alone would not deliver the high revenue growth that would fuel further expansion of the power sector.

The British-controlled Municipal Council identified industrial electricity as a promising growth sector. In his December 1907 report, Aldridge noted that the sale of "electricity for power purposes" grew by 114 percent to 141,146 units (kWh) and stated: "The development of the power load during the daytime, when the lighting demand is comparatively small, is to be encouraged by all possible means, because of the important part it plays in improving the load factor of the station, which in turn indirectly assists in cheapening the cost of production."[15] Aldridge had every reason to be optimistic. In 1907, the number of electric elevators in Shanghai's Financial District more than doubled within a year from ten to twenty-three. The harbors in the Eastern and Central districts also employed electric hoists to handle cargo. A larger generator was installed following the expansion of the Fearon Road Power Station, which gave an additional boost to the Shanghai Tramways. Electric motors seemed to be opening up new revenue streams.

Aldridge's 1907 report outlined the most ambitious expansion of Shanghai's electrical power infrastructure and signaled the rising importance of industrial power use. The Municipal Council chose to build the new power station at Yangshupu—the center of Shanghai's textile industries. Located on the eastern edge of the International Settlement and along the Huangpu River, the Riverside Power Station at Yangshupu had a superior location compared to the Fearon Road Power Station, which had "no conveniences for coal and a very poor supply of water for condensing purposes. The barging of coal, especially in winter, is a difficulty and all coal had to be carried into the works in baskets and already 300 baskets come in every day."[16] The Municipal Council was optimistic about the growth prospects, which is evident from Aldridge's description that the "station should be laid out to accommodate 5,000 kilowatts in plant, but so arranged to permit indefinite extensions as required."[17]

Aldridge engaged the services of Arthur H. Preece—the son of William Preece, the engineer-in-chief of the British post office who was himself a globe-trotting partner of the electrical and telecommunications firm Preece & Cardew. Projected power demand was so high that Preece proposed installing two turbo-generators of 1,500 kW each, instead of 1,000 kW each as previously planned. Shanghai's load factor was comparable to Coventry and West Ham in England, so there was no need to worry about overcapacity. Preece, who had managed electrical utilities in Calcutta, estimated that Shanghai Power would recoup the cost within two and a half years. Shanghai's power grid relied on high-voltage overhead lines, which were cheaper than the partially underground system in Calcutta.[18]

Outside of Shanghai, China's earliest power stations started out with electric lighting as their core business and then began promoting the use of electric motors. The first power stations established by Chinese businessmen followed the "government-supervised, merchant-managed" model that emerged from the self-strengthening movement. In 1890, Huang Bingchang, a Chinese businessman living in the United States, sought approval to build the Canton Electric Light Company. Zhang Zhidong, the governor-general of Guangdong and Guangxi and a leading self-strengthener, duly approved his request. When Huang's illumination company failed in 1905, the Qing government acquired it and merged it with another foreign-owned power station to form Guangzhou Electric Company. Of the thirty-three power stations founded before 1911 solely on Chinese capital, twenty-seven registered themselves as *diandeng chang* (illumination companies).[19] The remaining six referred to themselves as *dianqi gongsi* (power companies) or *fadian chang* (power stations), but they were reincorporated from other *diandeng chang*. One example of an early adopter who saw the potential for electricity to become a general-purpose technology was located in Wuhu, Anhui Province. Wu Xingzhou from Hefei amassed some wealth from selling sundry goods. He had set up Mingyuan Electric Light Company in Wuhu, Anhui Province, in 1906. Plowing the profits from electric lighting, Wu Xingzhou went on to build a flour mill, matchstick factory, machine works, telephone exchange, and other industries.[20] The diversification of electric companies catalyzed industrialization in places beyond Shanghai as well.

Energy Revolution in Textile Production

Back in Shanghai, cotton mills became the largest consumer of electrical power within a few years of the completion of the Riverside Power Station in 1912. Even with their wildly optimistic projections, Preece and Aldridge grossly underestimated Shanghai's power needs. Older cotton mills continued to generate their own electricity with in-house steam engines, as their old machines ran on a different power rating than that supplied through the grid. Japanese cotton mills, founded at the moment when Shanghai expanded its electrical power-generating capacity, capitalized on the new sources of cheap and abundant electricity. Only a few Chinese-owned cotton mills switched to purchasing electricity from the grid. Many of them, however, arrived on the scene a little too late and resorted to the preexisting practice of generating their own electricity. They either acquired obsolete machinery powered by steam engines or simply had their applications to connect to the grid rejected by the power company. Unequal access to electricity became a source of contention between Chinese and foreign textile mills.

The mechanization of cotton textile production in China was well underway even before the introduction of electrical power. By the late nineteenth century, the influx of machine-spun cotton yarn from India had displaced hand-spun cotton yarn in China. As Tomoko Shiroyama noted, the coarse and durable yarn was so popular that farmers in the cotton-producing region of the Lower Yangtze bought Indian yarn to weave cloth for sale.[21] At the same time, the reformers of the self-strengthening movement called for the establishment of cotton mills as part of an effort at import substitution. In 1878, the leading reformer Li Hongzhang founded the Mechanized Textile Bureau under the "government-supervised, merchant-managed" model. Equipped with 530 looms and 35,000 spindles imported from the United States, the first state-sponsored cotton mill began operations in 1889. In 1893, after a fire destroyed the machinery, official-industrialist Sheng Xuanhuai took over the cotton mill and incorporated the Huasheng Company.[22] One of its subsidiaries was Huaxin Spinning Mill founded by the Shanghai Circuit Intendant Gong Zhaohuan in 1888.[23] The Dasheng Number 1 Cotton Mill, which is the subject of Elisabeth Köll's study on regional enterprises, was founded in 1895 by scholar-official Zhang Jian under official patronage.[24] Between 1895 and 1914, Chinese, British, and Japanese businessmen established ten cotton-spinning mills in Shanghai, which led to a massive increase in power demand.

Existing power stations, however, did not generate enough power for industrial use. Even after its reincorporation in 1888, the peak instantaneous power output of Shanghai Power Station was 150 kW, which was less than half the needed power of the smallest cotton mill in Shanghai.[25] Tongchang Xieji, which was equipped with only 2,000 spindles, installed a 350 kW steam engine to drive its machinery.[26] Cotton mills farmed out the repair, maintenance, and operation of the steam engine to a mechanic known as a *laogui* (old devil). The cotton mill paid the laogui a monthly wage based on the number of reels installed.[27] Furthermore, the earliest generation of alternating current technology was not suited for industrial use. The Fearon Road Power Station generated a single-phase AC at 2,200 volts. Single-phase power systems deliver pulsating currents, which is incompatible with textile machines that ran at a constant rate. Single-phase AC systems also needed more conducting material, which made it costly to expand the electrical power network. The low-generating voltage meant that the power could only be transmitted for a short distance. The Fearon Road Power Station was too far from the cluster of cotton mills on Yangshupu Road for power transmission to be economical.

When the Yangshupu Power Station entered operations in 1912, the Shanghai Municipal Council began promoting the electrification of industries within the International Settlement. Hengfeng Cotton Mill, which had started out as Huaxin Spinning Mill, was one of the first cotton mills to acquire electricity from the grid. Its general manager, Nie Yuntai, was the grandson of the self-strengthening reformer Zeng Guofan. According to David Faure, his father, Nie Jigui, began accumulating shares in Huaxin after taking over as the circuit intendant of Shanghai in 1890 and acquired a two-thirds stake in the company by 1905.[28] The Nie family took control of Huaxin in 1909 when it was auctioned off. Nie Yuntai saw the transition away from steam engine to electrical power as a means to wrest operational control away from the laogui and improve product quality. The driving force for the looms became uneven when the temperature of the steam engine fluctuated, which made the yarn quality uneven. Nie eliminated the problem by acquiring fifteen electrical motors with a total power rating of 554 hp and Hengfeng's coarse yarn with a thread weight of 16s became the standard product for transactions in the Shanghai Cotton Exchange.[29]

Japanese cotton mills entered the scene just as the new power station came online and took advantage of this new mode of production. Established in 1914 as a joint venture between British, Japanese, and Chinese businessmen, the Shanghai Spinning Factory with about 30,000 spindles initially relied on a 3,200 kW steam engine to power its operations. As it tripled the production capacity and branched out into cloth weaving, Shanghai Spinning Factory began purchasing electricity from the Riverside Power Station.[30] Electricity was a major cost component. For example, a Japanese textile mill equipped with Toyota cotton looms spent about 1,200 sterling pounds on electricity, amounting to 2.8 percent of its total cost.[31] Japanese mill owners shifted production to Shanghai, not only to lower production cost and evade the threats of tariffs against Japanese goods,[32] but the availability of cheap electricity was also a contributing factor. Shanghai's electric tariff rate (2.5 pennies per unit) was 37.5 percent lower than that of Calcutta and comparable to London.[33] By 1914, power and heating overtook electric lighting as Shanghai Power Company's main revenue source, largely driven by demand from new Japanese cotton mills.[34]

Many existing cotton mills continued to use preexisting steam engines and generators and did not make the switch to purchasing electricity from the power company. According to the National Statistical Investigation of Electrical Power Industries collected by the National Construction Commission in 1929 and data compiled by the Union of Chinese Cotton Mills in 1933, all Chinese and foreign textile mills set up before 1915 continued to generate their own power.[35] The Sino-British Textile Bureau, cofounded by Jardine, Matheson & Company and local Chinese merchants in 1895, retained its 3,100 kW steam generator to supply its own electrical power well into the 1930s.[36] Up until 1929, a total of 600,000 spindles across China were still running on steam power. Zhang Wangliang, an electrical engineer who worked in cotton mills, noted that cotton mill owners knew that it was cheaper to purchase electricity from the power company but were reluctant to change for four reasons. First, an electrical motor had a lower power rating than the steam-powered machines but was more expensive. Second, the installation cost was high. Third, electric motors were more expensive to maintain. Finally, it was wasteful to abandon the flywheels, cables, and belts from the old machines.[37] Obsolete machinery persisted because it was still the most cost effective technology for small cotton mill owners grappling with capital shortage.

Cotton mill owners who purchased electricity from the power company had to contend with disruptions due to breakdowns and power shortages. The Riverside Power Station reported severe breakdowns within two years of its completion. In January 1915, the turbine blades of generators number 1 and 2 broke off. Generators number 3 and 4 burned out in September and October 1917. The power company alleviated the shortage by installing a fifth generator set with a 10,000 kW capacity. Due to World War I, the power company waited until 1919 to ship the damaged number 1 and 2 generators to Japan for repairs, while the delivery of two generator sets from Britain had to be delayed until 1920.[38] The power company was also unable to keep up with new demand. Eight more Chinese cotton mills founded between 1918 and 1921 with a total of 246,000 spindles established in-house electrical units to ensure energy self-sufficiency.[39] Hengfeng, one of the first cotton mills that connected to the power grid, expanded its operations after World War I to cater to increased demand. Hengfeng added 164 cloth weaving machines and more than 23,000 spindles when it opened a new plant in 1921. Seeing that the power station failed to meet additional demand, Hengfeng also installed 2,198 kW of generating capacity. Of the ten British and Japanese mills in Shanghai, only four of them installed their own generators or steam engines, while the rest received their electricity from the power grid. In comparison, sixteen out of twenty-eight Chinese-owned textile mills generated their own power.[40]

Chinese mill owners gauged their energy demand by referencing the experience of similar cotton mills in Japan. Both Japanese and Chinese mills manufactured coarse yarn with a thread density below 30s by spinning short-fiber raw cotton from India and China.[41] One of the most detailed estimates of energy input published in Chinese was based on the records of a single day's production by factory manager Sugimoto Aki of Minoshima Cotton Mill at Wakayama Prefecture, Japan. Sugimoto derived the energy efficiency with a few known variables—quantity of 20s, 30s, 32s, and 40s yarn produced and the total energy consumption of the day. He proceeded to subtract electrical energy used for worker dormitories and lighting and assumed that the rest of the electricity went toward powering the machinery. He then determined the average revolution per minute of the front rollers of the drawing frames based on all these variables. After finding out the electrical energy expended in the preliminary stages of drawing, Sugimoto went on to the calculations for the slubbing, intermediate, roving, spinning, and winding processes. The result showed that one bag of 20s cotton yarn (406 lbs) required 265 kWh of electricity, that of 30s yarn and 40s yarn being 397 kWh and 520 kWh respectively.[42] The estimate served as a reference for mill owners to select a generator with appropriate power ratings to run their operations. That said, the calculations from Sugimoto's energy accounting exercise could not be applied universally. Variations in cotton fiber length, roller rotation speed, machine component weight, and cotton yarn fineness meant that every cotton mill faced different energy needs.

Most notably, Chinese cotton with its shorter average fiber lengths required more energy to process.[43] Analyzing Sugimoto's calculations in relation to the information from the 1906 *International Library of Technology* textbook helps to explain why this is the case. The short length of the fibers required a shorter distance between the roller pairs. For comparison, the distance between the bottom rollers for American cotton was 1 ¼ inches, while that of Egyptian and Sea-Island cotton was 1 ⅜ inches. The roller setting for Chinese cotton was shorter at ⅞ to 1 ⅛ inches. Assuming that the weight of the components was the same, the rollers working on Chinese cotton would have to complete more revolutions per minute to spin the same length of yarn.

Many cotton mills continued to assume the risk of generating electricity in-house instead of purchasing power from the electric company. Cotton mills directly bore the fuel and maintenance cost. Coal prices in Shanghai were double that in Calcutta, most likely the result of Japanese conglomerate Mitsui Bussan cornering the coal supply.[44] Production had to be put on hold when the generators were shut down for routine maintenance, and electrical units were prone to accidents. In May 1926, the electrical unit in the Japanese-owned Xihe Cotton Mill blew up, killing one and injuring four others. The explosion was so strong that it burned through five-inch-thick steel plates. Two Japanese technicians had their fingers severed during the accident but managed to survive.[45]

Electricity as a Bargaining Chip

In the summer of 1925, cotton mill owners who were increasingly relying on recently expanded power stations for electricity realized that the new technology came with political risks. On May 30, 1925, Shanghai municipal police shot and killed a number of demonstrators who protested the ill-treatment of Chinese workers in Japanese-owned textile mills. Known as the May Thirtieth Movement, this incident is remembered as an anti-imperialist labor movement. It also marked the first instance when electrical supply was used as a bargaining chip in a political crisis. Cotton mills at the center of the labor unrest saw their operations grind to a halt when the Shanghai Municipal Council imposed a power blockade on the city's industries. Chinese-owned cotton mills that maintained the older practice of generating their own electricity, however, were able to partially continue their operations amid this political crisis.

The Shanghai Municipal Council quickly secured the public utilities upon the first signs of unrest. Shortly after the declaration of a state of emergency on June 1, 1925, an American naval party moved in to protect the Riverside Power Station and waterworks the next day. Thirteen hundred Chinese power station workers joined the general strike on June 4. They were among the 200,000 workers across two hundred enterprises who participated in a months-long stoppage in the summer of 1925.[46] The Electricity Department hired one hundred Russian workers as strikebreakers to keep this essential service operational. Chief engineer Aldridge intended to retain thirty of these workers and fire forty-six striking Chinese workers to "form a nucleus that can be depended upon in the event of future industrial action."[47] By June 29, 1925, the Electricity Department decided to discontinue power to all consumers except food producers on July 7 at noon, citing labor shortages. Harumi Goto-Shibata pointed out that the British minister in Beijing, C. M. Palairet, used the termination of electrical power to pressure Chinese authorities into dealing with agitators hoping that it would resolve the strike.[48] The Chinese in the International Settlement sought the assistance of the Italian consul-general de Rossi, whose empty threats directed at the Municipal Council were duly ignored.[49]

The power blockade severely crippled Shanghai's cotton industry. Two months of work stoppages caused Chinese cotton mill owners to sustain a loss of 3.26 million Chinese dollars. Five Chinese cotton mills equipped with their own generators resumed production in July and sold about 285 bags of coarse yarn per week. But with 500,000 spindles taken offline, production capacity was reduced by 1,000 bags per day.[50] Japanese-owned mills bore the brunt of the losses, and between July 16 and 27, Hsu Yuan, the Jiangsu provincial commissioner for Foreign Affairs in Shanghai, negotiated with Japanese consul-general Yada Shichitaro to settle the strike.[51] As the Chinese and Japanese came closer to breaking the deadlock, Japanese mill owners pushed for the resumption of electrical power before the workers returned to work. The Electricity Department, however, adhered to its policy of supplying electricity only when the strike came to a complete end and rejected the request from the Japanese councilor S. Sakuragi.[52] On August 11, Japanese mill owners and Chinese workers arrived at a compromise. One of the six conditions was "that those mills having electrical power are to resume operation at once. In those which were dependent upon the Electricity Department for power, workers should resume work as soon as power is furnished in them."[53] The inclusion of this provision was meant to prevent workers from collecting wages when the mills were not yet operational.

The British insisted on using their leverage on their control over electrical power to delay the resolution of the strike. Chief engineer Aldridge once again turned down the Japanese request to resume power on August 13, stating that "until the departmental employees have resumed duty upon the Council's own terms, it is impossible to restore supply."[54] At this point, the British sensed that a labor dispute that resulted from Japanese ill-treatment of Chinese workers was turning into an anti-British boycott. The Chinese and Japanese had settled their differences, but the standoff within British industries was not over. In a memorandum dated August 21 from the British consul-general to Shanghai Sidney Barton addressed to the British legation in Beijing, Barton rejected any arrangements that "enabled the Japanese mills to reopen while the British mills were closed and a purely anti-British strike maintained." In this document, he concluded by calling the power supply a "most effective weapon," which "may yet be retained for use by increasing the volunteer labour sufficiently to supply the Japanese mills, while refusing to supply the Chinese mills until normal conditions return." He also scornfully dismissed the threats from the Japanese Mill Owners' Association to boycott the Municipal Electricity Department and set up their own plant and reiterated his point that "the incident of May 30th arose from the protection offered by the Municipal Council to Japanese interests against the anti-foreign and particularly anti-Japanese agitation, arising out of the Japanese mill strike."[55] Hsu Yuan, who had resolved the conflict between the Japanese mill owners and Chinese workers, offered to persuade the Chinese electrical workers to return to work unconditionally on August 26, but the Electricity Department insisted on the termination of two hundred Chinese workers and the retention of the Russian strikebreakers.[56]

Within days of Barton's saber-rattling, it became clear to the British councilors that the strategy of holding up the electrical supply was no longer tenable. Council Chairman Stirling Fessenden warned that the discontinuation of the power supply had caused the Chinese and other foreign powers in Shanghai to see the Municipal Council as an agency that furthered British interests. He concluded: "Any impression therefore that the Council was acting in the interests of any one national would undoubtedly be read as a further argument by the Powers interested for revision of the present form of Municipal administration. He therefore suggested that a reply should be forwarded to the Commissioner for Foreign Affairs to the effect that when all the workers required have returned to duty, the Council will restore the power supply."[57] The British backed down and withdrew their stipulation for the firing of two hundred electrical workers but conveyed their intention to discharge the workers whose services may no longer be needed. The British-led Municipal Council resumed electrical supply on September 8, a full eighteen days before persuading the workers to resume work at the British mills.[58]

Shanghai's cotton industries experienced a sharp downturn after the May Thirtieth Incident. Japanese yarn flooded the Chinese market when Shanghai's cotton mills ceased production. Distributors even removed the Japanese brand labels and marketed them as Chinese-made products to get around the boycott of Japanese goods. A number of Japanese mills closed down in the aftermath of May Thirtieth. The number of cotton mills fell from twenty-six to twenty-two in 1925, then rebounded to twenty-four in 1927.[59] Between the Guomindang's campaign in Shanghai in April 1927 and early 1929, no new cotton mills were founded in Shanghai. Frequent strikes crippled the transportation networks and hampered further expansion of the Riverside Power Station. Despite the sharp downturn in Shanghai's cotton industry, power demand from other industries bolstered the revenue of the Electricity Department.

In the wake of industrial unrest, the Shanghai Municipal Council made plans to sell the Electricity Department, even though it contributed most of the government's revenue. A. W. Burkill, who chaired the Electricity Special Committee, explained that the sale would ensure that the provision of electrical power would be "regarded purely as an industrial concern and beyond the scope of political influence."[60] Burkill saw the corporatization of electrical power as a way to maintain cheap and abundant electricity in Shanghai and dispel the distrust stemming from the Municipal Council's handling of the May Thirtieth Incident. The huge growth potential of Shanghai's power market meant that the Municipal Council was able to sell the Electricity Department at a good price. By the late 1920s, the Shanghai Power Station was selling more power units and achieved a higher load factor than the Manchester Power Station in Britain. On April 1929, the Municipal Council passed a resolution approving the sale of the Electricity Department to American-owned Electric Bond & Shares Company Limited for 81 million taels of silver. With its financial future secured, the power company quickly completed the installation of a medium-pressure boiler and raised the power station's capacity to 161,000 kW.[61] Renamed the Shanghai Power Company, it was poised to enter a new period of accelerated growth.

Four new cotton mills in Shanghai that opened up in 1929 became potential customers for the newly incorporated power company. Shenxin Number 8 mill and Xiefeng Yiji did not adopt the self-supplied power model and chose to purchase electricity from the grid.[62] In an advertisement published in the 1932 edition of the *China Cotton Journal*, the Shanghai Power Company proclaimed that it supplied Shanghai's industries with "more than 193,000 h.p. of electricity," and it was "the only company in Shanghai that was capable of supplying such a [large amount of] electrical power."[63] Outside the International Settlements, cotton mills also began to acquire electricity from the grid. Hengda Cotton Mill at Pudong made this transition. It had relied on a 500 kW generator at its facility because there was no power station in the vicinity when it started operations. When Pudong Electrical Power was founded in 1932, Hengda purchased power directly from the local company and stopped generating power in-house.[64]

Not every new cotton mill followed the example of Shenxin Number 8 and Xiefeng Yiji. Longmao Cotton Mill, established in 1929 by a certain Mr. Yang, acquired the assets of Japanese-owned Tōka (Donghua) Number 1 mill that had been founded in 1920.[65] Built during a time when the Riverside Power Station was unable to keep up with power demand, Tōka Number 1 supplied its own electricity with a 1,250 kW generator with an output voltage and frequency of 600 V and 60 Hz.[66] Shanghai Power Company supplied a three-phase alternating current supplied at 220 V and 50 Hz, which was incompatible with Longmao's machinery. Yang raised six million Chinese dollars to replace the equipment to match the cost efficiency of other cotton mills that took advantage of cheap electrical power from the grid. Before the equipment upgrade was complete, Shanghai suffered massive aerial bombardment when Japanese and Chinese forces clashed between January 28 and March 3, 1932. Situated near the conflict zone, Longmao suffered massive losses and liquidated all assets in 1932 upon bankruptcy.[67] Longmao's collapse illustrated how undercapitalization forced small Chinese-owned textile mills to adopt obsolete technology to save on equipment cost, thereby condemning them to less efficient modes of production.

An investigation sponsored by the Institute of Pacific Relations published in 1933 revealed the energy inefficiencies within the cotton textile industries. Cotton milling consumed the largest share of electrical power, but its ratio of economic value to energy consumption was low. Cotton spinning and weaving consumed 34.3 percent of Shanghai's power output, while it accounted for 28.5 percent of the total value of Shanghai's industrial output. Shanghai's cotton industries took charge of the entire process of cleaning raw cotton, turning raw cotton into workable yarn, and weaving the thread into cloth. This vertical integration, according to Carles Broggi, sheltered the firms from price volatility by "producing more sophisticated products with less competition and higher prices."[68] The clothing industry, too, was an equally energy-intensive sector, as it used 5.52 percent of the city's total energy output but only contributed to 3.57 percent of industrial production value.[69] Despite the increased abundance of electrical power in Shanghai, many cotton mills continued to generate their own electricity. By 1933, self-supplied power accounted for 12.5 percent of motive force in Shanghai's cotton mills. The practice persisted not only because older mills continued to power their aged equipment with their own steam engines but also because mill owners remained skeptical about relying on a foreign-owned electric company to power their operations. The persistence of obsolete technology hampered the energy efficiency of Shanghai's cotton textile sector, while unequal access to electricity created resentment and distrust within Shanghai's business community.

Cotton mills outside Shanghai expanded their operations by relying on self-supplied power. The local power stations in the major cotton production centers of Wuxi, Changzhou, and Nantong suffered from chronic capital shortage, which limited their ability to acquire new generators to keep up with demand. With a combined capital of 403,000 Chinese dollars, the five power stations in Wuxi founded between 1921 and 1923 with generating capacity ranging from 10 to 100 kW operated old steam and gas engines. In comparison, the Zhenxin cotton mill at Wuxi founded in 1905 with 1.25 million dollars in capital was able to afford a 1,544 kW steam engine. Sixteen years later, Qingfeng cotton mill, which had double the amount of capital, also opted to install a 1,000 kW steam engine.[70]

Outside the Lower Yangtze region, cotton mills continued to supply their own electricity as power demand from the cotton mills far exceeded the capacity of the local power company. When Naigai set up operations in 1916, the Qingdao Elektrizitätswerk left behind by the Germans had been destroyed during the Japanese invasion in November 1914. The Japanese administrators completed the rehabilitation only on December 1919 after they installed a 1,200 kW boiler. The big four Japanese cotton mills (Naigai, Nissei, Fuji, and Nagasaki) in Qingdao had a combined generating capacity of 50,000 kW in 1929, which was ten times that of the Qingdao Power Station.[71] This was also true of Chinese-owned cotton mills in Tianjin (Hebei Province) and Wuchang (Hubei Province). By 1934, cotton mills accounted for 43 percent of the generating capacity and 58 percent of energy usage of all self-supplied power within China.

Energy shortage was a major constraint on Chinese cotton mills. One cotton mill addressed this problem by building its own power station to alleviate this problem. This was Dasheng Cotton Mill in Nantong, founded by Zhang Jian, a scholar-official who resigned from the Hanlin Academy and took advantage of the late Qing reforms to venture into industrial development. With 87,852 spindles, it was the largest cotton mill outside Shanghai. As Elisabeth Köll and Qin Shao have both argued, Zhang Jian's cotton mill became a patriarchal industry and a vehicle for regional self-governance.[72] The provision of electricity was crucial to Zhang Jian's vision of transforming Nantong into a "model county." Shao noted that Zhang drew up plans for a regional power station in 1920, just as his family branched out into other industries.[73]

Dasheng submitted an application to the National Construction Commission (NCC) of the Guomindang government in Nanjing to build its own power station in 1932. Dasheng needed at least 4,000 hp to run its 240 cloth weaving machines. As the steam engine deteriorated with age, mill owners made up for the power shortfall by installing a diesel generator. This, however, increased the fuel cost.[74] Dasheng "sought to economize the cost of electrical power" by building its own power station.[75] Huang Hui, a Cornell-educated engineer at the NCC, reviewed Dasheng's application. He agreed with their assessment and concluded in his inspection report, "Although the company is near the cotton-growing areas and acquired raw materials cheaply, its production costs were still high, making it unable to compete with Japanese cotton mills and new Chinese mills." Huang

Hui attributed the high costs to three reasons: (1) old equipment; (2) heavy interest burden; and (3) cost of maintaining the social enterprises.[76]

Dasheng pushed through with its plans to construct the power station, even as its revenue took a hit during the Great Depression. In 1932, the shareholders approved a funding proposal, in which Dasheng put up 30,000 Chinese dollars for start-up capital and mortgaged its assets to raise another 50,000 Chinese dollars. Dasheng petitioned the Construction Commission in March 1933 to obtain an operating license for the power station. To facilitate the transportation of electrical equipment and fuel, the new power plant was built along the Yangtze River and right next to the Number 2 cotton mill. Named the Tiansheng Port Power Station, it was equipped with one British-made boiler that generated twenty-five tons of steam every hour and two German-built generators from Allgemeine Elektricitäts-Gesellschaft (AEG) with a generating capacity of 750 kW.[77] The new power system connected Dasheng Number 1 mill at the outskirts of the city with the Number 2 mill located along the Yangtze. Electricity was transmitted along fifteen kilometers of 22 kV transmission lines, and factories along the transmission line connected themselves to the new power grid. Tiansheng Port Power Station also sold surplus power to Tongming Electric Company, which had served Nantong for three decades. Both companies maintained exclusive control over their franchise areas.

The regulatory authorities were satisfied with the performance of Dasheng's power station. Yun Zhen, the director of the Construction Commission's Electrical Bureau who took charge of electrical power infrastructure development during the war, only required minor revisions from Dasheng. When he inspected the power station with Huang Hui, Yun Zhen waived the stress test, as they found all the equipment to be in good condition. Huang Hui reported that the power station generated 80,000 kWh daily and achieved a load factor around 97 percent, meaning almost all the generated power was utilized. The Tiansheng Harbor Power Station did not undermine the operations of the preexisting Tongming Electric Company, and Tongming lowered the electric tariffs of its users in Nantong. Dasheng Cotton Mills paid 1.65 cents per kWh and sold its surplus power to Tongming at 2.65 cents per unit. Tongming drastically reduced the electric tariffs in Nantong making its electric tariff 25 percent that of the nearby silk production center of Huzhou.[78] As a result, silk weavers in Huzhou would point to the low tariffs in Nantong and accuse their local power station of price gouging.

THE ELECTRIFICATION OF cotton mills in the Lower Yangtze exemplifies the challenges facing China's industries as they pursued economic growth while being constrained by capital scarcity. The late Qing to Republican era saw the rapid expansion of China's industries. Tim Wright's study on China's coal industry before 1937 cites John Key Chang's findings that China's net industrial output grew at 9.4 percent per annum between 1912 and 1936.[79] Aggregate numbers only tell part of the story, however. The uneven access to electricity among cotton mills in Shanghai reveals how problems of "economies of speed" shaped China's early industries. Unable to purchase electricity from the grid, mill owners decided to install obsolete machines driven by steam engines or bear the cost of maintaining the factory's electrical units. Even after the corporatization of Shanghai Power Company in 1929 lowered the cost of purchasing power from the grid, many of Shanghai's cotton mills continued to rely on their in-house power supply. The persistence of the "self-supplied power" model had little to do with maintaining operational resilience but more to do with the lack of capital. As seen in the case of Longmao, this poorly capitalized firm lacked funds to replace the old machinery of its predecessor, which led to the persistence of obsolete technology. It had to sacrifice long-term sustainability in exchange for short-term economic gains. The electrification project by Dasheng in Nantong was truly exceptional and could not be replicated by other cotton mills that lacked the financial resources or political connections.

The struggle for electrical power in the transnational colony of Shanghai also shaped conceptions about economic sovereignty in modern China. In the absence of strong institutions, capitalists of different nationalities devised ways to defend their business interests. The Shanghai Municipal Council, which was dominated by the British, had taken over the provision of electricity to ensure the prosperity of the International Settlement and safeguard the interests of ratepayers of different nationalities. Ever since it began supplying power to Chinese, Japanese, and British cotton mills in 1914, the Electricity Department's impartiality had come under question. Chinese mill owners had their applications for electrical connection rejected and accused the council of granting preferential access to the Japanese and British. There were also conflicts among the non-Chinese capitalists. As we have seen, the British cut off electricity to the International Settlement with the twin objectives of forcing the Chinese and Japanese to negotiate a truce and retaliating against the Japanese for stirring up trouble that led to an anti-British boycott. The May Thirtieth Incident aroused the nationalist sentiments of both workers and capitalists alike.

Chinese intellectuals who came of age in the 1920s advocated the nationalization of electrical industries to address problems of uneven access, as they witnessed the inequalities created by a power sector dominated by foreign capitalists. Zheng Youkui, one of China's first generation of economic histo-rians, voiced support for this nationalistic approach to infrastructure develop-ment. In the conclusion of his study on China's pre-1949 economy, Zheng saw "an efficient, honest, and competent government" as the way to break the "log-jam" of China's socioeconomic problems and urged the state to use its coercive power to interfere with anything "ranging from consumers' taste to the direction of investments and capital formation."[80] The ensuing chapters discuss how eco-nomic nationalism inspired China's first generation of engineers to turn toward careers in public service, which in turn led to the rise of a highly interventionist developmental state following four decades of war and revolution.

The electrification of Shanghai's cotton mills within a compressed period of two decades marked the beginning of China's transformation into a contributor to the Great Acceleration of the Anthropocene. The Municipal Council bench-marked the Riverside Power Station's performance against five of Britain's largest power companies. By the second half of the 1920s, the number of units sold and load factor in Shanghai exceeded that of Manchester, Birmingham, Glasgow, Liverpool, and Sheffield. The factories clustered along Yangshupu Road simply bought up any additional power generated following each expansionary cycle. A process that had taken a century or so to unfold in Britain was completed within a decade in 1920s Shanghai. The growth screeched to a halt in 1941, however, after the Japanese takeover of Shanghai during the start of the Pacific War, but the foundations for accelerated development had been put in place.

The Lower Yangtze offered other models for economic growth. In the next chapter, the imperial powers that dominated the industries in Shanghai recede to the background in the silk-production center of Huzhou. Our focus is on the local elite who made use of the new energy technology to assert their control over the modes of production. Localism pervaded the energy politics of Huzhou, as its industrialists spearheaded the effort against the central government's attempt to regulate the power sector. While cotton milling fostered the rise of central-ized power systems, silk weaving led to the creation of highly localized power networks that catered to the energy demands of a small locale. The national economic fabric woven by the electrified silk loom looked quite different from that created from mechanized cotton spindles.

CHAPTER 2

Defending the Public Good

C HINA'S SILK INDUSTRY PLUNGED into crisis in the early twentieth century. Japanese mills introduced electrically driven jacquard looms modeled after French designs and mass-produced low-cost silk brocades with complex tapestry designs. Japanese machine-woven silk displaced the hand-woven crepe silk produced by its peasant households. In its heyday, Huzhou, a silk production center extending forty-three miles east to west and seventeen miles north to south on the southern shores of Lake Tai, boasted that "all under heaven are dressed in Huzhou silk." Its farmers enjoyed a high standard of living simply by planting one annual crop of rice and allocating most of their land to mulberry cultivation. To revive Huzhou's economy, its silk weavers would simply follow the example of Shanghai's cotton mills and mechanize their production process. This, however, was easier said than done.

The energy economics of silk weaving differed from that of cotton spinning. Purchasing electricity from centralized power networks made sense for cotton mills. In addition to saving on the maintenance cost of their power equipment, cotton mills were able to scale up their operations without being constrained by preexisting capacity. Silk filatures were smaller in scale. One coal-fired steam engine suited their operations perfectly, as it generated the motive force for their machinery and provided the boiling water used to immerse the cocoons. Silk filatures entered the scene before electrical industries. Italian merchants imported the first modern silk filatures to Shanghai in 1880, two years before the first power station was established. Up until 1931, all seventy-three silk filatures in Shanghai generated 1,637 kW of motive power in-house and did not purchase electricity from the power company.[1] In the previous chapter, we saw how the owner of Hengfeng cotton mill, Nie Yuntai, connected his mill to the grid to wrest operational control from the "old devil" (*laogui*) of the engine room. Liu Dajun's 1933 study showed that silk filatures paid the laogui wages of 0.55 silver taels per reel each month and 0.10 taels per reel per day for the coal. It also noted the wide variations in power efficiency, with one horsepower from a steam engine powering between two and forty reels. Using 1931 production figures, Liu estimated that operating one reel required forty *jin* of coal consumed each day, which meant that a quarter pound of coal had to be combusted to make one pound of silk.[2] This confirmed Tomoko Shiroyama's point that the Japanese silk industry became more competitive as it adapted European filatures to local conditions with silk factories having the flexibility to choose between coal and waterpower as their energy source.[3]

The electrification of Huzhou's silk industry in response to foreign competition set into motion a series of institutional changes that had broader economic and political implications. Huzhou silk merchants based locally and in nearby Jiaxing and Shanghai not only imported the electrical jacquard looms that gave their Japanese competitors an edge but also funded the development of Huzhou's electrical infrastructure. The electrical jacquard loom not only gave a new lease of life to Huzhou's economy but also drew Huzhou into broader regional and national political struggles.

A comparison between Huzhou and the nearby silk-production center Wuxi reveals how differing local power relations led to divergent pathways of industrial development. On a broad conceptual level, Huzhou's industrialists adopted "bourgeois practices" similar to those in Lynda S. Bell's study about the silk industry in early twentieth-century Wuxi by bringing "the peasant economy into contact with a new international system of mechanized silk production."[4] In the face of competition, Huzhou's elite fell back on the delicate power-sharing arrangements that had shaped their local politics and pooled their capital to invest in technology that boosted their productivity. The availability of capital allowed Woo-shing Electric in Huzhou to expand its generating capacity to 1,757 kW by 1932. Xue Shouxuan, who, according to Bell, monopolized Wuxi's silk cocoon markets and silk filatures, did not invest heavily in Wuxi's electric industry. While Wuxi's silk industry with fifty filatures and two hundred cocoon firms was larger than that of Huzhou, it did not have its own electrical industry by 1932. Wuxi's only surviving electric company, which was managed by an uncle of Xue's father-in-law, did not even generate its own electricity and resold power acquired from the state-owned Qishuyan Power Plant in Wujin. Its revenue was only 3 percent that of Woo-shing Electric.[5]

The electrification of silk weaving in Huzhou sparked a contestation of ideas about "saving the nation through industrialization" (*shiye jiuguo*). As the local elite used electricity as a tool to assert their dominance over the local economy, they vigorously resisted the national government's attempts to take over the electrical power sector by masterfully appropriating the state's rhetoric of "national salvation" and "public interest." By the 1920s and 1930s, Huzhou stood at the front line of a three-way struggle among power producers, consumers, and state regulators. In 1929, Huzhou's power plant managers Li Yanshi and Shen Sifang spearheaded a movement to resist the nationalization of China's electrical utilities. Six years later, they called on state regulators to intervene when their consumers accused them of price gouging. Amid this tug-of-war between private enterprise and the state, the Nanjing government encouraged the co-evolution of markets and bureaucratic institutions, which became a source of resilience that allowed the newly unified nation to respond to exogenous economic shocks and pursue economic growth during a time of political instability.

Saving Huzhou Silk

Huzhou's silk industry faced the onslaught of international competition between the 1890s and 1920s. In the absence of strong national institutions, the local elite seized the initiative to modernize its industries by securing funding and developing infrastructure. The patriotic movements in the aftermath of the May Fourth Movement aroused the nationalist sentiments of China's earliest engineering graduates, who in turn called on the central government to proactively regulate public utilities, or even seize control of them. These two visions of "saving the nation through industrialization" developed in parallel in the early years of the republic but began to collide in 1927, after the Guomindang regime concluded the Northern Expedition and built their capital in Nanjing. Conflicts arising from the electrification of Huzhou's silk industry reveal how the newly unified nation's dream of coordinating resource allocation for broad-based economic growth threatened the elite's desire for retaining autonomy over their local economy.

Huzhou's silk industry weathered numerous boom and bust cycles while it was swept up in the political turmoil of the late nineteenth century. International demand for Huzhou silk soared after the Qing's defeat in the Opium War. Between 1845 and 1859, Huzhou's raw silk export grew at 27.05 percent per annum.[6] The appearance of the first mechanical filatures began to depress demand for handspun silk (*tusi*) produced by peasant spinners in Huzhou. Compared to factory-made silk (*changsi*), handspun silk had inconsistent thread thickness, more impurities, and was more susceptible to breakage. Shiroyama points out that foreign-owned silk filatures were in fact joint ventures by compradors working for foreign firms and Chinese silk merchants. China's first mechanized filature, founded in Shanghai in 1863 by the British conglomerate Jardine,

Matheson, and Company and later renamed Ewo Filature, for example, had 60 percent Chinese capital. Huzhou silk merchants realigned their business model and profited from mechanized silk weaving. The first wholly Chinese-owned filature was founded in 1882 by Huang Zuoqing—a Huzhou merchant.[7] By 1895, factory-made silk commanded twice the price of handspun silk. Farmers stopped retaining their cocoons to reel their own silk and sold all their cocoons to cocoon houses. Silk firms in nearby Nanxun tried to match the consistency of machine-spun thread through rereeling. Despite these improvements, however, rereeled silk only achieved a market share of 15.96 percent.[8] According to Lilian Li, the percentage of Shanghai's white silk exports produced by steam filatures increased from 11.7 percent in 1895 to 98.5 percent in 1915.[9]

Huzhou's merchants, who participated in the global silk trade through their operations in Shanghai, recognized the urgency to mechanize silk production in their hometown. Huzhou crepe silk, marketed as *yangzhuangsu* (plain western silk), was suited to making fall and winter clothing and was highly prized in American and European markets.[10] By the 1910s, however, mechanically woven silk from Japan began crowding out Huzhou's main exports. A Huzhou silk firm responded to the competition by mass-producing silk fabric with ten jacquard looms imported from Japan in 1914 but was driven into bankruptcy in 1917 when Japanese merchants flooded Huzhou's silk market with a cheap mechanically woven cloth known as "wild chicken poplin" in 1916.

As the attempt to replicate Japanese production methods failed, Huzhou silk merchants designed a new product to open up new markets. Jin Lisheng, another Huzhou silk firm owner, made a last-ditch effort to save Huzhou's flagging silk industry by creating the prototype for *huasige* (Chinese silk poplin) with two manually operated looms. Cool to the touch and durable, huasige became popular in the tropical markets in Southeast Asia. Driven by increased demand, Huzhou's silk firms mechanized their operations to produce huasige, and three electrically powered silk filatures sprouted up in Huzhou in 1917.[11] The huasige emerged as a product of "national salvation through industrialization" during a time when the new Chinese nation had not yet been clearly defined.

After the founding of the republic in 1911, Huzhou was renamed Wuxing County. Huzhou gentry based in Shanghai promoted the electrification of their hometown and established in 1914 the six-hundred-square-meter Wooshing Illumination, which was equipped with a 100 hp (73.5 kW) British-made steam engine and a 68.8 kVA three-phase alternating current generator. Its co-founders—Wang Yimei, a director on the board of the Association of Cocoon Producers of Jiangsu, Zhejiang, and Anhui; and Guan Zhiqing, the comprador

for Siemens—raised 50,000 silver dollars to finance the construction of Huzhou's first power station. In the beginning, the power station only supplied
electricity to light the streets of the county seat from 5 pm to midnight. Much
like what was happening in Shanghai, the electrical industry expanded in anticipation of demand from the local textile industry. In 1917, the shareholders
raised another 70,000 silver dollars to build a new power station at the northern gate right next to Huzhou's largest silk filatures.[12] Around that time, three
of Huzhou's largest silk filatures and weavers, Guangyi, Lisheng, and Dachang,
had planned to install diesel generators in their factories.[13] The power company, however, "forcefully dissuaded" all three silk makers from proceeding
with their plans. Dachang's owner, Niu Jiechen, acceded to the power station's
request. Together with the power station's co-owner, Guan Zhiqing, Niu negotiated the purchase agreement with Siemens for the turbine. Dachang paid
a fixed amount of six hundred silver dollars each month for electrical power,
which provided a stable source of revenue to support the power station's expansion.[14] The 600 hp boiler-turbine installed in 1921 gave a huge boost to
Huzhou's silk industry, and silk factories enjoyed lower electrical tariffs as a result of their preferential customer status. Woo-shing Illumination changed its
name to Woo-shing Electric Company in 1923 to reflect its new core business
interest. At the peak of the boom in 1925, Huzhou had more than fifty silk factories with 5,000 electrical looms and produced 530,000 bolts of silk annually.

Coming of Age in a Revolution

Just as Huzhou's industrialists preserved their economic dominance by financing the construction of the local electrical infrastructure, a competing vision
for electrification led by a strong national government began emerging from the
nationalist movements of the 1910s. Sun Yat-sen, the leader of the Chinese Nationalist Party who had supported the armed uprising that overthrew the Qing,
called for the transformation of electrical power into a public good by nationalizing or regulating the industry. In 1919, Sun laid out his industrial policy, the
"Principle of the People's Livelihood," in which he called for all natural resources
and social goods to be turned over to the state so that "profits accrued by these
can be applied by the state for public purpose."[15] Sun's ideas entered the scene,
just as China's first generation of engineers was swept up in the revolutionary
fervor of the 1911 Revolution and May Fourth Movement in 1919.

   Two of China's first electrical engineering graduates from the class of 1921
found themselves on opposite sides of the nationalization versus privatization

debate. They were Shen Sifang, who would later become Woo-shing Electric Company's chief engineer, and Yun Zhen, an influential engineer-bureaucrat within the Guomindang regime. Both men graduated from the Government Institute of Technology, which had been founded as the Nanyang Public School in 1896, and later reorganized as Jiaotong University in 1921. Both belonged to the first cohort of electrical engineering students admitted in 1918 and served as co-secretaries of the university's first student council established in 1919.[16] S. R. Sheldon, who graduated from the University of Wisconsin in 1894 and served as the secretary of the forestry fund in Shanghai, was the program's inaugural dean.[17] These engineering students received a broad-based education in the four-year curriculum. They studied Chinese classics, Western literature, and foreign languages, in addition to courses in science and engineering. They also participated in nationwide demonstrations to protest China's mistreatment in Versailles in May 1919.

Shen and Yun, the two student council co-secretaries, viewed campus activism skeptically. Shen's writings offer few clues about his views on the student movement. In 1919, he published an essay in the college's engineering bulletin to call for the widespread use of electrical irrigation to lighten the farmers' physical burden.[18] Yun Zhen, alternately, positioned himself as a pragmatic reformer, much along the lines of the Changzhou School of Confucian practical statecraft from his birthplace of Wujin in Jiangsu Province.[19] As a student leader, Yun observed how campus walkouts failed to garner public support, as student leaders focused exclusively on political and diplomatic issues of little interest outside elite circles and failed to address the practical economic concerns of workers and merchants. He called on his fellow leaders to teach as volunteers in schools to equip workers and peasants with practical skills and engage in community service to help people in need, instead of rousing the crowd with sensational talk of "national subjugation."[20]

Engineering graduates looked toward the German-born American professor of electrical engineering at Union College, Charles Proteus Steinmetz (1865–1923), as their role model. Student publications and the commercial press retold the story of how Steinmetz, the son of a railroad worker and born with a hunched back, overcame great adversity to become a leading authority in electrical engineering.[21] A biography published in the religious journal *Mingdeng* in 1921 emphasized how Steinmetz was a hunchback with an angelic mind, seeming to suggest that China could break free from its servitude to the imperial powers and join the ranks of advanced industrialized nations through the mastery of science and technology. Chinese students latched on to Steinmetz's socialist leanings. Ronald Kline pointed out that Steinmetz became a socialist in 1883, after he was drawn to Etienne Cabet's ideas, which "outlined a centrally planned, democratic, egalitarian nation-state with no money, private property, or police." Steinmetz later fled the University of Breslau when Bismarck cracked down on the socialists in 1888.[22] Chinese writers emphasized Steinmetz's "hatred for the tyranny of capitalists."[23]

Upon their graduation in 1921, China's first cohort of electrical engineers faced the harsh reality of securing employment in an industry still in its infancy. Yun joined the bureaucracy of the new national government in Nanjing in 1928 after shuttling between temporary positions in the United States and China.[24] Shen, however, sought employment from small power stations in Zhejiang and Jiangsu owned and operated by local gentry. The idealism of their youth had faded into the background.

Yun crossed paths with several revolutionaries from the Young China Association when he pursued graduate studies at the University of Wisconsin at Madison with financial sponsorship from his uncle and during the early years of his engineering career upon his return to China. At Madison, Yun shared a room with fellow Young China Association member Fang Dongmei and continued to monitor political developments in China. In 1922, Yun left for an internship at Westinghouse in Pittsburgh, even before completing his master's thesis. Eager to reunite with his wife, Yun returned to China in 1923 and taught at the Zhejiang Industrial School at Hangzhou. In addition to teaching, Yun began organizing the Chinese Engineering Society and was acquainted with Tian Han, the playwright and activist who would write the lyrics to the "March of the Volunteers"—the song that would become the national anthem of the People's Republic of China. Yun worked as the chief engineer for a cotton mill in Zhengzhou, Henan, but had to flee when rival warlord armies clashed in Henan during the fall of 1924. Upon returning to his hometown, Changzhou, in 1925, Yun Zhen met Yun Daiying, an early leader of the Communist Party, who suggested Yun take up a teaching position at Sun Yat-sen's revolutionary base in Guangzhou. After objections from his elder brother, Yun decided against it. He instead secured a teaching position at Southeast University at Nanjing through another Young China Association member and arrived in Nanjing in 1925, just a few months before Sun Yat-sen's death.

Yun was away from China when Sun Yat-sen's successor Chiang Kai-shek staged the Northern Expedition to subjugate regional warlords between 1926 and 1928. The warlord regime that controlled Nanjing appointed Yun as an expert and sent him as part of Jiangsu Province's delegation to the Philadelphia

Sesquicentennial Exposition in 1926. Yun later remained in the United States after the Philly Expo to receive advanced training at the Dwight P. Robinson Company for one year.

Shen, however, stayed in the Lower Yangtze after completing his education and promoted electrical irrigation through his work in the power company. He saw the electrification of agriculture as the key to boosting China's national economy and made the first attempts to introduce electricity to rural areas in the Lower Yangtze. As the power line engineer for Qishuyan Zhenhua Power Station in Changzhou, Shen extended the power lines to nearby villages and introduced electrical water pumps to alleviate water shortages. He conducted a trial by connecting two 24-hp engines to the transmission lines to power pumps that irrigated 2,000 *mu* (329.5 acres) of rice fields. These experimental fields reported a bumper rice harvest. Within three years, the land area under electrical irrigation increased to 40,000 mu. Shen estimated that the cost of irrigating one mu of land each year came to 0.60 Chinese silver dollars, which was much less than that of rearing a water buffalo. Furthermore, because the power company owned the engines and pumps, farmers did not have to bear the cost of equipment.[25] The power company now appeared to take over one of the traditional roles of the gentry—the distribution of water resources.

Li Yanshi, who worked alongside Shen Sifang at Zhenhua, hailed from the ranks of the gentry. A native of Huzhou, Li studied German in Qingdao when he was fifteen and pursued a degree in anatomy at the University of Vienna in 1910, before he transferred to Berlin University to study law in 1911. He received his law degree in 1916 and returned to China to teach German at Qingdao University, Peking University, and Catholic University of Peking. He left his teaching post around 1924 after being appointed director of Zhenhua. Li supported Shen's efforts to introduce electrical irrigation to the fields around Changzhou. He resigned his post at Zhenhua in 1925. After working briefly at Ziliujing in Sichuan, Li returned to the Lower Yangtze as the director of the Nanjing Power Works. He relinquished this post in 1926 after being elected onto the board of directors at Woo-shing Electric in his hometown.[26] Shen was then called on to inspect the power equipment at Nanjing in January 1927 after repeated reports of generator breakdowns and excessively low transmission voltage across the system.

Zhenhua Electric and Nanjing Power Works, where Shen and Li had spent the early part of their careers, became targets of state acquisition. After capturing Nanjing from the warlord armies of Sun Chuanfang in March 1927, Chiang Kai-shek established the Nationalist government in Nanjing the following month. In February 1928, the Guomindang regime founded the National Construction Commission (NCC) to coordinate the development of transportation, hydrology, forestry, mining, and other modes of economic development. Yun returned to Nanjing in 1929, shortly after Chiang wrapped up the Northern Expedition upon securing the allegiances of major warlords. Zhang Renjie, one of the four leaders of the Nationalist Party and incidentally a descendant of one of the four prominent patriarchs of Huzhou, appointed Yun as an executive secretary at the Communications Bureau of the National Construction Commission.[27] In that year, the NCC placed Zhenhua Electric and Nanjing Power Works under the direct administration of the central government. In 1930, Yun took up the post of deputy director of the Electrical Industry Bureau. Sun Yat-sen's vision of nationalizing the electrical industries for the greater good was no longer just an idea on paper but was now a policy actively pursued by the Nanjing government.[28] With that, the two leading graduates from the class of 1921, who pursued diverging career paths, found themselves on opposite sides of the privatization versus nationalization debate.

Shen anticipated the Nanjing government's nationalization of the electric company in the new national capital. He had inspected the Nanjing Power Works in the early months of 1927 and detailed the rusty boilers and unrepairable generator blades in his report. Rampant electric theft had also caused the transmission voltage to drop to one third of the voltage setting.[29] Shen appeared hopeful that the nationalization would provide a necessary intervention to transform a failing entity into a model power station for the newly unified nation. The rehabilitation of Nanjing Power Works marked the first step of the Nationalist government's scheme to achieve "national, political, and ideological unity" through the transformation of Nanjing into a national capital that was clean, efficient, and offered the newest technology.[30]

While they thought that the nationalization of Nanjing's electric company was acceptable, Shen and Li accused the state of stealing the fruits of their labor by acquiring Zhenhua. The electrical irrigation program introduced in 1924 fit perfectly with the Guomindang regime's modernization programs, and Woo-shing Electric was reaping dividends from their efforts to electrify silk production. Having studied the silk industry in Japan, Shen and Li envisioned the introduction of the refrigeration of silk worm eggs, electrically powered incubators, electric driers, temperature control for silkworm breeding, and even electrical insect killers.[31] Shen and Li saw the need to mobilize private operators to guard against further state encroachment and prevent Woo-shing Electric from becoming the next acquisition target.

The revolutionary discourse in the early years of the republic opened up two separate paths to saving China through industrialization. The Guomindang regime's assertion of control over the commercial, financial, and industrial centers within its realm of control in the Lower Yangtze placed it on a collision course with the local elite. What was the role of the national government? How much autonomy should be granted to the local industries? The privatization versus nationalization debate that would unfold over the next five years continued to offer conflicting answers to these unresolvable questions.

Saving Private Interests

The central government's vision of "national salvation through industrialization" came into direct conflict with the economic interests of the local elite. The Nanjing government set out to fulfill Sun Yat-sen's plan to place the electrical power sector under state ownership, which allowed the state to allocate power to industries of national strategic interest. In response, power plant owners portrayed the NCC's attempts to consolidate the fragmented electrical industries as a covert operation to confiscate their assets and undermine local economic influence. Appropriating Sun Yat-sen's ambiguous revolutionary ideals, private power plant owners not only forced the central government to halt the nationalization project, but they influenced policy makers to devise a light-touch regulatory framework. Private power plant operators continued the push for greater deregulation by pointing to the experience of electrical utilities in Europe and America. They positioned themselves as protectors of the public interest by keeping electrical tariffs low by using market forces and pressing for safeguards for property rights.

The NCC's first statistical investigation of China's electrical industries painted a dismal picture. With no budget to conduct on-the-ground surveys across China, it pieced together fragments of data gleaned from provincial economic construction bureaus, registration materials transferred from the Ministry of Communications, and articles published in magazines and journals.[32] The investigation concluded that a small number of foreign-owned power stations dominated the largest electrical power markets in major cities, while hundreds of Chinese-owned power stations that served small county seats teetered on the brink of collapse. Small privately owned power stations lacked the capital to expand their capacity, as revenue from electric lighting and power for smallholding industries barely covered operating costs. All in all, 523 Chinese-owned power stations accounted for a mere 24.7 percent of the nation's generating capacity. Their market share was smaller than the thirty-five foreign-owned power stations, which amounted to 32.7 percent. One hundred and forty-nine factories installed their own power generators, which constituted the remaining 36.9 percent.[33]

Power stations across China acquired generators from different countries at different times, which led to the lack of uniformity of industrial standards. This piecemeal development resulted in a diverse fuel mix. In 1929, 57 percent of electricity was generated by coal-fired steam turbines, whereas 36 percent was from diesel. Diesel generators offered a quick fix for power stations to meet surging power demand quickly. After the Nanjing Power Works came under state control, the Electrical Bureau instantly relieved the power shortage by installing two 137 kW diesel generators, while waiting for steam turbines to be shipped from England and the pipes from the river to the power station to be laid.[34] Furthermore, diesel generators, unlike steam turbines, did not require water, so the power station did not have to compete with the agrarian economy for water access. There was also no nationwide voltage standard. The NCC only obtained alternating current frequency settings for 155 power plants across China—ninety-three were set at 50 Hz, while sixty-two at 60 Hz.[35] Even within Shanghai, the seven franchise areas under the control of two foreign-owned and five Chinese-owned power stations had different voltage and frequency settings. The International Settlement ran on 220 V/50 Hz, while the French Concession was set at 100 V/50 Hz. The hodgepodge nature of China's electrical industry reflected the fragmentary nature of China's national polity.

The Nanjing government lacked the resources to impose uniform standards. According to laws passed in 1929 and 1930, power stations, state-owned or private, were required to obtain an operating license with a franchise term of thirty years, and their rate schedules had to be approved by the commission.[36] The national government, however, failed to secure the compliance of power plant owners even within its sphere of influence in the Lower Yangtze region. The administrative hassle of overseeing a large number of small power plants littered across Jiangsu, Zhejiang, and Anhui Provinces proved to be daunting. In Zhejiang Province alone, out of the 112 private power plants with an average capacity of 77.5 kW, only twelve had registered with the NCC by December 1929.[37] The Electric Utility Regulation Board, which would only be established in 1932, was severely understaffed and its eleven commissioners had to cover the entire nation.

Right around 1929, private power plant operations began petitioning the government to impose a light-touch regulatory framework. Within a year of the NCC's founding, central, provincial, and municipal governments acquired Dayouli in Hangzhou, Zhenhua in Changzhou, and Pulin in Kaifeng. Private power plant operators denounced these takeovers as "confiscation" by state authorities. Private power plant owners leapt into action upon hearing that the provincial construction commission planned to acquire the local power station at Tai County in Jiangsu for one third of its asset value.[38] On July 19, 1929, representatives from thirty-eight power stations descended on Nanjing to deliver the petition to the Guomindang Party headquarters, Executive Yuan, Legislative Yuan, and the Construction Commission.[39] In the memorandum submitted to the central party leadership, the petitioners accused the party of violating company law by imposing arbitrary charges and intervening in personnel matters.

Power plant owners pleaded with the Nanjing government to safeguard their property rights by invoking Sun Yat-sen's writings. They pointed to specific passages in Sun's writings to assert that "the government should take over the public utilities, if the financial resources of the people proved to be insufficient to manage them properly," but "had no right to take over businesses that had already been established by its people." The state's takeover of Chinese-owned power plants thus amounted to covert plunder (*bianxiang lueduo*).[40]

The petitioning movement coincided with the drafting of the regulatory framework for China's electrical industry. Between July and December 1929, the signatories maintained a presence in Nanjing to push for six demands: (1) uphold the supervisory guidelines approved by Chiang Kai-shek in the second National Congress of the Guomindang; (2) ensure that regulators are limited to oversight and disallowed from direct management; (3) prevent takeover from provincial and municipal governments; (4) bar municipal governments from collecting franchise fees; (5) safeguard assets of investor-owned power stations and return nationalized power stations to the owners; and (6) waive the reapplication of operation licenses.[41] The collective action of the power plant owners led to the founding of the Association of Private Chinese Electric Corporations (*Quanguo minying dianye lianhehui*) in December 1929. The association listed a membership of 120 power stations from ten provinces, thereby making its claim to be an organization with national standing.

The association's name also exemplified how business owners appropriated the bureaucratic categories defined by the Guomindang regime to secure their business interests. The term *minying* literally means "operated by the people" and was used to denote entities whose capital came from the people rather than the government. By collectively referring to themselves as minying entities, power station owners imbued new meaning to the term. Unlike the terms *shangban* (merchant-run) or *siying* (privately run), which carried the connotations of profiteering, the term minying conveyed the idea that a power station was not just another business but an enterprise that served the public interest.[42]

The petitioners not only were granted five out of the six demands, but they also compelled the Nanjing government to draft regulations that sheltered them from competition. Passed by the Legislative Yuan in December 1929, the "Privately Owned Public Utility Regulation Law" required electrical utilities to comply with standards for financial reporting and safety but relieved their tax burden and safeguarded their profits. Power companies bore a tax burden of less than 1 percent of revenue. The only power company subject to a royalty tax was the foreign-owned Shanghai Power Company, which handed 1 to 5 percent of gross revenue to the Shanghai Municipal Council.

The Electric Service Code passed in 1931 further enshrined the principles of ensuring the economic viability of power plants large or small. Power stations with generating capacity higher than 10,000 kW were classified as Category 1, 1,000 to 10,000 kW as Category 2, 100 kW to 1,000 kW as Category 3, and lower than 100 kW as Category 4. Power stations with greater capacity had to supply electricity for more hours and meet higher service standards. For example, Category 1 power stations must supply electricity around the clock, whereas Category 4 power stations were required to provide power for six hours each day. The policy makers spelled out the rationale in the draft of the bill:

> From the users' perspective, a continuous supply of electricity day and night is optimal. From the power stations' point of view, it is most economical to operate when there is demand and cut off power when there is none. In places served by small power stations, it is difficult to maintain daytime electrical power and keep the lights on late at night. If the regulations are too stringent, these power stations will incur losses. This is not the purpose of this legislation.[43]

This four-tier categorization system benefited small and medium power companies like Woo-shing.

The NCC would later elaborate on the rationale in 1936 at the World Power Conference in Washington, DC. It explained that the policy makers hoped to "attract capital to this field" with "a much more liberal rate of return," which meant that "if an electric utility manages itself in an efficient manner and charges its customers with reasonable rates, it is entitled to enjoy a rate of return on the capital as high as 25 per cent."[44]

The tacit approval of the Guomindang leadership could have led to this highly favorable outcome. The inaugural issue of the association's journal reveals the political connections of these power plant owners. The list of authors who penned dedications for the inaugural issue of the association's journal reads like a who's who of Republican China. They included T. V. Soong (Song Ziwen), the brother of Madame Chiang and minister of finance, and Lin Sen, then vice president of the Legislative Yuan.[45] Sun Yat-sen's "Last Testament" graced the inside cover. The association's vice chairman, Guo Zhicheng, who was then general manager of Dazhao Electric Company in Zhenjiang and a former army general, alluded to Chiang Kai-shek's policy of pacifying the interior before resisting foreign aggression in the preface to the inaugural issue. Guo accused "outside threats" of undermining the accomplishments of the Chinese-owned electrical power sector.[46] He was not referring to foreign competitors, but warlords who refused to pay their electric tariffs, thereby hinting that the Nanjing government would be regarded as another extractive warlord regime if it pressed forward with its nationalization plans.

Shen Sifang and Li Yanshi, who had seen the Zhenhua Power Company taken over by the state, continued to press for deregulation even after the association's initial victory. Shen signed the petition on behalf of Woo-shing Electric and neighboring Jiaxing Electrical Company. He was also elected as a member of the executive committee along with Li. Laying out the association's vision for the next twenty years, Shen Sifang argued that nationalization was detrimental to the long-term development of the power sector. He noted, "While it might have been necessary for the government to take over one or two power stations and administer it as a model for others," state acquisition, however, "discouraged investment in the electrical power sector and made it more difficult for private power plant owners to secure loans."[47]

Basking in the success of their petitioning efforts, Shen Sifang and Li Yanshi represented China as official delegates to the Second World Power Conference in Berlin in 1930. They toured the power stations of Manila and Singapore during their stopovers on their journey to Berlin. They joined more than 3,900 delegates from thirty-four nations at the Berlin conference and enjoyed a rich conference program that featured 392 papers covering about eighteen topics. Following the conference, they took part in a tour of the Ruhr Valley, where they visited the manufacturing facilities of Siemens and Brown Boveri & Company and saw generators several hundred times the size of those used in Huzhou roll off the assembly line. After returning from the conference, Shen and Li chastised the Nanjing government for failing to pay enough attention to the largest professional conference for the electrical industries in the world. Two Chinese engineers residing in Germany and one division chief from the Ministry of Commerce were roped in to bolster China's ranks. The British, Americans, French, and Japanese sent hundreds of delegates. The Chinese delegation had sent only fifteen members.[48]

Shen and Li continued to press for greater protections of property rights from the Nanjing government upon their return. In their joint report to the association's members, both men noted that the nationalization versus privatization debate had emerged after the end of World War I but claimed that most of the conference delegates representing electrical corporations in Western Europe and America favored privatization. Li Yanshi, who received his law training in Germany, pointed out that public-private partnerships with the municipal authority taking a minority stake was the norm in Germany's electrical industries. He believed that this arrangement was viable because the German constitution safeguarded property rights, which ensured that private industries would receive a fair share for services rendered to the state. He reaffirmed this view after a conversation with a German electrical expert who had surveyed China's electrical industry in the 1920s, and who told him that public-private partnership would not work in China due to the lack of proper governance and political instability. Li then concluded that privatization was the optimal path for China's electrical industry.[49] Li also criticized the Nanjing government's onerous registration requirements. For example, it was a requirement for Chinese power plant owners to renew their operating licenses every thirty years. According to Li, Britain had long since abolished a similar requirement.[50]

Drawing on Shen and Li's expedition report, the association continued to push a simplistic argument that government-run power stations were less cost efficient than private ones. In a speech titled "What Is the Purpose of National Salvation through Industrialization?" at the 1933 annual meeting at Hankou, Li Yanshi delivered his most scathing criticism on nationalization. He claimed that private power stations were better able to preserve their capital, as they optimized their returns to investments and did not succumb to political pressure. In the same year, Dong Shu published electric tariff figures showing that municipal power plants in the United States charged 36.6 cents per unit of electricity, whereas private power plants set the price at 25.7 cents per unit, and in places such as Arizona, the price per unit of municipal power plants was five times that of privately run utilities.[51] These simplistic arguments ignored factors that might influence the pricing, such as variations in power consumption patterns, fuel costs, and scales of production. Private power plant operators cherry-picked their cases to reinforce their argument that nationalizing the power sector meant that consumers would have to bear higher costs.

The association's argument that state-owned power companies were less cost effective than private ones was contrary to the situation on the ground. Power stations administered by the national government charged a lower tariff than private power stations. With a generating capacity of 17,100 kW, Qishuyan Power Station, formerly known as Zhenhua Power Station, charged an average of 2.13 cents per kWh for bulk power supply. Tiansheng Port Power Station operated by Dasheng Cotton Mill in Nantong was six times larger but charged 2.65 cents per kWh for bulk power sold to a neighboring power station.[52] Smallholding power stations also enjoyed monopolistic rents, as the state guaranteed not only exclusive rights in their franchise areas but also a generous rate of return.

Chinese electric corporations in the Lower Yangtze became the greatest beneficiaries of the state's retreat from nationalization. Zhenhua and Nanjing Power Works became the only ones acquired by the Nanjing government. Between 1928 and 1937, no additional power stations came under government control. In addition to having a lack of funds in the national treasury, the Nanjing government aborted its nationalization plans due to the political backlash from the local gentry who saw the state takeover of local power stations as an affront to their authority. Mines and factories with self-supplied electricity also built power lines to sell excess power. One example was Changxing Coal located within the boundaries of modern-day Huzhou established in 1931. It started out as a power plant to fuel mining operations but built transmission lines to nearby towns to power indoor and street lighting.[53] Between 1928 and 1932, the electrical industries in the Lower Yangtze provinces of Jiangsu, Zhejiang, and Anhui reported an increase in the capital stock by 44.53 percent, 75.71 percent, and 6.35 percent, respectively. The number of privately run power stations in Zhejiang remained unchanged at 110. Despite being sheltered from competition, market share of Chinese utilities inched up slightly from 37.9 percent in 1932 to 40.5 percent in 1934.[54] The "Rules of Standardization of Frequency and Voltages" barely had an impact. Four years after the NCC mandated 220/380 V and 50 Hz as the mains voltage and frequency, the adoption rate inched up from 81.6 to 82.4 percent in 1934.[55] Even if the power companies violated the service code, they faced a maximum fine of one thousand Chinese dollars, which amounted to a fraction of a medium-sized power company's daily revenue.

The light-touch regulatory framework limited the enforcement powers of the Electric Utility Regulation Board. State regulators lacked the capacity to prosecute electrical theft or help power companies recover arrears of electrical charges, and power companies were unable to depend on the state to protect their property rights. Two years after railing against the evils of nationalization, Li Yanshi and Shen Sifang sought state intervention when their largest customers refused to pay their bills.

Restoring State Control

As the high tide of laissez-faire economics receded, state regulators seized on the opportunity to impose some semblance of order in the fragmented electric marketplace. Having fueled its growth through the sale of electricity to silk firms, Woo-shing Electric was reeling from the effects of the sharp downturn in silk exports during the Great Depression. A long-standing dispute between Woo-shing Power and Huzhou's silk weavers finally came to a head in August 1935. The trade association representing the local silk industry called on members to withhold payments to the power company. Exploiting loopholes in the service code, Woo-shing Electric retaliated by cutting off power to these local silk firms. The Electric Utility Regulation Board resolved the dispute by imposing a standardized pricing system to remove confusions arising from private discount agreements. Developments in China mirrored the trends unfolding across the world during the Great Depression, with central governments taking more proactive roles in infrastructure development. The program of the Third World Power Conference in Washington, DC, in 1936, which stood in stark contrast to the Second World Power Conference in Berlin, reflected this shift in mentality. As early as 1929, its organizers called for the return to a "general type of program which had characterized the First World Power Conference," which focused on national policy and planning issues.[56] The program directors saw "public education and discussion" as the key to understanding fundamental policy questions surrounding electrical infrastructure development in each nation and placed greater emphasis on topics such as national and regional planning, rationalization of gas and electric distribution, and public regulation. While private entrepreneurs made up most of China's delegation in 1930, government officials from the Electricity Utility Regulation Board, some of whom mediated the Huzhou power dispute, served as spokespersons for China's electrical industries in Washington, DC, in 1936.

Owners of Huzhou's largest silk mills, who had provided capital to the power station and served as its largest customers as early as 1917, felt betrayed when they learned that the power company had granted a deeper discount on their electric tariffs to newly established rice mills. The Confederation of Machine-Woven Silk Producers led by Huzhou's largest silk firm Dachang aired their grievances on the front page of the local newspaper on August 6, 1935. It alleged that the silk industry's support for electrification has gone largely unrewarded. Dachang's owner, Niu Jiechen, had supported the power company's expansion in 1921 but never received the preferential discounts as promised. Falling silk prices between 1927 and 1930 pushed Huzhou's silk industry to the brink of collapse. Niu pleaded for lower electric tariffs. In 1929, Woo-shing Electric and Dachang, the largest silk mill at Huzhou, agreed to a "discount of 12.5 per cent of the nominal rate" on the condition that silk mills "installed [the] newest machinery purchased by the [power] company," which complied with electrical power standards.[57] Two years later, Niu Jiechen learned that the power company had offered a 50 percent discount to newly established rice mills. As silk output began falling, Woo-shing Electric looked toward the eleven rice mills in Huzhou to soak up excess supply. On average, one rice mill consumed 16,723 kWh per month—four times the mean of one silk firm.[58] Woo-shing Electric tried to placate the silk industry by offering them the discounted eight cents per kWh rate granted to rice millers.

Li Yanshi and Shen Sifang, who had once accused the Nanjing government of profiteering from state acquisition, came under attack from their consumers for predatory pricing. The silk weavers wanted the power company to honor their preferential customer status, which meant that silk firms should pay seven cents per kWh and receive rebates for overcharged amounts backdated to 1929. On August 7, Woo-shing Electric flatly rejected these demands and published a fifteen-point rebuttal to the trade association's statement, as "charges if left unanswered could create serious misunderstanding." The power company denied Dachang's claims for privileged status. There was no written proof that the power company recognized the silk weavers as a "preferential customer." The co-founder, Guan Zhiqing, who had made the verbal agreement with Niu Jiechen, was long dead. The silk firms rightly pointed out that Huzhou charged a higher tariff. At eight cents per kWh, the rate was three times that of Nantong. The power company defended its pricing policy by pointing out that it faced higher costs on generating equipment and interest on outstanding loans.[59] Furthermore, the power company incurred high transaction costs, as it had to go through the trouble of collecting small sums of monthly dues from more than one hundred silk firms. To add insult to injury, a number of large silk firms stole electricity by making illegal connections to the grid.

Woo-shing emphatically refuted allegations of fraud and accounting irregularities. They accused silk manufacturers of spreading misinformation. Diesel prices did not fall by 40 percent as claimed. They also denied inflating meter readings by 20 percent during routine meter changes or confiscating the safety deposit, claiming that the silk manufacturers had conflated the refundable deposit with the nonrefundable meter maintenance fee. The electric company also maintained that they were simply following common practice by collecting a streetlight fee from all its residents.[60]

Consumers also seized on this moment to voice their concerns over the power system's safety record. In April 1935, an electrical system malfunction set off a fire at Dachang. The electric company and silk factory could not agree who was at fault. Woo-shing Electric insisted that the silk factory was negligent as it overloaded its internal circuits, but the silk factory blamed the fire on the shoddy connections between the grid and factory. Representatives of the silk industry contended that power outages had become much more frequent, to which the power company explained that they had to cut off power to carry out routine summer maintenance.

Woo-shing Electric then masterfully used the laws on the books to break the deadlock. It started out by invoking the rules that the NCC enacted to prevent power companies from falling into insolvency through cut-throat competition and overspending. According to Article 12 of "Regulations Governing Privately Owned Industries," electric tariffs could be reduced only when net profits exceeded actual capital receipts by one quarter.[61] Power companies were also required to place half of its profits into a reserve fund. Put simply, the financial difficulty not only made it impossible, but also illegal, to lower prices.

After ten days of loudhailer politics through the local media, the power company invoked an article from the Service Code and announced that it was cutting off power during the most productive hours of the silk firms. According to Article 42, Woo-shing Electric, which was classified as a Category 2 power company, was obligated to generate electricity for eighteen hours a day. The power company acknowledged that it "violated the regulations and voluntarily terminated twenty-four hour supply of electricity." It decided to cut off power from 6 am to noon each day, the time during which silk filatures went about most of their work. State regulators had to step in before the electrical blockade inflicted real damage on Huzhou's fragile economy.

When the power blockade took effect on August 15, the county government immediately summoned all parties involved to an emergency meeting. The former county head Dai Shixi chaired the meeting and put forth three suggestions: (1) silk weaving and machine weavers will be charged a discounted rate of 7.6 cents per kWh; (2) the silk industries should settle all outstanding payments by the end of September; and (3) the power company should complete the revision of its regulations as soon as possible.[62] The county government thought that the silk industry was concerned about price levels but sidestepped the "preferential status" question.

The dispute became an opportunity to regularize electric pricing. Zhu Dajing, a section chief from the Construction Commission, took charge of the investigation. He had dealt with safety complaints in the Hangzhou-Xiaoshan area in 1933 and traveled to Beijing with bureau chief Yun Zhen to settle arguments between the power company and streetcar operators in 1934.[63] The case in Huzhou was much more complex than these earlier complaints, as it involved longstanding local disputes. Zhu ultimately determined the private agreement between Dachang and Woo-shing Electric to be the crux of the problem. The NCC could not enforce a private contract, so both parties had to resolve their disagreements through a civil suit in the courts. He made it clear that ad-hoc agreements generated distrust and confusion between the producers and consumers but acknowledged that the electric company had the flexibility to grant greater discounts to its largest customers. One month after the dispute, the NCC proposed the implementation of a multitier pricing system that offered discounts to consumers who used more electricity. The first 50 kWh was priced at 7.5 cents per kWh. The rate for the next 100 kWh of energy was reduced by one cent. Energy in excess of 500 kWh cost 3.5 cents per kWh.[64] The policy had the combined effect of relieving the burden of the largest power consumers among Huzhou's silk industries and introducing some regularity to electrical pricing. These mediation efforts in addition to the promulgation of industry guidelines suggest a shift from the light-touch regulatory approach of the early years to one that was more interventionist.

The regulators working for the NCC had the opportunity to compare China's regulatory framework with other nations when they attended the Third World Power Conference in Washington, DC, in 1936. Chinese authorities appeared to guarantee an excessively high rate of return and tended to side with power companies during pricing disputes. The report by John E. Zimmerman, the president of United Gas Company in Philadelphia, identified "reasonableness of utility rates" as the guiding principle for rate regulations. State commissions have held 7 to 8 percent return on the value of property at the time of the rate of inquiry to be reasonable, but that "more than 8 percent would justify rate reduction."[65] By contrast, power stations in China were obligated to reduce rates and expand capacity only after crossing the "25 percent rate of return on capital" threshold. State-commission regulators in the United States also had price reductions that had saved ratepayers millions of dollars every year. Zimmerman claimed that state commissions had been "generally favorable to utility consumers" and one state commission even asserted that "90 to 95 percent of its decisions were in favor of the public."[66] In their report to the Third World Power Conference, the NCC's delegates made no mention of pricing disputes happening in China and simply outlined the responsibility of the central and local governments in electric utility regulation. It was not known if the American examples presented at the conference shaped

Chinese regulators' perception of what constituted an acceptable rate of return. The Chinese engineer-bureaucrats who reported on the progress of China's electrical industry modestly stated that it was "still in its infancy," and only a handful of publicly owned power stations "can be considered as satisfactory."[67]

The Capital Power Works in Nanjing and Qishuyan in Changzhou faced greater financial pressure than their privately owned peers. Unlike the privately owned electric companies that could postpone capital acquisition until they achieved high profitability, state-owned plants had to expand their operations even though they were unprofitable and were obligated to maintain low tariffs. The generating capacity of the Capital Power Station in Nanjing quintupled within a decade of the state takeover and its electric tariff was half that of private power companies in the region. Its boilers and turbines were even put up as collateral for loans to stave off bankruptcy. The Qishuyan Power Plant in Changzhou maintained a load factor of 58 percent, comparable to the Shanghai Power Company, but faced a liquidity crunch as ratepayers failed to keep up their payments. Despite its financial struggles, it went ahead with plans to construct a secondary plant in Shanghai. In the end, state-owned power stations succumbed to the same market forces that led to diminishing profit margins across the fragmented electrical power sector.

THE ELECTRIFICATION OF Huzhou's silk industry illustrates how the everyday politics of power consumption shaped the dynamics of nation-building during China's transition from empire to republic. Huzhou's gentry used the electrically powered jacquard loom as a tool to reassert their economic and political dominance. Failing to secure political influence through the constitutional reforms of the late Qing and support of the Nationalist Party in the 1910s, the gentry of the Lower Yangtze focused on protecting their business interests. Facing competition from mechanically woven silk from foreign silk firms, the Huzhou gentry embarked on the mechanization of silk fiber and cloth production to save themselves from financial ruin. Having secured the market without assistance from national political institutions, Huzhou's industrialists resisted the intervention of the Guomindang regime that was headquartered in the Lower Yangtze after 1927. They seized the initiative to shape the regulatory framework through an artful combination of resisting nationalization and appropriating the rhetoric of national salvation, thereby reshaping state-gentry relations.

Well aware of its limited capacity, the Nanjing government scaled back its ambitious plans for nationalization and opted instead for incremental reforms through broad policy directives, which in turn provided favorable conditions for power producers, consumers, and regulators to adapt to a market that was constantly in flux. The NCC's regulatory approach to the electrical industries reflected what Yuen Yuen Ang described as "the three key problems of adaptation inherent in the processes of variation, selection, and niche creation."[68] The electrical service code enacted in 1931 can be seen as an example of a "franchising mode of decentralization." The widespread compliance to registration and financial reporting requirements by 1934 suggested that private power plant owners recognized the importance of these regulations in legitimizing their operations. The Nanjing government also relied on market forces to eliminate weak players. The regulators used the four-tier categorization system as a tool to focus their supervisory efforts on large power stations that had a greater impact on industrial output. They flexibly adjusted the definition of success for their bureaucratic agents, once they recognized the impossibility of nationalizing the entire power sector. Instead of ordering the shutdown of failing power stations, the NCC stealthily expanded the franchise area of Qishuyan Power Station into areas with power demand. The Capital Power Works served as a benchmark, which pressured private operators to lower prices and improve system reliability.

While the adaptability of the Guomindang regime inspired institutional resilience, it led to systematic weaknesses across the electrical power sector. Regulators allowed smallholding power stations to operate with no interventions, as long as they turned in their paperwork and did not cause any accidents. In their bid to guarantee high rates of return and lessen debt burden, regulators disincentivized power stations from undertaking capital expansion. The gentry who owned these smallholding power stations distributed their profits (if any) as dividends, instead of investing in new generators and extending power networks. The residual inefficiency resulting from the persistence of smallholding power plants constrained industrial growth in the Lower Yangtze.

The electrification of Huzhou's silk industry also reveals how global economics penetrated local politics. Industrialists and government regulators alike looked toward the experiences of other industrialized powers for shortcuts to accelerated economic growth. The Huzhou gentry pioneered the mechanization of silk filatures by importing machinery and replicating techniques of their foreign rivals. This not only quickly reversed the declining fortune of the local silk industry but also catalyzed the expansion of Huzhou's electrical infrastructure. Shen Sifang and Li Yanshi of Huzhou's power station also encouraged the local silk industry to emulate the Japanese example of electrifying multiple processes between silkworm breeding to silk weaving. The high growth potential of Huzhou's electric market motivated Shen and Li to participate in the petitioning effort against nationalization. The Association of Chinese Electric Utilities pushed for greater deregulation by drawing on the information gathered from its participation in the 1930 World Power Conference in Berlin. It based the argument that privatization leads to lower electric tariffs on simplistic comparisons of price data between American and British power companies. It also glossed over the fact that municipal governments played a coordinating role in power distribution. While Shen and Li returned from the 1930 Berlin conference fully convinced that privatization would lead to higher demand and lower electric tariffs, Chinese government officials who presented papers at the 1936 Washington conference learned how state-led electrical infrastructure projects facilitated the efficient allocation of natural resources.

The electrification of silk production in Huzhou suggests the existence of a different possibility for China's transition from the organic to the carbon economy. Silk is a product of an organic economy. Unlike cotton mills that could increase their output by importing fiber from abroad, the scale of the silk industry in Huzhou and elsewhere was ultimately constrained by the growth cycle of the locally grown silkworm. Woo-shing Electric accelerated the processing of the raw materials of Huzhou's organic economy but did little to increase the output of raw silk. Like the hundreds of small power stations littered around Jiangsu and Zhejiang, Woo-shing Electric remained an isolated electric utility serving a small local area. Even then, smallholding power stations could have developed into hundreds of community microgrids scattered across the Lower Yangtze and might have had the potential to serve as the foundation for a decentralized power distribution system. Unlike macrogrids that suffer from power loss when they transmit electricity over long distances, small-scale power networks reduce energy wastage by responding dynamically to shifts in local demand. The destruction unleashed by the Japanese invasion in 1937 closed off this alternative model. The fire and fury unleashed by wartime mobilization forced the Japanese and Guomindang regimes to place electrical industries under centralized state control.

In his opening address to the 1936 Third World Power Conference titled "Power and Culture," Lewis Mumford seemed to have foretold the tectonic shift of China's electrical power sector. The Chinese delegates present at Constitution Hall must have heard Mumford utter these words:

> One of our American historians, Henry Adams, predicted that during our generation the world would change from a mechanical to an electrical phase, with a vast increase in energy and an acceleration of all social

processes. He pointed out that this might mean either the speedy destruction of our civilization, or the building up of a new world on a different basis.[69]

One year after attending the conference in Washington, DC, these Chinese engineer-bureaucrats would devote themselves to building a wartime power network under the threat of enemy bombardment. The Japanese and Guomindang regimes would reconfigure China's energy geography as they harnessed her resources for all-out war.

CHAPTER 3

Unleashing Fire and Fury

IN THE SPRING OF 1936, Sun Yun-suan, a twenty-three-year-old electrical engineer with the Lianyungang Railroad Bureau Power Station in Jiangsu Province, caught the attention of Yun Zhen, the director of the Electrical Bureau, with a passionate essay calling for the nationalization of the electrical industries. A graduate of the Sino-Russian Industrial University in Harbin, this young engineer-bureaucrat would later rise through the ranks of the Republic of China government in Taiwan and would serve as premier from 1978 to 1984. In the spring of 1938, as Japanese troops advanced upstream on the Yangtze River toward the provisional capital Wuhan, Sun and other engineers of the National Resources Commission (NRC) were tasked with transporting hundreds of tons of power equipment into the inland provinces to set up the electrical infrastructure that would power the defense industries. Sun rushed to Lianyungang in Jiangsu Province, 128 miles east of Xuzhou where Japanese and Chinese forces faced off from January to June 1938. He oversaw the transportation of two 500 kW boiler-turbines 1,200 miles west to Ziliujing in Sichuan. Facing the fire and fury unleashed by the Japanese Imperial Army backed by a massive industrial base, the electrification of China's interior provided the spark that kept the flame of Chinese resistance going. With the Japanese capturing 97 percent of the nation's generating capacity, engineer-bureaucrats who retreated to Southwest China shouldered the responsibility of building an electrical infrastructure with a resilience to withstand the vagaries of war.[1]

The 1941 map titled "The Final Days of the Anti-Japanese Resistance" sent a message showing that the efforts of engineer-bureaucrats serving the Guomindang regime were futile (see figure 3.1). It taunted senior Guomindang (Chinese Nationalist) party officials and capitalists for fleeing the destruction of Shanghai and Nanjing. It depicted Chiang as a coward hiding away in Southwest China as the Japanese conducted massive air raids over the wartime capital, Chongqing. In contrast, North China, which had been incorporated into the East Asia Co-Prosperity Sphere, was beaming with hope.

FIGURE 3.1. The Final Days of the Anti-Japanese Resistance. Propaganda map from 1941. In this map, an electrified North China radiates hope, while the lower Yangtze descends into darkness with Chinese capitalists and Chiang Kai-shek's Nationalist regime fleeing to safety in the Southwest. Source: P. J. Mode Collection of Persuasive Cartography, Cornell University Library.

The map also illustrates the reconfiguration of China's energy geography in the wake of the second Sino-Japanese War (1937–1945). The war led to a reversal of fortunes between the Lower Yangtze and North China, and cities like Shanghai and Nanjing, home to some of China's largest electrical utilities, suffered massive disruption during this time. The Japanese increased the capacity of North China's electrical industries by taking advantage of its proximity to major coal deposits, as they went about transforming North China into an industrial base for its defense industries. The engineer-bureaucrats, who retreated with Chiang to Southwest China, were building the electrical infrastructure from the ground up as Japanese bombs rained down on Chiang Kai-shek's inland bases.

This chapter traces the development of three wartime electrical power systems that emerged during the second Sino-Japanese War. They are the North China

Electric Corporation (Jp. *Kahoku dengyō kabushiki kaisha*) headquartered in Beiping; the Central China Waterworks and Electricity Company (Jp. *Kachu suiden kabushiki kaisha*) based in Shanghai; and the Lakeside Electrical Works, hastily constructed and completed in June 1939 in Kunming, Yunnan Province. It draws on archival records to build on earlier research completed by Tajima Toshio's research team at the University of Tokyo and the economic historian Kanemaru Yūichi.[2] All three cases not only illustrate how the exigencies of war forced China and Japan to bring the electrical industry under direct control but also examine the dynamics that pushed China toward a carbon-intensive path of development. By the end of the war, China's electrical industries were no longer concentrated in Shanghai but extended outward into other parts of the nation—a trend that continued after the founding of the People's Republic in 1949.

The mobilization for war caused both China and Japan to revive aborted plans to nationalize the electrical industries and catalyzed the growth of coal-fired power stations. The turn to coal happened in North China with its rich coal deposits and also in Southwest China with its coal deposits of low combustibility. The most obvious benefit of coal-fired power is that the cost per kWh for coal was much lower than for hydropower.[3] In addition, wartime regimes took advantage of the modularity, scalability, and calculability of the coal-fired power systems. Both Japanese occupation forces and the Guomindang regime went about vertically integrating coal mining with power generation toward different ends. The centralization of coal distribution allowed the Japanese to cut down on the cost of fuel transportation. They encouraged the expansion of North China's regional power system by generating electricity closer to the fuel source, while they strictly curtailed the supply of coal to the Lower Yangtze to save costs. In the case of Lakeside Electric Works in Kunming, the state acquisition of the coal mines was aimed at boosting and securing coal supply. Unlike wartime Germany and the United States, where, according to Thomas Hughes, the government funded the "development of power plants of unprecedented size," wartime electrification in China saw the concurrent expansion of centralized regional networks in North China and the growth of small coal-fired power systems that could be shut down and redeployed in response to the demands of warfare in Southwest China.[4]

Gearing up for Total Warfare

Years before the outbreak of the second Sino-Japanese War in July 1937, the national governments of China and Japan attempted to divert energy resources toward national defense by asserting direct control over the electrical industries.

In anticipation of all-out war against Japan, the Guomindang regime in Nanjing established the National Defense Planning Commission in 1932, which was then reorganized as the NRC in 1935. Subsumed under the Military Affairs Commission, the NRC spearheaded wartime industrial mobilization efforts.[5] Securing electricity for its defense industries was among the NRC's top priorities. In the meantime, the Japanese government began pushing for the nationalization of its electrical utilities, while its national policy companies surveyed the electrical infrastructure of major cities across China. The resistance to nationalization in China and Japan broke down when armies from both countries clashed after July 1937.

The Guomindang regime's engineer-bureaucrats had always been aware that the low output of China's electrical industries severely limited its industrial capability and placed it at a disadvantage against Japan. All of Japan's cities and more than 80 percent of its countryside were electrified.[6] At 1,670 million kWh, the annual electrical output of China's electrical industry in 1929 was only 15.8 percent that of Japan.[7] Chen Zhongxi, the director of the NRC's Electrical Power Bureau, explained the implications of this energy gap by converting electrical energy output into man-hour units. Taking 170 kWh to be equivalent to the labor output of one worker in one year, Japan's production capacity in 1929 would be equivalent to 62.11 million workers per year, whereas China's was 9.8 million.[8] The NRC foresaw the loss of the coastal cities to the Japanese in the event of war. Planning for the scenario of the nation's electrical power capacity falling into Japanese hands, the Ministry of Economic Affairs presented a five-year plan in 1936, outlining plans to construct state-sponsored heavy industries inland.[9] As revealed later in this chapter, the Guomindang regime executed too little of the plan too late.

Even before the Nanjing government began planning for the eventuality of war, private and state-backed Japanese companies had already plotted the incorporation of China's power sector into their imperialist enterprise. As early as 1935, the economic investigation unit of the Southern Manchurian Railway Company completed a 185-page report containing a comprehensive list of power stations in China. Designated as a top-secret document, its table of contents offered a chilling preview of the Japanese path of invasion, starting with Beiping and Tianjin in the north, sweeping across the North China plain, then south toward the Yangtze Delta.[10]

The Japanese investigators built on the National Construction Commission's statistical survey of China's electrical industries and gathered additional operational details from a few important power stations. The 1935 report reprinted figures on the capital and generating capacity of individual power stations from public Chinese government records. It included other information withheld from Chinese state regulators. The entry about Beiping Chinese Merchants' Electric Company revealed a shocking level of specificity. It provided a complete inventory of the six turbines, listing the date of purchase, power output, country of origin, and manufacturer. Data on daily coal consumption and coal price and the layout of the power station's facilities were also included.[11] Such technical details facilitated the subsequent Japanese takeover. Ide Taijiro, a Japanese engineer who visited North China in 1938, noticed that Chinese power stations seldom retained drawings of their power equipment, as their foreign suppliers took charge of all the repair work.[12] The Japanese ended up taking all the power stations listed in the report with the exception of those in Yunnan and Guangxi.

The survey provided much-needed information for Japanese power companies to make inroads into China's power sector. In the spring of 1936, the five largest electrical power corporations of Japan, namely Tokyo Electric Light Company, Nippon Electric Company, Ujikawa Electric, Daidō Electric, and Tōhō Electric, set aside their business rivalries and conducted a joint investigation in North China. As early as August 1936, Kōchū Kōshi, a subsidiary of the Southern Manchurian Railway Company whose name translates into Revive China Company, joined hands with the Tianjin Municipal Government to establish the Tianjin Power Company, which would become the largest power station in North China. On December 8, 1936, the Big Five founded the North China Electrical Industries Corporation (Jp. *Hoku-shi denryoku kōgyō kabushiki kaisha*), with a start-up capital of 5 million yen evenly shared among these five companies.[13]

While the Big Five plotted their expansion into the China market, they learned about the Japanese government's plans to nationalize the electrical industries. As was the case in China, Japanese power companies accused the government of attacking private property rights. In December 1936, representatives of the Japan Electric Association argued that Nazi Germany halted the nationalization of electrical power and recognized that privatization led to lower electrical tariffs.[14] Laura Elizabeth Hein noted that Tanomogi Keikichi, the communications minister of the Hirota cabinet, offered a compromise in January 1937 that would have allowed the Japanese government to "gain managerial control without seizing ownership." Even with support from Suzuki Teiichi, a lieutenant-general in the Japanese Imperial Army in charge of economic planning, Tanomogi failed to garner enough support for the bill in the Diet and had to withdraw it.[15] Opposition to nationalization would soften after the war with China broke out in July 1937, and the nationalization bill was subsequently passed in March 1938.

The Guomindang leadership recognized the Tanomogi plan as a precursor to military aggression. In an editorial for *Sweat Blood Monthly*, the official mouthpiece for the Guomindang in Shanghai, Liang Mu wrote, "The nationalization of Japan's electrical-power industries would allow the military to wield supreme authority over the Japanese economy. The military would no longer be restrained by corporations and political parties."[16] The author urged the Nanjing government to respond by integrating the development of the electrical power sector with China's defense industries. Unknown to Liang Mu, the Japanese were hatching a plot to transform North China into their military-industrial base by harnessing its energy resources.

Electrification of Japanese-Controlled North China

A plan couched by economic planners in the Japanese client state of Manchukuo outlined a strategy to nationalize North China's electrical industries. Dated July 1937 but possibly drafted earlier, the "North China Economic Construction Plan" by the Manchukuo Ministry of Industries proposed to "establish North China into a military base" for the Japanese "within a short period of time." It called for "first and foremost the expansion and strengthening of electrical power corporations to obtain abundant and cheap electrical power."[17] First, the Japanese would establish electrical power bureaus directly administered by the Ministry of Industries. These power bureaus would be responsible for the maintenance, operation, and expansion of all power-generating facilities and distribution lines. Next, government agencies would sponsor the development of large-scale power stations that would drive down the average cost of power. The Japanese planners believed that they could "ignore the small power stations, as they will not be able to withstand the competition from the new [state-owned] power station and either merge with us or exit the market."[18] The Japanese also had to reverse a long-standing policy of "hydropower first, thermal power second" and aggressively exploit the coal deposits in North China. The state-owned power corporation would build large power stations (larger than 1,000 kW) near the major coal deposits and channel the electricity to urban centers with high energy demand.

The old capital, Beijing, renamed Beiping in 1927, saw its electrical power system reorganized under the North China Industrial Development Plan devised by the Japanese. In December 1937, the Japanese placed Beiping Chinese Merchants' Electric Light Company, which was then the largest power company in North China, under direct administration of its collaborator regime—the

Provisional Government of the Republic of China.[19] The North China Development Company, a Japanese national policy company, funded the purchase of its power equipment and materials.[20]

Private power corporations began supporting the takeover of North China's electrical utilities by the Japanese national policy companies and the military, as they recognized the alignment of corporate and state interests. With the exception of Ujikawa Electric, Japan's Big Five power companies sent engineers to North China. The North China Expedition Force divided the engineering contingent into five groups and sent the crew from Tokyo Electric Light Company along the Beiping-Wuhan railway, Nippon Electric Company along the Tianjin-Nanjing railroad, Tōhō Electric Company along Beiping-Suiyuan railroad, Daidō Electric to Taiyuan, and South Chosun Consolidated Power to other parts of Shanxi Province.[21] The engineering contingent served as key members of the military's logistical team. Private power corporations recognized the sacrifices of their engineers who placed their lives on the line. An entry in the company history of Tokyo Electric Company reads: "Due to supply restrictions, we had to rush old equipment from our Japanese homeland to places in need. We pushed ourselves to the limit to complete this painstaking endeavor. In the process, two members of our expedition team were killed in the line of duty."[22]

The Japanese seized control of the electrical utilities in North China through a combination of military action and strategic acquisitions. Between July 1937 and March 1939, engineering contingents attached to the Japanese military took over 15,594 kW of power-generating capacity mostly in medium-sized cities in the provinces of Hebei, Henan, Shandong, and Shanxi. The Tokyo Electric Light Company also raised the generating capacity of Shijiazhuang in Hebei, Zhangde, and Xinxiang in Henan by 4,500 kW.[23] By March 1939, the Japanese inherited 64,984 kW of generating capacity in North China largely intact.

Japanese electrical company executives, who initially resisted nationalization, came around to recognize the merits of having a national policy company coordinate the development of electrical infrastructure. Naitō Kumaki, a well-regarded China expert and vice president of Nippon Electric, voiced his ambivalence about the role of national policy companies such as Kochū Kōshi in electrification projects. He was of the opinion that government-backed corporations should limit themselves to public infrastructure projects and not interfere with the private market. Naitō, however, was hopeful that the national policy companies would speed up Japanese visions of remaking North China in the image of Manchukuo. He envisioned that electrical power alliances in North China would "spring into action swiftly to provide capital, technology and experience, and work hard to restore, consolidate and expand the power facilities."[24] Sogō Shinji, the director of Southern Manchurian Railroad, listed electrical power development as the first concrete achievement of the economic development plan.

Soon after the Japanese power companies worked alongside the military to take over power stations in North China, the once-aborted plan for nationalizing Japan's electrical industry was revived. In March 1938, the Diet approved the Konoe cabinet's plan to consolidate thirty-three power-generating companies into a special nationalized corporation named Hassōden that sold power to the transmission and distribution companies.[25] The organizational logic of Hassōden emerged from the cooperation between private enterprise and state policy companies that took place on the North China battlefield. The state coordinated fuel distribution and consumption, while power companies continued to generate income through the sale of electricity. In April 1938, the Big Five Japanese companies renamed their company in China the East Asian Electrical Industries Corporation (*Tōa denryoku kōgyō kabushiki kaisha*) to reflect its mission of building a new East Asian economic order. Fifteen more Japanese power companies took up a stake in this regional power company and increased its paid-up capital to 30 million yen.[26]

Having put in place an overarching corporate structure, the Japanese went about forming Sino-Japanese joint ventures to consolidate their grip on the North China power market. In a typical joint venture, the Chinese and Japanese each controlled half of the shares, while Chinese and Japanese officials shared equal representation on the management board.

The Tianjin Power Company, founded with joint Chinese and Japanese capital before the war, served as a springboard for collaboration in the electrical power sector. Naitō Kumaki envisioned Tianjin as the foundation on which the Japanese would build their electrical utilities in North China. The Japanese concession in Tianjin had already been home to Japanese cotton textile mills and iron works. The 30,000 kW of generating capacity that came online around February 1938 further consolidated Japanese dominance over North China's industries. Naitō projected Tianjin's average instantaneous power consumption in 1942 at 25,000 kW and expressed confidence that Tianjin's power company could keep up with demand.[27] In addition to the Tianjin Power Company, the Japanese also established other joint venture entities in North China, namely the East Hebei Electric Shareholding Company, Qilu Electric Shareholding, and Zhifu Electric Shareholding Company. Each corporation had a clearly delineated franchise area. The East Asian Electric Corporation bought up twenty-three large power stations previously owned by Chinese businessmen across North China.[28] For Chinese businessmen who had struggled to keep their smallholding power companies solvent, the Japanese capital injection threw them a lifeline.

On February 1, 1940, the Japanese founded North China Electric Company (Jp. *Kahoku dengyō kabushiki kaisha*), which became the first macroregional power corporation in North China. It directly managed power companies in Beiping, Tianjin, and Tangshan of Eastern Hebei, as well as power stations in Henan and Shandong under military administration. The minister of justice of Wang Kemin's Provisional Government, Zhu Shen, a graduate of Tokyo Imperial University, became its president. The vice president was none other than Naitō Kumaki. Co-owned by national policy companies and jointly managed by Japanese and Chinese business representatives, the new conglomerate mirrored Hassōden on the Japanese home base and Manchuria Electric in Manchukuo.

With paid-up capital of 177 million yen and operations spread out across Beiping, Tianjin, Tangshan, Jinan, Zhifu, Shimen, Kaifeng, and Xuzhou, North China Electric implemented prewar plans for large-scale power stations and long-distance power transmission lines in Hebei, Henan, and Shanxi.[29] Naitō advocated the construction of power stations near the coal mines and transmission of electrical power into the cities with high-voltage lines. This allowed the power company to cut down on the cost of transporting coal. The Japanese started by building the long-distance power transmission lines. One 77,000 V high-voltage trunk line between Tianjin and the port of Tanggu was completed in 1940, while another was completed between Tianjin and Beiping in 1942. In the meantime, they expanded the capacity of the power station at the Kailuan coal mines in Tangshan and completed the construction of the high-voltage power lines from Tangshan to Tianjin, Tanggu, and Beiping in 1944. Tajima Toshio aptly described this wartime power grid as the "prototype for the modern day Beijing-Tianjin-Tangshan power network."[30]

While fuel was abundant, the electrical industry in North China suffered from a chronic shortage of generating capacity. After taking over the 32,000 kW Shijingshan Power Station that supplied most of Beiping's electricity, the Japanese realized that the instantaneous power output of 18,000 kW was only a fraction of the listed power rating. They managed to increase it to 21,100 kW in May 1940 after some minor repairs.[31] Installing newly built generators from Japan would take several years. In the meantime, the additional generating capacity would have to come from somewhere within China. As the Japanese extended their control into Central and South China, some of the generating equipment in the Lower Yangtze would be transferred northward to alleviate this shortage. Ultimately, the rise of North China came at the expense of the Lower Yangtze.

Generating Shanghai's Electrical Crisis

The Japanese faced an energy crisis of their own making as they extended beyond the North China plain into Central and South China. They faced three major obstacles as they attempted to take over the power industries in the Lower Yangtze. First, the battle for Shanghai and Nanjing severely destroyed the electrical infrastructure of both cities. Second, even after rehabilitating the transmission lines and generating equipment, the Japanese-owned macroregional power company encountered resistance from the foreign concessions and its collaborator regime—the Reformed Government of the Republic of China. Finally, the high cost of transporting coal from North China to the Yangtze Delta made it expensive for the Japanese to provision fuel for the electrical utilities in Shanghai and neighboring cities. Naitō and Oshikawa Ichiro, the North China chief economic investigator for the Southern Manchurian Railway Company, offered a "rationalistic approach" to alleviate the problem—curtail coal supplies to Shanghai and transfer idle equipment northward. The Japanese were thus unable to exploit the industrial capacity of the Lower Yangtze, even after occupying the most electrified region in China, with 45.3 percent of the generating capacity in China proper spread out over the provinces of Jiangsu, Anhui, and Zhejiang.[32]

Several months after capturing Shanghai, the Japanese began planning the establishment of the Central China Waterworks and Electricity Company (Jp. *Kachu suiden kabushiki kaisha*), but soon realized the difficulty of managing the electrical power market in the Yangtze Delta. A team of twelve power company executives mainly from western Japan made a whirlwind tour of the Lower Yangtze between January 18 and February 6, 1938, shortly after the Japanese captured Shanghai in November 1937 and Nanjing in December 1937.[33] For reasons unknown, the "China expert" of Japan's electrical industries, Naitō Kumaki, did not participate in this expedition. Okabe Eiichi from Tokyo Electric Light Company, who happened to be the oldest delegate, assumed Naitō's duties. Other members received very little advance notice. Tsuji Hideo, a section chief from Ujikawa Electric, had five days to prepare for the trip. On January 17, 1938, Tsuji left Ōsaka for Nagasaki, hopped onto a ferry and linked up with the survey team in Shanghai three days later. The Japanese felt overwhelmed by the work required to rehabilitate the electrical power sector in the Lower Yangtze. Reporting his findings over lunch at Daitō Electric in Ōsaka, Tsuji's fellow delegate Ishikawa Yoshijiro exasperatingly asked: "What is to be done about Central China?"[34]

Shanghai was supposed to be the grand prize of the East China campaign. Ishikawa noted that the level of electrification in Shanghai's foreign concessions was comparable to any large Japanese city. Ishikawa's home base of Kyoto had a generating capacity of 100,000 kW, which supplied electricity to 1.107 million people spread out over 111 square miles, whereas power stations in Shanghai's foreign concessions had a combined capacity of 141,180 kW and transmitted power to 1.5 million people concentrated in 12.8 square miles.[35] In 1938, the Japanese were unable to seize control of the electrical infrastructure under American and French control in the foreign concessions as they sought to avoid confrontation with the Western powers.

The Japanese took on the task of restoring public services in 332 square miles of urban sprawl outside Shanghai's foreign concessions, which had mostly been damaged during the severe fighting in the Battle of Shanghai. Ishikawa felt an overwhelming sense of hopelessness when he surveyed the service areas of Zhabei Power Company in the north and Chinese Electric Company in the Old Chinese City in the south. Ishikawa reported that 70 percent of the residences and businesses were burned to the ground. This destruction reminded him of the destruction of working-class neighborhoods in eastern Tokyo during the 1923 Kanto Earthquake. The situation in Nanjing and Hangzhou was equally dismal. The retreating Guomindang forces also blew up a 15,000 kW power plant in Hangzhou and created one million yen's worth of damage.[36] What would it take to restore Shanghai and Nanjing to their former glory?

Despite the daunting nature of this task, Ishikawa rejected any recommendations to abandon the Lower Yangtze, as the British, Americans, and French in Shanghai's foreign concessions would move in to exploit the power vacuum. He suggested that:

> The five provinces of North China must be formed into a single bloc with Japan. We should however think about how to extend our influence into Central China, take a strong position on matters with those foreign countries, and think if they will do anything to diminish our stature. . . . By leaving Central China alone, pro-Japanese sentiment cannot develop. If the Chinese do not feel affinity to the Japanese, we will lose the objective of bringing peace to East Asia.[37]

His superior Tsuji Hideo voiced similar sentiments. In his private report, Tsuji pointed out that the American-owned Shanghai Power Company "exploited the desperation of the Chinese-owned power stations outside the concessions and seized control of [Shanghai's] power supply." Tsuji added, "As our military forces seized control of the power station in Nanjing, [Japan's] power companies must extend their control in the Yangtze Delta as soon as possible."[38]

The Japanese had little choice but to take on its responsibility of restoring order to the Lower Yangtze region. On June 30, 1938, the Japanese merged five electric companies and two water companies outside Shanghai's foreign concessions to form the Central China Waterworks and Electricity Company. Its executives unilaterally declared that the company would enjoy privileged status under the Japanese-backed Reformed Government of the Republic of China. The Reformed Government then declared it illegal for non-Japanese companies to supply electricity outside the foreign concessions and banned companies from generating their own electricity. It also waived the tariffs on heavy equipment, registration fees, and national taxes for Central China Water and Electric.[39] The Ministry of Industries of the Reformed Government and the commander of the Central China Expeditionary Army, Harada Kumakichi, ratified the agreement on August 9, 1938.

Economic historian Kanemaru Yūichi argued that Central China Water and Electric lived up to its promise of restoring electrical power to the Yangtze Delta, but later scaled back its operations due to severe constraints in capital, skilled labor, and fuel. Kanemaru cites company records from Manchuria Electric, which mention Japanese plans to install a 4,000 kW generator at Zhabei, two sets of 5,000 kW at Pudong, and a 30,000 kW generator at the Chinese Electric Company in the Old Chinese City. The Japanese also rebuilt the Capital Power Station in Nanjing, a state-owned entity under the Guomindang regime. They, however, struggled to find workers to operate the power plant, and Lieutenant-General Nakashima Kesago, who was implicated in the Nanjing massacre, admitted in his diaries that "many technicians and electricians were killed, as the army entered and raided the city."[40] Kanemaru concluded that right up until December 1941, Central China Water and Electric repaired most of the power lines damaged by war in Jiangsu Province. Following the attack on Pearl Harbor and the full escalation of the Pacific War, curtailment of coal shipments and a shortage of raw materials made it increasingly difficult to sustain the electrical infrastructure in the Yangtze Delta.[41]

Archival documents written between August 1938 and December 1941 largely support Kanemaru's findings but also reveal that the Japanese curtailed their recovery operations to conserve resources for their war effort. Restriction orders made it difficult for the electrical industries in the Yangtze Delta to receive much-needed machinery and fuel. In December 1938, the minister of industries for the Reformed Government, Wang Zihui, relayed orders from the Central China Expeditionary Army banning the transportation of metals and metallic products, coal, textiles, and alcohol from the Japanese-occupied areas into Shanghai's foreign concessions.[42] This gave the Japanese military absolute control over the electrical equipment and fuel flowing into Shanghai. The Japanese occupation forces also disrupted the American and French power stations in the concessions by throttling the flow of coal from North China.

The Japanese also deployed underutilized generating equipment in Shanghai to North China. They hollowed out the Chinese Merchants' Electric Company owned by Green Gang boss Du Yuesheng. The targeted destruction of Du Yuesheng's power company was also politically motivated. Ishikawa had singled out Du Yuesheng as an example of Chinese obsession with "securing promotions and getting rich" (Ch. *shengguan facai*).[43] Ishikawa saw Du as a typical unethical businessman who accumulated ill-gotten riches by colluding with the Guomindang leadership. Shortly after Du fled to Hong Kong, the Japanese moved one set of 3,200 kW generating equipment and another set of 6,400 kW to the Shijingshan Power Station and Steelworks in the western suburbs of Beiping. The remaining set of 6,400 kW generating equipment went to the Boshan Power Plant in Shandong Province. The Chinese Electric Company later attempted to recover the equipment by filing a restitution request with the Guomindang government after the war ended, claiming that generating equipment with a total rating of 16,000 kW was "immediately seized by the enemy," possibly months after the fall of Shanghai.[44] Transferring the underutilized equipment from Shanghai saved the Japanese the hassle of transporting refurbished generators from Japan. The Old Chinese city sustained such terrible damage that demand for electricity was anemic. Industrial activity barely recovered during the years of Japanese occupation, leading Central China Water and Electric to shelve its plans to construct a 30,000 kW plant to serve this area.[45]

Forced to make do with scarce resources, Central China Water and Electric balanced its budget by withholding payments to the collaborator regime in Shanghai. Its insistence on privileges for fee waivers antagonized the leaders of the Reformed Government of Shanghai, and a dispute over public service remuneration broke out between the power corporation and the collaborator regime in March 1939. Power companies under its control had signed an agreement to pay 1 percent of its electric light revenue and 1.5 percent of electrical power revenue to offset the cost of using roads and waterways. As the parent company, Central China Water and Electric was liable for these fees. The Municipal

Government withheld electric and water tariffs until the power corporation met its payment obligations. The Japanese-backed power corporation objected to these demands by pointing to the tax waiver clause and reasoned that such extraneous charges hampered its abilities to carry out its duties as a public service provider.[46] After months of wrangling, company executives finally admitted to the Municipal Government's Finance Department in May 1940 that the company was unable to meet these payments due to mounting losses.

The Japanese also leveraged their control of the coal supply to strengthen their influence on municipal politics in the foreign concessions. Japan's Ministry of Commerce and Industry started imposing coal rationing in May 1940.[47] As the war dragged on, the Japanese began cutting back coal supplies to the American and French power companies. The foreign concessions, which had been enjoying a stable power supply all this while, began experiencing power shortages. In that month, the Shanghai Municipal Council proposed implementing daylight savings time in the summer months to reduce power demand for illumination. Deputy Secretary T. K. Ho indicated that shops had already dialed their clocks one hour ahead to capitalize on longer daylight hours.[48] With electric light accounting for 11 percent of power consumption, such measures did little to save Shanghai from blackout.[49] On March 3, 1941, the American-owned Shanghai Power Company was left with three months of fuel oil in its reserves, forcing it to submit a proposal to the Shanghai Municipal Council calling for a curtailment of power consumption. It noted that it had "no assurance that it will now be able to secure fuel from sources other than Kailan consisting of 25,000 tons of coal monthly or approximately half of current requirements."[50]

Japanese council members made it clear that the Japanese government was unable to ease restrictions on coal and suggested that fellow councilors seek alternatives. At a meeting on March 5, 1941, convened to discuss the power crisis, the Japanese councilor Okamoto Issaku fired the first salvo stating that "some form of restriction of supply is a necessary precaution and that restriction on the basis of monthly consumption since January 1, 1940, is reasonable." Council chairman W. J. Keswick, the *tai-pan* of the Jardine Matheson conglomerate, agreed with Okamoto.[51] Well aware that the Japanese would not agree to increased coal shipments, council members changed tack and scrambled to locate alternative sources of coal. Minutes before nightfall, the council announced the first power restriction orders in the history of the concessions, calling for a 25 percent reduction in power usage. With the blessing of another Japanese council member Hanawa Yutaro, then director of the Shanghai branch of the Mitsui Corporation, the council authorized Shanghai Power to negotiate a shipment of 150,000 tons of coal from Calcutta to Shanghai. Crunching the numbers in the conference room, the Chinese accountant Xi Yulin estimated that the coal would cost 300 Chinese dollars per ton, thirty times the cost of coal from North China.[52]

Shanghai's Chinese businesses saw these power curtailment measures as an energy blockade directed at non-Japanese businesses. Chinese businessmen alleged in a *Shenbao* editorial dated March 15, 1941, that Japanese companies abused their privileges and did not comply with power reduction orders. Speaking on behalf of the Chinese business community, the *Shenbao* editorial stated that "the Japanese hold the key to Shanghai's electrical power. If the Japanese authorities can ship more coal from North China to Shanghai, problems with electrical power can be immediately resolved."[53] In other words, this was a manufactured crisis.

The Japanese pressed the council to double down on measures to curb power use. After weeks of negotiation, the council learned that the emergency coal supply from Calcutta had to be delivered in at least eight shipments. Hanawa from Mitsui Company, which had controlled Shanghai's coal supply, cautioned against taking up the charters.[54] By June 1941, Shanghai Power began planning to work with 35,000 tons per month—70 percent of the minimum required amount. Okamoto assured the council that the Japanese would provide at least 25,000 tons of coal to Shanghai every month.[55] Shanghai Power had to work within the limits spelled out by the Japanese.

After the attack on Pearl Harbor in December 1941, the Japanese assumed direct control over Shanghai's electrical infrastructure. The American-owned Shanghai Power Company came under Japanese military administration. The Japanese also claimed majority representation on the Municipal Council. Okazaki Katsuo, the Japanese consul-general of Shanghai, assumed the chairmanship. Okazaki explained that "since Utilities are essential to public safety and the defense of the area under the Japanese, the Japanese authorities cannot countenance the control of public utilities by enemy firms at the outbreak of the Pacific War."[56] The Japanese military imposed further restrictions on the movement of goods in Shanghai. In April 1942, businesses in Shanghai had to obtain permission from the Kōain (East Asia Development Board) to use, manufacture, or sell thirteen categories of materiel, including iron and steel products, nonferrous metals, and coal.[57]

Shanghai's power crisis worsened after Japanese administrators assumed total control and imposed stricter regulations to prevent Shanghai from draining precious fuel and mineral resources away from the Japanese war effort. Following the March 1941 restriction orders, Shanghai Power reduced its power output by 45 percent. The city imposed a ban on advertising lights and neon lamps but kept the streetlights on for safety.[58] Its output would fall even lower. To plug a massive deficit, the Japanese councilors ordered an increase in electric tariffs. Okazaki pointed out that the Shanghai Power Company incurred monthly losses of 12 million Chinese dollars per month, as the price of coal increased five or six times during the war. The losses of Central China Water and Electric widened further after taking over Shanghai Power. Okazaki secured the new consul-general's full backing to push through a 200 percent increase in electrical tariffs. Apart from one Chinese councilor who protested that "the revised tariffs [had] been forced down the Council's throat without an opportunity being afforded to give the matter full consideration," the other councilors endorsed the plan and called the price increase "justifiable."[59] The non-Japanese councilors knew that the tariff revision had the full backing of the Japanese imperial government and that any resistance was futile.

As the largest consumer of electricity in China, Shanghai bore the brunt of power conservation campaigns in the Japanese-controlled regions. After the abolition of the Municipal Council in July 1943, the Shanghai Municipal Government began administering the former foreign concessions. The Bureau of Social Welfare of the Shanghai Municipal Government gathered the owners of the most energy-intensive industries in Shanghai and formed an Industrial Fuel Economy Committee to discuss strategies to lower coal, electric, and gas allotments in November 1943.[60] On the last day of 1943, Shanghai mayor Chen Gongbo mandated shops to switch off electrical lights before 7 pm and restaurants and amusement facilities to end business before 11 pm daily.[61]

With power curtailment campaigns achieving limited savings, the Japanese turned their attention to improving Shanghai Power's fuel efficiency. The Industrial Survey Department ordered the power company to tighten controls in the boiler rooms, noting that "it is essential that boiler attendants are chosen who have a high standard of intelligence and experience." The authors scrutinized every step of the power-generation process and suggested cleaning and descaling boilers, preheating boiler water, unclogging the injector nozzles of the boiler feed pumps, and moderating the chimney draft to ensure that there was enough air to support combustion. It also recommended damping fine coal with 10 percent water to prevent the "loss of fine coal being sucked up the chimney before complete combustion and also [to] keep the coal from dropping through the grate."[62] Most important, the Industrial Survey Department called for the monitoring of boiler room workers to ensure that they were not making illicit profits by selling coal ash to charcoal briquette factories. Combusting every ounce of coal would put an end to their profiteering.

Business associations clamored for exceptions as the fuel crisis worsened. The military administration rejected most of these requests but made allowances for security reasons. Through personal connections with the Power Conservation Committee, two pharmaceutical companies in Shanghai had their excess usage penalty waived in February 1944, when they went over their power allotment to increase serum production during a meningitis outbreak. The Miscellaneous Grain Association also successfully petitioned for the restoration of electrical power for ten of their members in November 1944. Its president told the military authorities that although their members did not process grain for the military, flour products from puffed rice, broad beans, tapioca, and peas "fed the masses that cannot afford to buy rice." Furthermore, the ten shops only used 1,452 kWh of electricity per month, which was less than a medium-sized factory.[63]

By the end of the eight-year Japanese occupation, chronic power shortages had badly crippled Shanghai's industries. The high cost of transporting coal to the Yangtze Delta from North China forced the Japanese to cut back on coal shipments. By July 1944, Shanghai Power made do with 29,000 tons of coal, nearly 40 percent less than its prewar levels. It had to meet the target of sharply reducing monthly coal consumption by another 7,500 tons by August 1944.[64] Shanghai's electrical industry, which included the largest power company in East Asia before the war, was falling into a death spiral. Taking stock of the damage at the end of the war in 1945, the Provisional Senate noted that the total generating capacity of Shanghai's Big Five power companies fell from the prewar levels of 260,000 kW to less than 140,000 kW.[65] The Chinese Electric Company, Western Shanghai Power Company, and Pudong Electric simply resold electricity acquired from Shanghai Power, as their generating capacity fell to zero during the war.[66] As discussed in chapter 6, the wartime energy crisis destabilized Shanghai's economy and ultimately contributed to the collapse of the Guomindang regime in 1949.

Keep the Fire Going

What was happening in Southwest China under the Nationalists? The war broke out a year after China's economic planners had completed drafting the five-year plan to develop state-run industries inland. The engineer-bureaucrats retreated to the provinces of Sichuan, Yunnan, and Guizhou, which had little to no power-generating capacity (see table 3.1). They had to familiarize themselves with the natural and human environment of these frontier regions and build the electrical

TABLE 3.1. Electrical industries under Japanese and Guomindang control, circa 1939

Provinces under Japanese control	Number of power plants	Investment ('000 yuan)	Generating capacity (kW)
Jiangsu	107	47,818	125,740
Anhui	25	1,828	4,644
Zhejiang	109	13,057	30,908
Fujian	29	5,457	11,555
Guangdong	35	9,622	36,060
Jiangxi	13	1,115	3,792
Hubei	18	5,192	20,427
Shanxi	6	1,530	5,572
Henan	7	1,430	2,110
Shandong	23	5,038	52,044
Hebei	17	16,528	44,079
Cha'har	1	278	335
Suiyuan	2	591	608
Hunan*	12	2,479	7,074
Total			344,948

Provinces under GMD control	Number of power plants	Investment ('000 yuan)	Generating capacity (kW)
Yunnan	3	1610	1,614
Guizhou	2	110	165
Shaanxi	1	300	709
Tibet	1	83	100
Sichuan	22	4685	5,172
Xikang	1	21	25
Gansu	3	44	141
Ningxia	1	100	100
			10,872

NOTE: Figures are from 1935 national survey completed by the National Resources Commission. Hunan is partially under Guomindang control. Source: Data from "Chen Zhongxi, "Dianqi shiye yu dianqi shiye jianshe," *Ziyuan weiyuanhui jikan.*

infrastructure from the ground up with bits and pieces of salvaged equipment. The need for cheap and abundant electricity to fuel the Chinese war effort likewise prompted the engineer-bureaucrats to build a coal-dependent energy infrastructure in a region that had poor quality coal and vast hydropower potential.

Engineer-bureaucrats, who had fulfilled both civilian and military duties before the war, promptly transitioned to wartime mobilization. Yun Zhen, the director of the NRC's Electrical Power Bureau, was promoted to major-general of the National Revolutionary Army. He was no stranger to the military, having served as the academic director of the Academy of Military Communications in Shanghai. For a monthly stipend of 200 Chinese dollars, Yun taught soldiers and officials of the Nationalist Army how to use radio equipment.[67] The military authority conferred by the state did little, however, to help Yun Zhen complete the mission of evacuating the electrical industries into inland provinces. He watched helplessly as power-generating equipment of the state-owned Nanjing Power Company fell into Japanese hands. The banks forbade the NRC from removing the generators, which had been put up as collateral for loans. During the battles to defend the Lower Yangtze between August and November 1937, only three power stations in Shanghai and one each in Suzhou, Changshu, and Hangzhou complied with orders to provide daily operational updates to the NRC.[68] With the Japanese swiftly advancing into Central China, Yun's colleagues became hard-pressed to act on earlier plans to build power grids in Southwest China that would provide the driving force for makeshift defense industries.

Even before the retreat into Southwest China, the NRC had already decided to prioritize the construction of coal-fired power stations. Huang Hui, a Cornell-trained engineer, concluded that the geology and hydrology of Southwest China, a region with more than half of China's hydroelectric potential, was ill-suited for hydropower. Huang identified four major obstacles:

1. Our nation has no large waterfalls. There are many small waterfalls in the Southwest, and they are mainly suited for small and medium hydro-electric plants . . .
2. In our nation, the forests are underdeveloped, the rivers are untamed, and there are few natural lakes, causing huge fluctuations in flow volume. If we construct reservoirs, there is a danger of flooding too much arable land. It is more economical to complement hydropower with fossil-fuel power.
3. There are huge seasonal fluctuations in the flow volume of rivers. The water level in the wet season can be 20 or even 30 meters higher than that of the dry season. This is unseen in Europe and America . . .

4. The weather in North China is harsh. The rivers are frozen for long periods. The water carries huge amounts of sediments. The loess level is thick. It is difficult to find solid rock [on which to build dams]. There were few problems like this in the South.[69]

All these factors hampered the development of hydropower in Kunming, Yunnan's provincial capital. Shilongba, China's first hydropower station, was built in the western hills of Kunming in 1910 and had a paltry generating capacity of 2,000 kW. Liu Jinyu, a physics graduate from the University of Paris, took on the task of building Kunming's electrical power infrastructure. He learned that during the dry months from November to April, the water level in the rivers and lakes dropped sharply and the hydroelectric power plant only ran at half of its full capacity.[70] There was no further hydropower development after the completion of Shilongba in 1910. This left the NRC with only one option: to transform water power into firepower. Pump the water from Kunming Lake into a boiler and use the steam to spin a turbine! Even with scarce coal deposits, it was more viable to build a new coal-fired power plant than to upgrade the existing hydropower facilities.

The same geological and climate factors that hindered hydropower development also delayed the construction of the Kunming Lakeside Electrical Works. Fang Gang, a young civil engineer, identified a plot for the power station six miles west of Kunming in Shizui Village of Majiezi township, which sat next to a huge body of water and was situated on a major transportation route for coal.[71] The on-site survey in 1936 failed to take into consideration the soil conditions, and the construction crew later realized that the soil could not support the weight of the power-generating equipment and absorb the vibration from the rotating turbines. Liu Jinyu had to bring in pile workers from Xiashesi in southern Hunan to lay the power station's foundations. When the construction team arrived in May 1938, they ran into Kunming's rainy season when poorly paved roads turned into mud tracks, delaying construction until October 1938. The shortage of basic construction equipment, such as saw blades and hammers, also caused further delays.[72]

Meanwhile, Yun Zhen and his colleagues, who had fled from Nanjing, coordinated the transportation of power-generating equipment to Kunming.[73] The NRC only managed to salvage two sets of generating equipment, however. The first came from the Xiangtan Power Station in Hunan. Workers strapped the 2,000 kW turbine weighing three hundred tons across twelve boats and moved it downstream along the Xiang River to Qiyang in western Hunan Province. Upon reaching the first destination, Yun and Huang received orders to transfer the turbine to Liuzhou in Guangxi Province. Laborers from Kunming then hauled the turbine to northern Vietnam and loaded it onto a freight train that ran along the Hai Phong–Kunming Railroad.[74] The second set, built by Babcock and Siemens, was on a ship from Singapore to Shanghai. The NRC had it diverted to Hong Kong and sent to Kunming via the Haiphong-Kunming Railroad.[75] Besides the Kunming Lakeside Electrical Works, the NRC managed the construction of six power stations across Sichuan, Gansu, Yunnan, and Guizhou. All in all, it dismantled 2,200 tons of power-generating equipment and shipped them over more than a thousand of miles of rivers and roads, before reassembling them in these remote inland provinces.[76]

To secure the fuel supply for the newly constructed power station, the NRC acquired the Mingliang coal mines located fifty-six miles northwest of Kunming in September 1939. Opened in 1902, Mingliang became the largest coal producer in Yunnan before the war by supplying the local handicraft industry, tin smelters, and major railroads. Its lignite coal was well suited to British-built boiler-turbines. Under the takeover agreement, the government provided 2.2 million Chinese National Currency (CNC) dollars of capital, while private shareholders offered 0.6 million CNC dollars of equity. The newly incorporated Mingliang Holdings Company appointed Yun Zhen as managing director and remodeled itself after other state-owned defense industries.[77] The mine hired more workers and increased output from prewar levels of 9,220 tons to 34,036 tons in 1940.[78] The power station had to make do with this low-quality coal. Lu Yuezhang noted that the lignite coal deposits in Kaiyuan, which was 136 miles south of Kunming, had a water content of 23 to 32 percent. Its combustibility ratio was much lower than that extracted by Japanese-owned Fushun colliery in Manchuria.[79] The engineer-bureaucrats remained unperturbed by the poor quality of coal and pointed to the Soviet example of boosting electrical output by burning low-quality peat.

Kunming Lakeside Electric Works demonstrated great resilience during the war. With a generating capacity of 4,336 kW, it supplied 3 million kWh of electricity annually to 520 users, including some of the most important defense industries in Kunming, through forty-three miles of high-voltage transmission lines and fourteen miles of low-voltage distribution lines.[80] Liu Jinyu reported that the power station recorded only one total blackout in 1940 lasting for two hours—the result of an air raid. There were about three minor power stoppages each month during routine maintenance, and it recovered from nine power failures caused by lightning strikes in the rainy season.[81] The engineer-bureaucrats also began preparing for the eventuality that the power station and other factories on the shores of Kunming Lake would become prominent targets for aerial bombardment.

As early as April 1940, the NRC considered relocating part of the Kunming Lakeside Electrical Works into the mountainous hinterland of Kunming to evade air raids and moving it closer to its fuel source. Liu Jinyu identified a site on the boundary of Songming and Yiliang counties, about fifty miles northeast of Kunming. It was much closer to the Mingliang coal mines than the previous location, which helped reduce fuel transportation costs. Even better, the high mountains in all four directions shielded the new power station from air raids.[82] The Penshuidong Power Plant was built here and would get its name from a blowhole at the foot of the western cliffs where it is located.[83]

The air raid on Kunming in October 1940 after the Japanese invasion of French Indo-China forced the engineer-bureaucrats to implement their contingency plans. Two armament factories had already relocated into the caves near Penshuidong and needed a power supply. After completing a topographical survey in December 1940, Tao Lizhong, a civil engineer with Kunming Electric, accompanied Liu on a visit to the armament factories in the caves to study how they adapted to the mountainous terrain. The site visit alleviated Tao's concerns. He found a cave eighteen feet wide and twenty-four feet high with just enough space to accommodate the turbines. Engineers erected a building outside the mouth of the cave and placed the boilers there. The road between the cave and the boiler plant was only nine feet wide, thus reducing the area exposed to aerial bombardment. The engineers constructed a protective wall between the cave and adjacent building, which had a small tunnel that would allow only one person to crawl through at a time. This helped shield the turbines from any explosion outside the cave.[84]

Squeezing a power station into a space that doubled as a bombproof bunker tested the limits of the NRC's civil engineering capabilities. Besides limiting damage from air strikes, the engineers devised innovative solutions to pack bulky turbines into tight spaces and address problems with ash removal, ventilation, and noise reduction.[85] The cave turned out to have certain advantages. Instead of building an ash basement, the engineers adopted the sluice conveying method, which uses cold water from the condensers to flush the coal ash away from the cave. The building design saved a lot of cement and put the water source to good use. It was also not necessary to construct a steel-and-concrete sub-base to stabilize the turbines, as the mountain rock was strong enough to absorb their vibration. Powered by a 150 hp "run-of-river" installation, the power plant did not need to be equipped with an external diesel generator.

The Kunming Lakeside Electric Works and the Penshuidong Power Plant
survived the onslaught of aerial bombardment during the war. With the destruction of Shilongba Dam during two raids in October 1941 and January 1942, they
supplied most of Kunming's electricity.[86] The defense industries ran their own
generators to make up for shortfall. The NRC had its hands full maintaining the
existing electrical infrastructure, and plans for hydropower development within
Yunnan Province did not get off the ground until after the war. The NRC also
would not execute plans to dam the Praying Mantis River in Northern Yunnan
that they had drawn up in 1944.[87] Packing a coal-fired power station inside a
cave was challenging, and building a hydropower dam with bombs raining down
from time to time was impossible.

THE TRANSITION TOWARD a carbon-intensive developmental model emerged
as a result of scarcity during the second Sino-Japanese War. The Japanese and
Guomindang regimes not only turned to coal as a primary fuel source for power
generation but also centralized its distribution and consumption. The outcome
was independent of the abundance of coal within the areas under their control. The practices of generating electricity close to fuel sources in North China
and burning peat with low carbon content in Southwest China bore striking
parallels with the GOELRO (State Commission for Electrification) plan the
Soviet Union launched in 1920. At the same time, capitalist forces were at play.
Japanese electric companies capitalized on the war in North China and pooled
their capital and human resources to support the takeover of the electrical power
infrastructure. Collaborator regimes backed the formation of Sino-Japanese
power companies, as they sought to stabilize the urban economy by ensuring
a steady supply of electricity. The alignment of corporate and state interests facilitated the Japanese takeover of the electrical utilities in North China, which
in turn generated the economies of speed that paved the way for accelerated
growth of long-distance power transmission networks. The centralization of coal
distribution and redeployment of power equipment from the Lower Yangtze to
coal-abundant North China also reconfigured China's energy geography. These
changes had a lasting impact. North China's share of the nation's generating
capacity increased from 8.3 percent in 1936 to 15.8 percent in 1947.[88] And, as will
be shown in chapters 6 and 7, industries in Shanghai would continue to suffer
from chronic shortages in fuel and generating supply stemming from this earlier
disruption.

Subsequent regimes that ruled over China bore the cost of accelerated
growth and consolidation in China's electrical industries between 1937 and

1945. Power-generating equipment and transmission lines hastily built during the war were prone to breakdowns. Wartime electrical systems became sources of residual inefficiencies. As will be discussed in Chapter 6, the North China regional grid, once the shining example of Japanese wartime electrification, would prove to be a weak link that the Communists exploited during the Chinese Civil War (1946–1949).

Having experienced the trauma of losing 97 percent of China's generating capacity to the Japanese and seeing their enemies harness energy resources on Chinese soil to wage war on China, the NRC's engineer-bureaucrats developed an energy security outlook that viewed foreign interference with great suspicion. The self-reliant approach to infrastructure development brought the Guomindang regime into conflict with its allies. The next chapter explores in greater detail how the NRC's engineer-bureaucrats kept their American allies at arm's length to prevent American electrical equipment industries from imposing their standards on the Chinese electrical power sector. These engineer-bureaucrats pursued the relentless defense of economic sovereignty at the cost of alienating its global trade partners.

The developments outlined in this chapter also paved the way for China's transformation into a contributor to the Great Acceleration. The mobilization for war forced competing regimes that ruled over China to ramp up electrical power output. The generating capacity of North China increased by 89.6 percent during the war years, a growth rate that would not be achieved in subsequent years. It set also into motion the shift of industrial activity closer to coal deposits in North China. Inland provinces in Southwest China, which had a miniscule electrical output to begin with, saw the rapid introduction of electricity, as generating capacity in Sichuan, Yunnan, and Guizhou increased by 4.74 times, 5.18 times, and 11.36 times respectively.[89] The scale of the electrical industries in the inland provinces remained small despite the wartime build-up. The building blocks for a carbon-driven economy had now been put in place within China's interior. Using the limited supply of electrical power, the engineer-bureaucrats erected makeshift production facilities in Southwest China. Cut off from the global supply chain, they designed and built generators, transformers, and wires—electrical components that would transform energy in fossil fuels into motive force that would, in turn, unleash greater fire and fury.

CHAPTER 4

Dawning of the Copper Age

I N 1944, THE NRC'S Central Electrical Manufacturing Works received a long list of supplies from the US Army's field artillery unit that was deployed to Kunming. The barracks for the enlisted men alone needed to be equipped with eleven ceiling sockets; two drop sockets; four convenience outlets; one 30A knife switch; thirteen conical reflectors; two 15A porcelain fuse boxes; fifty porcelain insulators; 260 feet of no. 14 AWG and no. 18 and 100 feet of no. 8 AWG weatherproof copper wire; ten feet of flexible rubber-filled cords; two porcelain wire holders; ten 50 W and three 40 W incandescent lamps; and a miscellany of wood screws and porcelain cleats. Even more electrical fittings had to be installed in the recreation hall, washroom, kitchen, officers' barrack, and mess hall. The barracks also had to be connected to the local power grid with a number of transformers.[1] When the Guomindang regime retreated to Southwest China six years previous, everything on this supply list had to be shipped in from abroad. Cut off from the global supply network, Chinese engineer-bureaucrats spent those six years toiling away to jumpstart a domestic electrical equipment industry to sustain the war effort. Their hard work had just begun to pay off.

The electrical equipment industry found its wartime base in Kunming after a series of twists and turns. The proximity to known copper deposits was a key contributing factor. Having failed to evacuate the power-generating equipment from the capital, Nanjing, Yun Zhen, the engineer-bureaucrat who had been conferred a military rank of major-general during the war, took on the task of making the much-needed electrical components. He surveyed the Dongchuan copper mines in Yunnan on horseback in 1938. These mines had supplied the copper for coinage since the Qianlong era (1736–1796), they also had supplied the ore for the smelting plant that made high-grade copper for electrical components during the War of Anti-Japanese Resistance.[2] If the Bronze Age symbolized the maturity of early Chinese civilization, the period between 1938 and 1944 marked the dawn of China's "copper age," as the survival of the Chinese nation-state hinged on its ability to manufacture vital electrical components for arms manufacturing and military communications depicted in the advertisement in figure 4.1.

Eight years of ceaseless warfare totally transformed China's electrical equipment industry. Before 1937, China's electrical power sector had expanded in a piecemeal fashion. Power companies installed a wide array of electrical equipment, which led to a total lack of standardization. There was also little incentive to invest in electrical equipment manufacturing in China. Producing small quantities of components of varying specifications would have increased average production costs. At the same time, manufacturers in Britain, Germany, and the United States were facing overproduction and looked toward China as a market to dump their excess inventory. Wartime mobilization caused China to be cut off from the global supply chain, prompting engineer-bureaucrats to simultaneously pursue two different paths toward self-reliance—one being technology transfer with established manufacturers in the United States and Europe, and the other state-sponsored applied research. As Grace Shen noted in the case of Chinese geology, Chinese wartime science could not be dismissed as "too makeshift, too isolated, and too ephemeral to have had a significant impact on global developments."[3] Similarly in the case of electrical equipment manufacturing, engineer-bureaucrats conducted applied industrial research for viable substitutes amid wartime scarcity, as they charted the path toward self-reliance. Seeing their work as integral to the quest for energy independence and the defense of economic sovereignty, these engineer-bureaucrats vigorously resisted attempts by American electrical industries to impose their industrial standards on China. This came through most clearly during negotiations over the terms of the technology transfer agreement between the NRC and its American partner, Westinghouse.

Electrical equipment manufacturing grew from a neglected sector into the lynchpin of "Free China's" defense industries during the war. The state-run Central Electrical Manufacturing Works (hereafter Electrical Works) started with the manufacturing of wires, radio tubes, and field telephones—vital components for military communications. Judging electrical equipment manufacturing to be more important than power generation, Yun Zhen relinquished his portfolio as the director of the Electric Power Bureau to focus on his duties as the general manager of the Electrical Works.[4] Explaining his decision to a reporter with the *Ta Kung Pao* (L'impartial) newspaper in July 1948, Yun said, "Before our electrical equipment industry is developed, any increase in electrical power will only enlarge foreign exchange disbursement and drain the wealth of our nation. Those who are far-sighted will try to devote some energy to break this deadlock."[5] Yun had every reason to be proud of his wartime achievements.

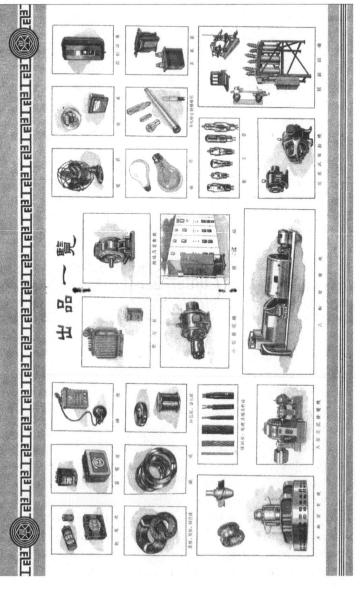

FIGURE 4.1. *Products of the Central Electrical Manufacturing Works.* With the exception of the light bulb and electric fan, every other component featured in this catalogue had to be imported prior to 1939. Source: Academia Sinica, Institute of Modern History Archives, Taipei, Taiwan.

By the end of the war, the NRC created a state monopoly on electrical equipment and took a vital step toward achieving the goal of nationalizing the electrical power sector. Besides facilitating a homegrown electrical equipment sector, the NRC also enjoyed greater success in promoting national industrial standards. Back in 1931, the NRC decided on a 220 V/50 Hz setting for mains electric power, as it used less copper than 110 V/60 Hz. Wartime scarcity reinforced the need to administer these predetermined standards, causing Yun and his colleagues to resist lobbying efforts by private American manufacturers to adopt the 110 V/60 Hz standard commonly adopted in the United States.

Circuitous Path to Self-Reliance

Since the establishment of China's first power station in 1882, almost all power generation and transmission equipment had to be imported—even basic components like wires. According to Yun Zhen's 1939 report to the NRC, "Between 1932 and 1936, China imported 7 million CNC dollars' worth of copper wire, 710,000 CNC dollars of radio tubes, and more than 2 million CNC dollars of telephones annually."[6] Before the Japanese invasion in July 1937, there were approximately two hundred electrical equipment factories in China, assembling small appliances like light bulbs and electric fans, with 159 of them located in Shanghai. Severe capital shortage rendered most of these factories dormant, resulting in low output and poor product quality.[7] Shimizu Dōzō of the Asia Development Board (Jp. *Kōain*) offered an unflattering assessment of China's light bulb industry. He reported that Shanghai's factories manufactured sixteen million light bulbs annually between 1924 and 1934, amounting to one light bulb for every twenty-five people living in China and catering to 40 percent of domestic demand. Consumers avoided Chinese-made light bulbs, as good and faulty ones were sold together.[8] The same could be said for Chinese-made batteries, which had a shelf life of less than six months compared to more than two years for foreign batteries.

Imported machinery largely fueled China's industrial expansion before 1937. Chiah-Sing Ho's study based on maritime customs figures noted that machine imports surged between 1928 and 1932 during the first years of the Nanjing decade. Prime movers accounted for between 16.26 and 22.76 percent of total imports. More than 10,000 diesel engines were installed across the country. As mentioned in chapter 2, diesel generators allowed small power companies like Woo-shing Electric to quickly increase their capacity. The mastermind of Woo-shing's expansion was a comprador for Siemens AG in China. By 1932,

Germany overtook the United Kingdom as the largest exporter of electric generators. Three years later, Japan took over as the second largest exporter. On the eve of the Japanese invasion, Germany, Japan, and Britain accounted for 34.08 percent, 32.40 percent, and 15.19 percent of imported power generators, respectively. The United States was a distant fourth.[9]

The low prices of electrical equipment in China suggested that foreign manufacturers were dumping their excess inventory into China's market. Zhang Chenghu, the NRC's engineer who eventually negotiated the technology transfer agreement for wire manufacturing, was surprised to learn that electric wire not only accounted for 49 percent of the value of all imported electrical equipment, but also that its retail price in China was 5.0 percent to 26.7 percent lower than that in the United States. He concluded that "this overwhelming force of international competition made it difficult for any newly established domestic wire manufacturing facility to retain a foothold, as it would face many operational uncertainties."[10]

With small power companies acquiring equipment from the lowest bidder, state regulators achieved little progress in their drive for voltage standardization. China's prewar situation resembled the United States's, Britain's, and Germany's right after World War I. Thomas Hughes argued that the "inertia, divisive tendencies, and parochialism" only began to break down after 1926 with the appearance of large blocks of newly generated power, such as the Pennsylvania–New Jersey Interconnection, the Bayernwerk in Bavaria, and the National Grid in Britain.[11] There were no standardized guidelines to speak of in the United States. In 1927, R. E. Argersinger of the American Institute of Electrical Engineers (AIEE) lamented the unwillingness of American power companies to adhere to a single standard, even though they knew it would be cheaper for electrical equipment manufacturers to produce a smaller range of products at larger quantities.[12] In China's case, as discussed in chapter 2, private power operators banded together to resist state regulation. The National Voltage Standards, promulgated by the Nanjing government in 1931, resulted from a compromise between state regulators and private power companies. The National Construction Commission (NCC) decided on the service voltage of 220/440 V for direct current, and 220 V, 220/380 V, or 220/440 V at 50 cycles per second (single or three phase) for alternating current, as these settings were the most common. The distribution voltages were a mish-mash of all available settings used in the United States and Europe. The intermediate distribution voltage of 2,200 V, 2,200/3,800 V, and 6,600 V was in common usage in the United States, while the high-voltage transmission lines came from the European standards.[13] Recognizing the impossibility of enforcing a single set of industrial standards that had been imported from another country, the NCC exempted power companies established before 1931 from adhering to its standards.

The mobilization for war forced the Chinese power sector out of foreign dependency and created the urgency for greater standardization. Anticipating the fall of coastal cities to Japan, the Ministry of Economic Affairs passed a three-year plan in 1936 to build state-sponsored heavy industries in the inland provinces. The electrical equipment industry only received a paltry allocation of 15 million CNC dollars (5.5 percent) out of the total 271.2 million CNC dollars budget and lost out to other high-profile industries such as steel and oil refining.[14] Tasked with jump-starting the electrical equipment industry on a shoestring budget, the engineer-bureaucrats relied on technology transfer agreements to quickly put in place the necessary manufacturing processes. With their supervisors busy transporting power equipment away from the path of destruction, the NRC's junior engineer-bureaucrats were left on their own to negotiate favorable terms of trade with prospective foreign partners. Within a year and moments before the Japanese invasion, they secured three technology transfer agreements. The NRC came to acquire equipment and manufacturing processes for wires from the British, radio tubes from the Americans, and field telephones from the Germans.

With no time to develop production processes and machinery on their own, the NRC settled on a pragmatic policy of "importing foreign technology as far as possible."[15] Yun Zhen delegated the negotiation with foreign electrical industries to his fellow alumni from the Shanghai Jiaotong University (formerly known as the Government Institute of Technology of the Communications Ministry).[16] Meanwhile, Yun coordinated the transfer of power-generating equipment and construction of new power lines for newly established defense industries.

Upon their arrival in Europe and America, these young engineers leveraged their personal connections to establish contact with prospective partners. Zhang Chenghu, a Jiaotong University alumnus like Yun Zhen and who received his master's degree in electrical engineering from the University of Manchester, took charge of the technology transfer agreement for wires and cables. Having specialized in radio transmitters, he returned to China to work for the International Radio Station in Shanghai. His task of securing a technology transfer agreement with wire manufacturers had no bearing on his specialization, however, and Zhang prepared himself for the mission by visiting major electrical appliance retailers and wholesalers in Shanghai and studying tariff records. Zhang set his sights on three of the world's largest cable manufacturing firms—British

Insulated Cables Company, Callender's Cable and Construction Company, and Henley's Telegraph Works Company, Ltd.

Zhang asked the director of acquisitions for the British Boxer Indemnity Committee, Wang Jingchun, to introduce him to the general manager of the British Insulated Cables Company (BICC).[17] Zhang remained committed to the objective of self-reliance throughout the negotiation process and rejected the original proposal, in which the British would supply both the wire drawing machines and ready-made copper rods, because such an arrangement meant that "the control of materials remained in foreign hands." Furthermore, it would be impossible to transport copper rods into China during the war. He instead asked the British to furnish the Chinese with metal rollers, so that the NRC's factories could make the copper rods on their own.[18] The British agreed to the terms. They dispatched one British technician and one workman to China to help with the start-up and offered to train three Chinese workers in London each year. Upon the commencement of operations, the Electrical Works would give BICC 2 percent of its revenue as remuneration.[19]

The technology transfer process for radio tubes followed a similar pattern. Zhu Qiqing, another Jiaotong University alumnus, was sent to the United States. Unlike Zhang, however, Zhu had an advanced degree in radio communications and was operating within his area of specialization. He, unfortunately, had no contacts within the industry. Arriving in New York in the winter of 1936, Zhang contacted two Chinese interns at Arcturus Radio Tube Company in Newark, New Jersey, and the International Telephone & Telegraph Company (IT&T).[20] Zhu finally got his foot in the door at Radio Corporation of America (RCA). Talks quickly collapsed in January 1937, however, when both parties failed to agree on the licensing fee. Zhu then turned to the interns, who introduced him to RCA's competitor Arcturus Radio Tubes. The NRC purchased the circuit drawings and process specifications for a number of vacuum tubes for a mere 10,000 US dollars. Zhu sat down with the two engineering interns and drafted an inventory for a vacuum tube factory in China and then purchased the necessary equipment with the remaining budget.[21]

This barebones agreement provided sufficient information for the Chinese to produce vacuum tubes on their own. Workers from light bulb factories had transferrable skills that could be applied to vacuum tube manufacturing, such as the drawing out and cutting of tungsten wires, welding, glassmaking, and the pumping of air from glass encasing. After Shanghai fell to the Japanese in November 1937, one hundred glassworkers and technicians followed Feng Jiazheng, the general manager of the NRC's light bulb factory, on the inland retreat. Another fifty female apprentices joined the operations at the new inland production base. Using the circuit drawing designs and process specifications acquired from Arcturus, the light bulb factory launched the first "made-in-China" vacuum tubes shortly after its debut of the first "made-in-China" fluorescent lamps in July 1938.[22] Following the conclusion of the technology agreement with Siemens Halske for field telephones in 1937, Electrical Works was now in the position to manufacture three key components of military communications equipment for the National Revolutionary Army led by Chiang Kai-shek.[23]

Preoccupied with the evacuation of power-generating equipment, the NRC could not spare anyone else to acquire power-generating technology from abroad. Yun Zhen placed the existing state-owned electrical equipment factories under the control of the de facto signal equipment unit of the Nationalist Army. Established as a small classroom workshop at Shanghai Jiaotong University in 1927, the subunit made shortwave military radios for the National Revolutionary Army. After taking over this workshop in 1929, the NCC transformed it into the first state-owned electrical equipment plant. This makeshift plant supplied batteries and radios to the military and conducted research on electric motors and transformers. Just two days before the Battle of Shanghai broke out, this subunit withdrew with the Nationalist forces to the provisional capital, Wuhan. As attempts to hold off the Japanese at Wuhan became a lost cause, it retreated inland to Hunan in February 1938 and merged with Central Electrical Manufacturing Works one month later.[24] Yun Zhen tasked the unit with producing batteries, electric motors, and transformers, largely by imitating the designs of electrical equipment that had been imported to China. As detailed later in this chapter, this unit developed into a center for applied research during the war, as it was not bound by stipulations from technology transfer agreements and had the flexibility to devise improvised product designs.

The electrical equipment industry first retreated to Xiangtan, Hunan, and then subsequently to Kunming, Yunnan. One full year before the outbreak of the second Sino-Japanese War, Yun Zhen, along with the general managers of Central Steel Works and Central Machine Works, surveyed Jiangxi and Hunan in central China along major tributaries of the Yangtze River in July 1936. Two months later, the three managers decided to build factories in Xiashesi in Xiangtan City, Hunan Province. Xiashesi was about forty miles south of Hunan's provincial capital, Changsha—a Nationalist Army's stronghold that fell to the Japanese only in late 1944. With an area of 9,430 *mu* (1,553 acres), the plot on the Xiang River's north bank had two key advantages. Able to withstand 8,000 to 10,000 pounds of pressure per square foot, the soil could support the heavy weight of machinery and buildings of the electrical, machine, and steel works and, at an elevation of between three hundred and six hundred feet, the hilly terrain offered protection from ground forces and air strikes.[25]

The construction of the Electrical Works in Xiangtan served as a blueprint for subsequent wartime industrial complexes. As the leading electrical engineer serving the Guomindang government, Yun Zhen led the advance party of electrical engineers to Xiangtan to build the power plant that would be foundational to subsequent industrial development.[26] Electricity enabled the provision of tap water and gas as well as the construction of railroads and docks, giving rise to the infrastructure that would support the operations of the steel factory, machine works, and electrical equipment plant. For about a year, the new industrial base in Xiashesi appeared far enough away from enemy bombardment. It was more than eight hundred miles from the nearest coastal industrial base. That changed with the fall of the provisional capital, Wuhan, in October 1938, and the NRC followed the same planning procedure as when they uprooted these industries based in Xiashesi to Kunming.

As the Guomindang regime retreated from Wuhan to Chongqing, the state began to recognize the strategic importance of electrical equipment. Article 5 of the mobilization order issued by the Ministry of Economic Affairs on October 1938 stated:

> In order to meet the demands of extraordinary times, the Ministry of Economic Affairs, with permission granted from the Executive Yuan, will subject these industries under direct government control or manage them as public-private partnerships: (1) the mining of strategic minerals; (2) production of armaments and military goods; and (3) electrical equipment industries.[27]

In addition, the Guomindang regime offered military protection and financial capital to struggling Chinese enterprises, if they retreated inland with them to Western Hunan, Sichuan, Guangxi, Sichuan, and Yunnan.

Within a month of the Japanese capture of Wuhan, the Electrical Works closed down all manufacturing operations in Xiangtan and withdrew into Guilin and Kunming in Southwest China. Only ten workers stayed behind to guard its empty facilities and administer a training facility for apprentices.[28] The Electrical Works splintered into four different subunits, which produced wires, vacuum tubes, telephones, and electrical machines and spread out their operations in Kunming, Guilin, Chongqing, and Guiyang to reduce the risk of having the entire industry taken out during a single strike.[29]

Kunming emerged as the headquarters for the electrical industry for two key reasons: its connection to overland routes, and its proximity to copper deposits. As long as French Indo-China did not fall into Japanese hands, the NRC was able to transport materiel through the Yunnan-Vietnam railway. The first batch of machinery for the wire factory arrived at the northern Vietnamese port of Haiphong and was transferred to Kunming via rail.[30] Even after rail access was cut off, Kunming remained the first major stop inside China on the Burma Road and served as the first landing point for aircraft transporting supplies flying over the Himalayan Hump.

More important, Yunnan contained known sources of high-quality copper—a vital raw material for electrical equipment. Recounting his experiences in Yunnan in his speech about copper conservancy in 1951, Yun Zhen recounted his observations in 1938 and extolled the high quality of ore from the Dongchuan copper mines:

> During that time, Dongchuan produced 400 to 500 tons of copper ore, which were sent to Kunming and made into electrolytic copper. According to geologists, there is a copper belt 300 to 400 kilometers long, stretching from Dongchuan through Jinsha River [literally: Gold Dust River] toward the Tianbao Mountains of Xikang Province [today: Tibetan Autonomous Region of Sichuan]. The copper on the surface had been exploited during the Qing dynasty. The prospects of the underground deposits were promising. This is high quality ore. We can extract 4 *jin* of copper for every 100 *jin*. There are some smaller copper mines in the Northeast and Central China, but they only had a yield of 0.5 percent.[31]

He then added that the region experienced severe deforestation, as trees were cut down for fuel. But even then, the mines remained productive.

Various pieces of the electrical equipment industry finally came together on July 1, 1939. The vacuum tube and light bulb production team arrived safely in Guilin from Xiangtan in April 1939. By May 1939, the wire factory started producing bare copper wire and stranded cables from imported metal rods. When the power station became fully operational in June 1939, the copper refinery began extracting high-grade copper, making it possible to produce copper wire domestically. The field telephone and electrical motor manufacturing units, however, were still on the run or waiting for the arrival of machinery.[32]

Financially, the Electrical Works was much like R. H. Tawney's description of a Chinese peasant—"standing up to its neck in water, such that a tiny ripple would drown it."[33] The four units of the Electrical Works reported 2.1 million

CNC dollars in operating revenue, but chalked up 2 million CNC dollars in operating costs. It shored up its balance sheet with a number of accounting maneuvers. It took in 250,000 CNC dollars by selling old equipment from its field telephone factory and through a foreign currency gain of 570,000 CNC dollars. The Number 4 factory in Guilin earned 80,000 CNC dollars by selling electricity generated by the low-capacity power generators that it produced. All said and done, it generated a respectable net income of 950,000 CNC dollars during the first year of operations.[34] The Electrical Works also qualified for low-interest loans. When the Ministry of Finance requested the "Big Four" banks (Bank of China, Central Bank of China, Farmers Bank of China, and Bank of Communications) provide low-interest loans to state-owned industries in 1940, the Electrical Works became the largest beneficiary, as it received six million out of the 44.3 million CNC dollars lent out by the Big Four banks under the funding package.[35] The funds gave a much-needed boost to the Electrical Works. Instead of worrying about financial insolvency, its managers could now focus on the primary mission of supplying the military with wires, batteries, radio sets, and electrical power devices—the little things that kept the forces moving.

The industrial complex built around the copper refinery, wire factory, and power station soon became an obvious bombing target. After the Japanese invasion of French Indo-China in October 1940, Kunming came within range of Japanese bombers. On October 17, 1940, at 2:05 pm, twenty-six Japanese heavy bombers released hundreds of bombs from 9,000 feet onto Majiezi. According to Yun Zhen's dispatch to Chongqing, the bombers flew along an axis formed by the chimneys of the copper-smelting plant and wire factory, while southwesterly winds blew some of bombs off course.[36] In order to minimize construction costs, the Electrical Works had housed the wire production facilities in steel-and-concrete structures, while it placed the production units for field telephones and other electrical equipment in thatch-roofed mud huts.[37] Many of these makeshift structures sustained severe damage or even burned to the ground.

The highly prized wire factory bore the brunt of the Japanese bombing expedition. Yun Zhen detailed the damage:

> Two bombs landed on the roof of the copper wire plant. Several of the concrete beams were shattered, and the walls on the East collapsed. The small, medium and large wire stranding machines sustained severe damage, all testing equipment destroyed, circuit boards, electric meters, water, gas, electric lights and all other electrical appliances in the factory broken beyond repair. The machines on the West were generally fine.[38]

One crippled technician failed to take cover and escape in time and died during the massive bombing raid. Yun estimated that the wire factory lost half of its manufacturing equipment, 60 percent of raw materials, and 5 percent of its finished products. The Electrical Works filed for 400,000 CNC dollars of compensation but never heard back from the military insurance authorities.[39] Meanwhile, it busied itself fleeing from the enemy and spreading out its operations, while also reinforcing factory buildings. The wire factory evacuated two kilometers west to Suyuan, while the electrical equipment plant was rebuilt with brick walls and a clay roof on the original site. The field telephone facility moved twenty kilometers southwest to Anning, a city at the end of the line of the Sichuan-Yunnan railroad.[40]

Kunming's industries struggled to recover in the months after the aerial bombardment. They were also isolated, as the Allies were tied down on the battlefield and unable to render aid. On March 12, 1941, the Electrical Works established ground rules for the rationing of electrical equipment after meeting with representatives from the NRC, Ministry of Communications, Ministry of War, and Ordnance Department.[41] For the next few months, the Electrical Works turned down requests from the military on technical grounds; and it would take a year before they could fulfill the military's orders for insulated wires.[42]

The Lend-Lease Act of 1941 provided much-needed financial relief to rebuild the electrical equipment industries. The United States offered a 100 million US dollar loan in 1941. In the following year, the British furnished a 100 million sterling pound (equivalent to 500 million US dollars) loan through the form of export credits.[43] The Electrical Works drafted a list of several hundred types of materials and equipment worth 598,180 US dollars and submitted their orders through the Universal Trading Company in April 1942.[44] It came as no surprise that the largest expense item was 426 tons of copper rods for wire drawing at the cost of 140,500 US dollars. While the wire factory was designed to produce 2,000 tons of bare copper wire annually, the copper mines of northern Yunnan at best yielded 500 tons of copper. Local sources of copper were limited, especially after Japanese air raids severely damaged the copper refinery.[45] The next largest expense item was 6,865 kilometers of multicrystal tungsten wires of various densities for the radio tube and light bulb factories.[46] In fact, 57.5 percent of its first major purchase under the Lend-Lease Act went toward replenishing raw materials and equipment for projects established through the technology transfer agreements.

Despite being cut off from transportation routes, the Electrical Works remained heavily dependent on foreign suppliers for raw materials and machine parts. They streamlined their acquisition process by dealing with four primary agents in New York, London, and New Delhi. These agents had substantial experience with the China market and shared personal connections with the NRC's engineer-bureaucrats. The New York office of the NRC liaised with Universal Trading Company and Wah Chang Trading Company, both of which were trading firms run by Chinese engineers. Li Guoqin (K. C. Li), a Chinese-American engineer of Hunanese ancestry who founded Wah Chang Trading, exported tungsten ore from China to the United States before the war. During the war, the NRC took over his tungsten ore business in China and appointed his company as the primary agent of electrical equipment from the United States.[47] The NRC's representatives in India liaised with the New Delhi branch of the Pekin Syndicate (listed as *Fu gongsi* in Chinese records). The Pekin Syndicate started out as an engineering firm incorporated in Shanghai and primarily sold materials and machinery acquired from the British. Another agent for the Electrical Works was the China Purchasing Agency in London. These agents in New York, London, and New Delhi received the purchase requests from the Electrical Works and took charge of securing the release of the goods from the suppliers, seeking approval for export, and shipping the goods to the NRC office in Calcutta.[48] The NRC's officers then reserved cargo space on a plane flying from Calcutta into Kunming through the Transportation Bureau and paid its freight charges in full.[49]

The acquisition process was rarely as smooth as described above. Even if the Electrical Works fully paid for their orders, there was no guarantee that they would receive them. American and British suppliers often unilaterally canceled the Electrical Works's orders, citing the lack of availability of transportation. Take the example of the massive April 1942 order placed with Universal Trading. Four months after the initiation of the purchase, the Lend-Lease Administration denied a request to ship the components stating that only "when air transportation from India to China can be guaranteed, can such an application be reconsidered."[50] Even when conditions became favorable, the purchased items trickled in slowly. In November 1942, the Electrical Works urgently needed bronze sheets, bare copper wires, and switchboard wires to produce a batch of field telephones. Chen Liangfu, the NRC representative in New York, pleaded with the Universal Trading Company to resubmit the priority status application to the Lend-Lease Administration. Chen wrote that "we have greatly reduced the above order to a very few items, weighing not over 2 tons, we feel quite hopeful that air-transport space from India to China could be arranged."[51] He even tallied the weight of each item precise to one pound in order to convince

American officers that the goods could be shipped safely to China without over-loading a plane.

Even if the Americans and British allowed the shipment to leave their borders, goods were lost in transit, leaving the Electrical Works to bear full losses. Transportation and insurance costs were high. British agents charged 20 percent of the total value of their orders for insurance and freight.[52] Insurance also offered no absolute protection against losses. When six items from a motor generator set and motor alternator set went missing from two shipments in December 1941 and January 1942, the British insurers denied these claims stating that the ships had already unloaded the goods at Rangoon and Karachi. They cited the condition that "the insurance terminates directly after the goods are discharged at the final port of discharge." The Electrical Works had no other choice but to take out a fresh line of credit to replace those missing items.[53]

The Allies also withheld war materiel that was in short supply. The British curbed the export of rubber—a material used for replaceable parts in machinery and insulation. With the fall of Malaya at the end of January 1942, the British lost a major source of rubber and maintained a tight grip on its inventory. The British held their rubber stocks in India. Prospective buyers of rubber sought approval from the master general of ordnance of the British Army. In September 1942, the Pekin Syndicate in New Delhi, the NRC's agent in India, reluctantly accepted an order for ten tons of rubber and demanded cash payment for the purchases.[54] After nearly one month of negotiations, it finally received approval from the British authorities and confirmed that the Chinese would be allowed to tap the British export credits for these goods.[55] Rubber was rarely used as an insulating material after 1942, and material shortages prompted the Electrical Works to experiment with alternatives such as cambric and insulating paint.[56]

Overcoming Scarcity through Applied Research

Following the lean months between October 1940 and February 1942, the Electrical Works developed an in-house research unit to devise workarounds in the face of material shortage. Yun Zhen fell back on his internship experience at Westinghouse in the 1920s, during which the company was still focused on applied research. As Thomas Lassman noted, Westinghouse housed its research facilities within the manufacturing arm and transitioned from applied research to fundamental research around 1935.[57] Yun established a technical office at the headquarters of the Electrical Works in Kunming to coordinate the research. He divided research staff into three main groups: chemicals, machinery, and

electrical power. Among the twelve areas for research, applied sciences for electrical equipment manufacturing, raw materials and viable substitutes, testing
and developing electrical gauges, and the design of machinery and tools to
make electrical equipment were the most important. Team leaders of each research group submitted monthly reports, all of which would be forwarded to
the general manager.[58] The technical office recruited engineers who graduated
from the top of the cohort and conscientiously followed the research agenda
assembled by Yun Zhen and the management of the Electrical Works. Much
like other research teams scattered across Southwest China, Yun Zhen's research
team worked in makeshift facilities. It became a highly productive unit and accounted for more than half of the patents filed in China during the war, and
their research outcomes were applied in arms manufacturing and refurbishment
of industrial machinery.

Researchers devised more efficient ways to process locally available raw materials. With constant delays to the shipment of tungsten from the United States,
metallurgists devised ways to extract high-purity tungsten from wolframite, an
iron manganese tungstate mineral abundant in China. Ding Chenwei, an assistant engineer with the Kunming copper refinery, consulted Smithells reference
book, which stated that existing extraction methods could only produce tungsten oxide of 85 percent purity. Ding presented a three-step process to obtain
high-purity tungsten. He boiled wolframite in sodium hydroxide and soaked the
end-product in ammonia solution to make ammonium paratungstanate, which
can be converted to tungsten (VI) oxide. He claimed to attain 99.98 percent purity in repeated experiments.[59] Insulating material was also another area of interest. Zhang Lianhua, a chemical engineer with the Central Radio Manufacturing
Works in Guilin, patented his method for making mica plates—an insulation
material that could withstand high temperatures. Zhang's recipe called for mica
minerals from Guangdong, shellac from Yunnan, cow-hide glue, which was "sold
everywhere," and alcohol, which was "available in all pharmacies."[60]

By the second half of 1943, electrical engineering accounted for more than
half of the technological improvements published in the NRC bulletin. In fact,
all the announcements about technological improvements in December 1943
came exclusively from the Electrical Works and Central Radio Manufacturing
Works: the invention of a short-wave radio that consumed less electrical power,
modifications to field telephones that made it difficult for the enemy to intercept
messages, replication of German headphones, novel manufacturing processes
for more durable copper oxide rectifiers, and new research for carbon products

offered promising improvements to electrical power transmission and military communications.[61]

As the only state-owned electrical equipment unit not constrained by any technology transfer agreements, the Number 4 factory became one of the largest contributors to industrial research. As mentioned earlier, it had humble beginnings as a small classroom workshop in Jiaotong University during the 1920s. Taking charge of the production of all electrical equipment except wires, vacuum tubes, light bulbs, telephones, and radios, it took on the bulk of the Electrical Works's product offerings. Headquartered in Kunming, the Number 4 factory opened two other branches in Guilin and Chongqing. Ge Zuhui, an engineer with the Guilin branch, claimed that this plant "was entirely designed and constructed by the engineers of our own country," and "all its products are in high demand."[62] The workforce of the Number 4 factory grew from about fifty workers during its inception to eight hundred in 1944.[63] From 1942 to the end of the war, the Electrical Works replicated different types of electric motors, power generators, and transformers and supplied these components to power stations throughout Southwest China.

Engineer-bureaucrats at the Electrical Works also developed battle damage repair protocols as they transported, reassembled, and refurbished power turbines during the retreat into Southwest China. The refurbishment of hydropower turbines allowed the NRC to complete small hydropower projects under conditions of extreme scarcity. Longxihe Power Station near Chongqing, Yaolong Electric Company in Kunming, and the Xiuwen Hydroelectric plant in Guizhou were among the beneficiaries of the refurbishment projects.[64] The retrofitting of a damaged power generator for the Longxihe Hydroelectric plant provides a glimpse into the operations of the Electrical Works. Zhang Chenghu's description of the project is worth quoting in full:

> The Longxihe Hydroelectric plant of the National Resources Commission purchased a 2500 kW (600 rpm) generator, the steel sheet stator was deformed and the generator coils were completely destroyed. It appointed the Electrical Works to repair and modify it into a 1700 kW 6900 V generator, to suit the needs of the Changshou hydroelectric facility. Our plant took charge of the redesign, manufactured the generator coils on our own, used mica insulating material and completed the project in a year. Apart from handing the production and installation of the steel stator to the Minsheng Machine Works in Chongqing, we designed and installed every

other component. The generator has been in use for nearly a year, and we have not heard of any malfunction. It is truly an accomplishment for the domestic electrical industries to complete the retrofitting of high voltage and high capacity power generators.[65]

The Electrical Works's efforts increased the generating capacity in areas under Guomindang control. Between 1938 and 1944, generating capacity of electric power plants in Southwest China nearly doubled from 35,405 kW to 70,017 kW, while power generated increased from 73,621,694 kWh to 174,229,500 kWh.[66] Considering that the NRC did not establish any technology transfer agreements with foreign enterprises, and that it was unfeasible to transport huge power turbines from abroad, increased electrical output came from the refurbishment of damaged electrical turbines and power generators built through imitation and improvisation.

Besides increasing energy availability, the applied research facilitated the design and manufacturing of electrical tools critical for arms production. Electrical welding accelerated the production process of weapons and enhanced their durability. Zhang Chenghu listed three types of welding machines as part of its new product offerings. Most significant, the Electrical Works supplied the 21st armament factory with three 10 kVA spot welders for the manufacturing of mortar tail fins on a trial basis. Encouraged by the positive preliminary results, the Electrical Works proceeded to manufacture 20 kVA spot welders.[67] The Number 4 factory, along with its counterpart the Electrical Works, became the focal point of China's defense industries. Its production units in Kunming, Chongqing, and Guilin went beyond manufacturing wires, light bulbs, and field telephones for soldiers stationed in the mountainous garrisons in Southwest China. They also supplied armament factories with new manufacturing tools, such as infrared heating tunnels, dielectric heating machines for metalwork, and x-ray devices for checking metal defects, which improved the quality of Chinese-made weapons.

For all its resourcefulness, the Electrical Works maintained a modest production scale. Output peaked in 1942 and 1943 but fell significantly after the Japanese captured Guilin during the Ichigo Offensive in 1944, which forced the Electrical Works to relocate their operations from Guilin to Chongqing and Kunming. Writing in 1945, Zhang Chenghu acknowledged that "the state-run factories have shown slight improvement in recent years, but technologically advanced factories are making rapid progress. In comparison, we are lagging even further behind, as our annual output is even lower than the daily or weekly production of European and American factories."[68] The output of these make-shift wartime factories remained miniscule compared to the great electrical conglomerates such as General Electric, Westinghouse, and Siemens.

Despite these difficulties, the Electrical Works strived for self-reliance, while continuing to explore partnerships with foreign suppliers. In 1944, the NRC transferred Yun from the Electrical Works in Kunming to its New York Office, where he assumed the post of technical director and took on the task of securing technology transfer from electrical equipment manufacturers in the United States. Yun spelled out his guiding principles regarding foreign acquisition of electrical equipment:

> Our nation only has finite resources, making it impossible to revive the industries and organize new enterprises of all types with our limited manpower and capital. We have to closely examine the capability of each industrial sector. Then, we can decide which items are in huge demand that have to be manufactured on our own, which items are needed in small quantities that are not worth the effort to produce domestically and should be acquired from abroad. In addition, there are also some commodities that have to be imported initially, but we will have to eventually learn to build on our own.[69]

In his dealings with Westinghouse, Yun remained focused on the ultimate objective of self-reliance and took an uncompromising stance over the license fee to maintain the cost effectiveness of the technology transfer agreement. Yun treated his negotiations with prospective American partners as a battle to defend China's economic sovereignty. Electrical equipment was the key to China's energy independence, and Yun would do everything in his power to prevent China's electrical sector from being beholden to American corporate interests.

Standardization and the Defense of Economic Sovereignty

Cooperation with American electrical equipment manufacturers presented certain risks to the NRC, as it potentially undermined its efforts toward standardization pursued during the War of Anti-Japanese Resistance. Wartime constraint increased the urgency for greater standardization. The Electrical Works not only became the dominant supplier of electrical components but also functioned as a de facto national standards institute for the power sector. While the NRC enforced the adoption of uniform voltage standards, the Electrical Works manufactured a limited range of products that complied with standard voltage settings and achieved significant cost savings. However, American industries and professional associations pressured the NRC to abandon existing voltage standards in exchange for technical assistance. Attempts by American industries to impose their standards on China provoked a nationalistic backlash from the NRC's engineer-bureaucrats. This forced the NRC into an uneasy compromise when they concluded their technology transfer agreement with Westinghouse. Their American partners agreed to comply with Chinese industrial standards but negated the value of any knowledge produced by their Chinese counterparts. They also undermined the NRC's bureaucratic authority by cutting private deals with political leaders of the Guomindang regime. The Americans renounced the unequal treaties in January 1943. Yun saw the technology transfer agreement with Westinghouse as a reminder that the Americans never regarded the Chinese as equals.

Wartime austerity led to greater compliance with China's voltage standards. During the height of the economic blockade, the engineer-bureaucrats revisited the four most commonly adopted standards around the world—the 1930 National Electrical Manufacturers Association (NEMA) Standards of the Americans, the 1939 British Standards (known as BS 77), the 1932 Verband der Elektrotechnik, Elektronik und Informationstechnik (VDE) standards of the Germans, and the International Electro-Technical Commission (IEC) Standards.[70] Wang Shoutai, whose brother Wang Shoujing served as director of the NRC's technical office in New York, noted that all four prevailing voltage standards involved a tradeoff between cost and safety:

> Germany is a country that lacks resources, so it pays more attention to the conservation of materials in the design of electrical equipment than the British and Americans. This is not to say that the British and Americans are wasteful. It is just that the British and Americans have more resources at hand and devote greater care to product safety. This is not a question of "right or wrong."[71]

Wang found Germany's situation most suitable for China. Voltage settings adopted under the IEC standards, which were similar to the German VDE standards, had most in common with existing power stations. Following the IEC standards led to savings in raw materials, which in turn relieved the financial burden on China's wartime economy.

When American suppliers pressured the Chinese to adopt a 110 V/60 Hz voltage and frequency setting, the NRC explained that its 1931 voltage standards were best suited to a resource-poor country like China. Yun Zhen drove home the point that supplying a 50 Hz triple phase alternating current at 220/380 V required less copper than 110 V. The shortage of copper weighed on Yun Zhen's mind, as Kunming's copper smelting plants failed to keep up with the demand for raw materials from his electrical equipment factories. Yun also warned his colleagues that China could not afford to adopt the laissez-faire approach of the United States. He noted that Westinghouse manufactured six hundred different types of electrical machines to fit the specifications of thirty-five different combinations of transmission voltages across the United States, thus driving up the average cost of production.[72] Unlike Westinghouse, the Electrical Works lacked the capital to produce such a wide range of products. Yun Zhen also wanted China to avoid the situation in Japan, where the eastern region adopted 50 Hz transmission frequency, and the western region 60 Hz.[73] The War of Anti-Japanese Resistance obliterated many small privately owned power plants that resisted state regulations, leading to greater uniformity across the nation's power sector, and cooperation with the Americans must not undermine the progress made thus far.

Despite these disagreements over standards, the NRC went ahead and pursued a partnership with Westinghouse to acquire manufacturing expertise for power generators. Salvaging existing equipment might suffice for the war effort but it was not a viable long-term strategy. Based in Pittsburgh, Pennsylvania, Westinghouse maintained cordial relations with China's wartime government in Chongqing and emerged as the most ideal partner. As early as November 1941, Westinghouse offered paid internships to two of the Electrical Works's engineers, Chu Yinghuang and Lin Jin.[74] Westinghouse presented itself as "one of the most experienced and progressive electrical equipment manufacturers," prepared to transfer the necessary technology to build generators, transformers, switchgears, and electric meters.[75]

As early as 1942, Westinghouse began pressing the NRC to abandon its prevailing standards and adopt the 110 V service voltage commonly used in the United States. At a time when he struggled to get China's fledgling electrical equipment industry off the ground, Yun Zhen received a letter from Westinghouse detailing the merits of the 110 V system. In his May 1942 letter, S. V. Falinksy from the Associated Companies Department listed five benefits:

1. Safety—115 volts-to-ground is less dangerous than 220 volts-to-ground.
2. Cost of equipment built for 220 volts-to-ground is higher, because more insulation is required. Some countries specify a ground wire to appliance; others specify a special metal shroud to protect the terminals, etc.

3. In appliances where a low wattage coil is desirable (say, a 50-watt warmer in percolator), the wire size is very small even for 115 volts—and for use on 220 volts, it is impractical due to extreme fineness of the wire.
4. If a 115–230 volt system should be accepted, existing 220 volt appliances and other equipment could be used by connecting across the 230-volt line.
5. Small orders of special voltage appliances do not receive adequate development and testing.[76]

Yun Zhen, who had just completed the relocation of the field telephone unit to the Qingying mountains and was begging the Americans to release emergency shipments, offered a terse reply. He reiterated the official stance: China stayed on the 220/380 V system simply because it was most commonly adopted and used less copper. He further explained that the European IEC standards allowed China to preserve most of its preexisting transmission networks. It also opened up more options for sources of electrical equipment. Yun noted that China dropped the 6,600 V setting as the intermediate distribution voltage commonly used in the United States and instead adopted the IEC standards of 6,000 V, as the "existing generating plant capacity using 6,600 volts is comparatively small," and a "6,600 volt plant can still be operated if we use transformers having a few more taps to conform to the new 6,000 volt system."[77]

The American Standards Association (ASA) and the American Institute of Electrical Engineers (AIEE) also weighed in on the matter. The ASA acknowledged the NRC's argument, stating that "the greater cost of distribution at the lower voltage and the fact that 230 volts is already generally used in China would justify the Chinese Commission in adopting the IEC standard of 230/400 volts." The professional associations cautioned the Chinese against the wholesale adoption of the European standards simply because the IEC "have an international standing in the engineering profession."[78] AIEE chairman of the power committee A. C. Monteith pointed out that 20 kVA three-phase distribution transformers, 800A, 1,000A, and 1,200A circuit breakers, were all not part of the American standards. Five out of the eight settings for circuit breaker capacity listed in the Chinese standards were also not used in the United States.[79] Put bluntly, if the Chinese insisted on maintaining their voltage standards, the Americans would not be able to assist them with wartime electrification and postwar reconstruction.

The NRC pleaded with their American counterparts to take into account China's local conditions. Writing to the ASA secretary P. G. Agnew, Chen Liangfu expressed his appreciation of the ASA's "unbiased viewpoint" and related a story to him: "A shoemaker once made many pairs of shoes according to his own size. When he was asked by a customer to offer some bigger sizes, he remarked: Can't you cut your feet to fit my shoes?"[80] He made his point with the example of 33,000 V generators, which were not popular in the United States and England, but "will be very convenient in China. In metropolitan areas and suburban districts, power could be distributed without any step-up transformers. It needs [to be] only stepped up once for long-distance transmission and stepped down once for industrial or domestic use."[81] NRC chairman Weng Wenhao learned that American manufacturers supplied Soviet Union with power generators in compliance with IEC standards. There was no reason why the Americans would be unable to do likewise for the Chinese.[82] The voltage standard was more than a technical problem. It was an issue of Chinese economic sovereignty. The NRC stood its ground and selected the specifications that best suited China's needs. It successfully defended its decision. The first set of voltage standards ratified by the People's Republic in 1954 was based on these wartime decisions.

Having forced the Americans to engage the Chinese on their own terms, the NRC opened negotiations for a technology transfer agreement with Westinghouse. In a March 1944 proposal submitted to the Chinese government, Westinghouse formulated the plan "after consultation with engineers of the Chinese government," and it aimed to develop "an electrical manufacturing industry carefully patterned to the specific requirements of China."[83] The company also added that, "contrary to ill-famed procedures of the past, where foreign capital installed subsidiary companies in Far Eastern countries under foreign directorship and carried off the profits and fruits of local labor, Westinghouse wishes its products to be manufactured in China by Chinese owned and operated factories."[84] The proposed plant employed 3,870 people and manufactured 1,950 carloads of motors, generators, transformers, switchgears, meters, porcelain, insulation, and varnishes with an estimated net sales billed of 16.5 million US dollars. Westinghouse pledged to train 221 management staff in its East Pittsburgh plant in the first five years. It projected an operating profit of 2.475 million US dollars and estimated that the manufacturing plant would eventually achieve a net sales billed of 27.32 million US dollars as more workers joined its ranks.[85] Westinghouse did not spell out the licensing fee in the proposal, and this became a sticking point in Yun Zhen's negotiations.

William Hunt, Westinghouse's agent in China, submitted the draft of the technology transfer agreement to Finance Minister H. H. Kung (Kong Xiangxi), who then forwarded it to Deputy Minister Weng Wenhao. Yun Zhen, who had completed his political training at the Guomindang Party School at Chongqing, met Weng to formulate the response.[86] Weng said to Yun, "Consult the experts in your factory and see if the products to be co-developed suit our needs. Make adjustments and bring the list to the United States. While you are negotiating the price with Westinghouse, approach Allis Chalmers and General Electric and report to me."[87] Yun Zhen departed from Kunming in August 1944 onboard a US Army transport plane bound for Calcutta, took a train to Karachi to catch his flight to Morocco, then finally completed his transatlantic journey to New York. Twenty-three years after leaving the University of Wisconsin, Yun Zhen finally returned to the United States in the fall of 1944.

At his office on 111 Broadway, in New York City, Yun Zhen spelled out the terms of engagement to R. D. McManigal, the manager of Westinghouse Associated Company Department. He listed the specifications for the transformers manufactured under the technology transfer agreement and stressed that the voltage ratings were "recommended by the electric utilities in China."[88] Even then, three vice presidents of Westinghouse insisted on having China switch to the 110 V/60 Hz system. They argued that the 60 Hz system was more efficient, as the generator completed 3,600 revolutions rather than 3,000 revolutions in a minute. Yun refused to give in on this point. He tried to gain leverage by concurrently negotiating with Allis Chalmers and General Electric. A clause stating that the technical adviser had to supply engineering information "according to the requirements of C.E.M.W [Central Electrical Manufacturing Works]" was present in the draft contracts with both Allis Chalmers and Westinghouse.[89]

The value and ownership of intellectual property ultimately became sticking points in the negotiations. Yun accused Westinghouse of charging exorbitant licensing fees and levying high royalty rates that subjected the Chinese to a long-term financial burden. Yun had set an upper limit of 3 million US dollars for the licensing fee, which was to be paid within the first ten years of the factory's operations. Hunt bypassed Yun and struck a deal with Finance Minister Kung, who agreed to pay the 3.4 million US dollars licensing fee up front in one lump sum. Hunt then showed up at Yun's hotel in Pittsburgh with Kung's letter. Yun had no choice but to follow his superior's orders. In his December 1944 letter to the NRC's US trading agent, Yun wrote, "As Dr. H. H. Kung has already given approval to this agreement, we shall highly appreciate if you would kindly arrange with the United States Import and Export Bank to have the above amount of US$3,400,000 made available to us through your office."[90] Yun also complained about Westinghouse's plan to impose a fixed 3 percent royalty on annual production starting in the tenth year. Responding to the protests,

Westinghouse gave a minor concession: "Beginning from the 11th year, 3 percent royalty will be paid for the annual production of the first US$10,000,000, 2½ percent for the second US$10,000,000 and 2 percent for anything above US$20,000,000." It also agreed that the NRC was also entitled to use all information "including subsisting patents" by paying 450,000 US dollars.[91]

Looking back decades later, Yun not only distrusted private enterprise but also felt betrayed by his superiors. Negotiations with GE and Allis Chalmers fell through, as they found it too risky to invest in China. By January 1945, Westinghouse was the only contender for the joint venture with the NRC.[92] Yun was also infuriated that Westinghouse claimed ownership over any intellectual property generated by their Chinese partners. Yun recalled that McManigal insisted on Westinghouse gaining free access to any product improvement, inventions, and patents contributed by their Chinese partners, since "the Chinese are inexperienced apprentices who cannot possibly have too many inventions and creations."[93] In his interview with the left-leaning reporter Xu Ying of *Ta Kung Pao*, Yun lamented, "Technical cooperation should be mutually beneficial, so the Americans should also pay for our patents. It would be our honor, even if they only give us a dollar."[94] In the end, the electrical industries on mainland China did not benefit much from this technology transfer agreement. The NRC failed to follow through on plans to send more than two hundred technicians for advanced training at Westinghouse, as the Electrical Works became preoccupied with the takeover of Japanese electrical equipment plants in the northeast provinces and Shanghai after the Japanese surrender in August 1945. Following the Communist takeover in 1949, the plan for the NRC and Westinghouse to jointly develop electrical equipment manufacturing facilities in mainland China was aborted.

EIGHT YEARS OF wartime mobilization broke down decades of foreign dependence. The Electrical Works headquartered in Kunming set into motion the dynamics for accelerated development. Its incorporation marked a sharp departure from the laissez-faire approach toward the electrical power sector before and during the Nanjing decade. Zhang Chenghu, who built the wire manufacturing facility from scratch and took over the development of electric motors, pointed to the Electrical Works as a success story for state intervention. Coordinated industrial planning led to the optimization of limited resources, reduction in wastage, and timely delivery of orders.[95] Just before the Japanese surrender in August 1945, Yun made the case for nationalization during his debate with pro-business, entrepreneur-turned-politicians Lu Zuofu and Zhang Jia-ao by citing the achievements of the state-run Electrical Works.[96] The wartime experience showed that China could not rely on private enterprise to fund the expansion of electrical infrastructure. Yun argued that profit-optimizing capitalists only cared about short-term gains and state-owned enterprises promoted long-term policies, as they undertook the risk to "purchase equipment before there is demand and reinvest newly incurred profits."[97] Yun's view stemmed from his cynical outlook on private enterprises and was reinforced by his bitter exchanges with Westinghouse's representatives.

Accelerated development came at a price. Having to contend with material shortage, the NRC's system builders sacrificed output efficiency to achieve increases in generating capacity. This meant using metals of lower quality for turbine fan blades, increasing the content of metals with poor conductivity in wires and cables, or using alternative materials with a shorter shelf life. These resulted in residual inefficiencies in the form of high transmission power losses or frequent breakdowns, the consequences of which will be elaborated further in chapters 6 and 7.

Technology transfer agreements concluded in the wake of and during the War of Anti-Japanese Resistance integrated China with the global marketplace on one hand, but provoked a nationalistic backlash when state-owned industries suspected foreign partners of profiteering, on the other. The first three technology transfer agreements for wires, radio tubes, and field telephones allowed the NRC to acquire valuable expertise to manufacture much-needed electrical equipment. They also provided little leeway for adaptation and improvisation and required the NRC to replenish materials according to specifications spelled out in the terms of exchange. The earlier experiences prompted Yun Zhen to take precautions to guard against the loss of autonomy that came with technology transfer during his negotiations with Westinghouse. Despite his best efforts, the NRC was forced to accept high licensing fees on unfavorable terms. Accusations of "unfair trading practices" lurked beneath the signed and sealed agreement between Westinghouse and the Chinese government. Both parties never fully resolved disagreements over the proprietorship of intellectual property and procedures for access to technical knowledge. The trade wars of the early twenty-first century revolved around the same issues.

By supplying China with the tools to harness its vast fossil fuel reserves, the wartime electrical equipment industry initiated the Great Acceleration ultimately launching China into the second stage of the Anthropocene. The intense mining for metals and coal came with the establishment of the electrified military-industrial complex, leading to the "rapid and pervasive shift in the human-environment relationship" that Will Steffen, Paul J. Crutzen, and John McNeill associated with the first stage of the Anthropocene. During the War of Anti-Japanese Resistance, the electrical equipment industry developed technologies that "represented new applications for fossil fuels," which would later fuel the high growth rates during the second stage of the Anthropocene.[98] The products that rolled off the floors of China's first domestic electrical equipment factories were ultimately used to unleash destruction on the battlefield in the age of total warfare. With the end of the armed conflict, the energy released during wartime mobilization was channeled toward the transformation of China's natural environment.

The retrofitting of the hydropower turbine at Longxihe Station was a precursor to the tumultuous postwar reconstruction. Some of the engineer-bureaucrats involved in this project received the opportunity for advanced training in the United States. The next chapter follows these trainees as they looked to master dam builders working for the federal government of the United States for inspiration and conjured dreamscapes of a powerful China that would successfully harness the massive power of its rivers.

CHAPTER 5

Turning the Tide

O N NOVEMBER 1944, the American master dam builder John Lucian Savage produced the blueprint for a massive hydropower station that straddled the Yangtze River. Standing 738 feet, the proposed dam would have been twelve feet taller than the Grand Coulee, which was then the highest dam in the world, but the electricity produced by its ninety-six mighty generators would be five times that of its American counterpart.[1] For the first stage of its development, Savage ordered "the excavation and concrete lining of all (twenty-six) tunnel and shaft systems" that could be either concentrated on one side of the river or divided between two sides as depicted in figure 5.1. He also called for the construction of earth and rock fill cofferdams, the completion of six tunnel systems for four main and two station service systems, and the installation of sixteen main power units.[2] Jack Savage, the "Billion Dollar Beaver" even quipped that the Yangtze Dam would make the Boulder Dam, which only had four diversion tunnels, "look like mud pie."[3] Savage, however, did not specify the amount of time needed to complete the project. The construction would presumably be quick; returns on investment would also be high. He projected that twenty years of electrical power revenue would easily cover the 653 million US dollars needed for stage one. To him, all that mattered was that the Yangtze Gorges Dam would be a "CLASSIC" that would "change China from a weak to a strong nation."[4]

Savage did not conjure the dreams of accelerated development on his own. The vital spark of imagination came from two young Chinese engineer-bureaucrats trained at the Tennessee Valley Authority (TVA) during the height of World War II. Sun Yun-suan, the future premier of Taiwan, and Zhang Guang-dou, the future vice-chancellor of Tsinghua University in Beijing and advocate for the Three Gorges Dam, were among thirty-five engineers dispatched to the United States for this advanced training. They survived an arduous 15,504-mile journey across four continents. On their way to the United States, both men fell ill with malaria as they moved through sub-Saharan Africa and received treatment at a US Army Hospital in Accra, Ghana.[5] The TVA, where they spent the

FIGURE 5.1. *Artist Impression of John Lucian Savage's Proposed Yangtze Gorges Dam at Yichang.* Source: John Lucian Savage Papers, box 4, American Heritage Center, University of Wyoming.

most formative years of their training, happened to list malaria eradication as one of its objectives.[6] Inspired by the federal government's achievements, Zhang and Sun introduced themselves to Savage and described China's vast potential for hydropower development to the American dam builder.

Chinese engineer-bureaucrats selectively appropriated parts of the TVA's management practices that would allow China's state agencies to strengthen their grip on the electrical power sector. Unlike David Ekbladh and Christopher Sneddon, whose work primarily discussed how the TVA served American national interests, this chapter focuses on how the Chinese adapted lessons learned from the Americans to address pressing needs of national reconstruction. The TVA's technical assistance, which in the words of Sneddon, worked as "a geopolitical tool within the apparatuses of the American state as it sought to extend its influence over the underdeveloped regions of the planet in the mid-twentieth century."[7] No doubt, the Americans won over the hearts and minds of the Chinese trainees by graciously welcoming them to work alongside them at TVA sites. The Chinese engineers showed less interest in the "public-private cooperation," which, according to Ekbladh, set the TVA apart from statist Nazi Germany, the Soviet Union, and imperial Japan.[8] Having spent their early careers forcing private operators to comply with state regulations with limited success, the Chinese trainees looked toward the TVA for a roadmap to accelerated economic development under the auspices of centralized state power.

The hubris of victory gave rise to a blank-slate mentality that took hold among the Chinese engineers and their American counterparts. Sheltered from the vagaries of war on the American home front, Chinese engineers like Zhang and Sun observed how an American federal government agency worked without interference from private enterprise to raise the living standards of its citizens in the rural South through integrative river basin developments. The Allied Victory strengthened their conviction about the viability of the TVA model in China. Their plans for accelerated development of hydropower came at a time, when the tide of World War II was turning in the favor of the Allied powers. Savage's Yangtze Gorges proposal exemplified this blank-slate approach. Their grandiose plans failed to consider basic factors such as geology, sedimentation, and, most important, budgetary constraints of a postwar state. The time and energy spent on the megadam project diverted valuable resources from smaller hydropower projects that could have been completed in a shorter timeframe and addressed power shortages. The defeat of the Guomindang regime in the Chinese Civil War (1946–1949) would dash the dreams of accelerated development.

Forging Sino-American Technological Diplomacy

The TVA, an agency established under Roosevelt's New Deal, provided an ideal training ground for China's engineer-bureaucrats. Its projects exemplified the centralized approach to infrastructure and industrial development that the NRC sought to emulate. While NRC's engineer-bureaucrats had considerable experience in salvaging electrical power systems, they lacked expertise in designing large-scale hydropower plants and managing long-distance transmission networks. Their training at the TVA led to a shift in their mindset from survival mode to long-term strategic thinking. The integrative river basin developments, which achieved multiple objectives of flood control, fertilizer production, land reclamation, and hydropower generation and distribution, inspired the Chinese engineers to outline ambitious plans for national reconstruction. The engineer-bureaucrats even visualized TVA project sites in Tennessee, Alabama, and Georgia as microcosms for large-scale power developments undertaken at the national level that would catapult China into the ranks of industrialized nations.

The TVA and China's Nationalist government learned that they shared a common vision of the government's role in industrial development, paving the way for the TVA's partnership with the NRC. The New Deal's opponents accused President Franklin Delano Roosevelt of using federally funded projects to secure absolute political power. Lilienthal learned about a speech that Hu Shi, then China's ambassador to the United States, delivered at the bicentennial conference of the University of Pennsylvania on September 19, 1940. Hu argued that there is no "danger of dictatorship" in the United States, citing the fact that the president "cannot carry his own Dutchess County during all the eight years of his administration," and "when the party in power controlled 80 per cent of the votes in both Houses of Congress, the President could not push through Congress some of the legislation which Congress did not like."[9] David Lilienthal, co-director and later first chairman of TVA, expressed his gratitude to Hu for his spirited defense of the federal government and offered to call on the ambassador at Washington, DC.[10] Before departing from his post in 1942, Hu spoke in Nashville but did not stop in Knoxville to bid farewell to Lilienthal. A year later, the Cornell-trained electrical engineer Huang Hui, who was visiting the TVA, presented Lilienthal with a fan with a picture of the goldfish on one side and an inscription written by Hu Shi on the other. Hu's chosen passage was from military strategist Sun Zi's essay "On Nature," which aptly summed up the objective of the technological diplomacy between the NRC and TVA: "If you let things multiply by themselves, will it not be better to apply your effort to transform them?"[11]

Back in Chongqing, the NRC began nominating candidates for advanced training in the United States. With electricity emerging as the main focus of applied science research during the war, more than one third of those selected (twelve out of thirty-five) came from the power generation and electrical equipment industries. The NRC carefully rationed limited training opportunities among its industries. The electrical equipment industry was limited to six nominees. Yun Zhen had already secured internships for Chu Yinghuang and Lin Jin to study power generators at Westinghouse. He assigned the four remaining slots to high priority areas and made a compelling case for each of his nominees. In his recommendation for Tang Mingqi, a twenty-nine-year-old deputy engineer with the Central Electrical Works who had worked at Shanghai Siemens and the Chinese-owned Huacheng Electrical Plant before joining the Electrical Works as a circuit designer, Yun drove home the point that transformers served as "the lifeline of electrical systems, and the quality of design and production quantity will determine the fate of our electrical power sector."[12] Besides vouching for Tang's work ethic, Yun stressed the importance of learning new production methods to overcome existing limitations. He acknowledged that the Electrical Works was capable of producing medium and small transformers on an ad-hoc basis, but "encounter[ed] considerable difficulty in manufacturing high-voltage and large transformers and the mass production of medium and small ones."[13]

Engineers outside the power generating and electrical equipment industries highlighted the connections between their areas of expertise and electricity. Wu Daogen, a specialist in alloy manufacturing, provides the most telling example. Wu listed eight different types of steel smelted in electrical furnaces in his essay before declaring that, "the furnace is at the heart of alloy manufacturing, so it is necessary to pay attention to the supply of electrical current, the circuitry, electro-thermal efficiency" among other things. He pointed out that the electrode was the heart of the furnace, as "it facilitates the transformation of electrical energy to heat, which allows the melting of metals and creation of alloys."[14] No amount of manual labor can transform iron ore into electrical steel. It was electricity that enabled vital chemical reactions in materials processing.

Successful candidates had distinguished themselves in the line of duty. Sun Yun-suan and Zhang Guangdou salvaged power generating equipment to build electrical power systems from the ground up with little logistical support in far-flung regions. Zhang, who had graduated from the University of California at Berkeley and Harvard University, rushed back to China when the second Sino-Japanese War broke out in 1937. The NRC ordered him to rehabilitate the Longxihe hydropower station, sixty miles northeast of the wartime capital Chongqing—the process of which was discussed in the previous chapter. Sun, who had studied with Soviet instructors at Harbin Engineering University and joined the NRC in 1936 after working as an engineer at the Lianyungang Power Station in Jiangsu, quickly earned Yun Zhen's trust. As seen in the opening paragraphs of chapter 3, Sun coordinated the evacuation of power generation equipment from Lianyungang to Ziliujing in Sichuan. Sun was later deployed to Xining in the sparsely populated Qinghai Province to build a coal-fired power station. As the electrification of Xining failed to stimulate the development of industry, the NRC's director, Weng Wenhao, transferred Sun to Tianshui in Gansu Province.[15]

Sun and Zhang delayed their departure for the United States for several months. Their colleagues began leaving in Chongqing and Kunming in batches in June 1942.[16] With the Pacific Ocean closed to civilian air traffic after the bombing of Pearl Harbor, the Chinese engineers flew across the Himalayan

Hump from Kunming to Calcutta and waited for weeks before they could board the short flying boats of the British Overseas Airways Corporation, which hopped from city to city along the Indian coastline, Arabian Gulf, and Tigris River until they reached Cairo. They headed into sub-Saharan Africa on US Army Air Transport, before flying to Belem in Brazil, where they boarded a ship to take them to Miami.[17] A typical journey involved 15,506 miles of flights with twenty-eight transfers. Air sickness, harsh weather, endless waits, and sleep deprivation were part and parcel of this tortuous trip. With no delays, the journey from China to the United States took about five weeks.[18] That was rarely the case.

Following the same travel itinerary as their colleagues who traveled before them, Sun and Zhang encountered numerous obstructions along the way. But they found meaning in the sights and sounds of their arduous journey. One of Sun's most memorable moments was the flight on a BOAC flying boat over the Suez Canal on the tenth leg of his journey.[19] It conjured his memories of toiling in the arid regions in Northwest China. Sun wrote in his diary on March 27, 1943:

> In this boundless sandy desert, where one can barely see people walking around, there are roads filled with vehicles and oil pipelines that run for hundreds of miles. Hundreds of fearless engineers toiled tirelessly and self-lessly, in order to make such a contribution to mankind. The crisis of our nation is deepening. We should be ashamed to death that so many of our young and accomplished engineers shake their heads incessantly at every mention of Northwest China.[20]

The sight of the canal and pipelines in the desert brought to mind the NRC's Yumen Oil Fields in the deserts of northwestern Gansu. As the superintendent of Tianshui Power Plant hundreds of miles east of Yumen, Sun knew too well the unwillingness of young engineers to take up hardship postings in Northwest China.

Sun and Zhang also fell ill with malaria, further delaying their journey. They ran straight into the North African campaign and could not fly farther west from Egypt, so they headed south toward modern-day Sudan, Chad, Nigeria, and Liberia on US Army Air Transport. The US Army transport planes ferrying Sun and Zhang could have played a part in speeding up the spread of malaria.[21] On April 18, 1943, while in Accra, Ghana, Zhang awoke to paroxysms of high temperature, chills, and profuse sweating. Five days later, Sun exhibited the same symptoms. Sun filled a gaping hole in his dairy between April 26 and May 3, 1943, with a phrase "Receiving treatment for malaria at a US military hospital at Accra."[22] They recovered by May 12, just in time for the next flight out from Accra to Natal in Brazil. On May 20, 1943, Zhang Guangdou established contact with the New York office of the NRC for the first time in the ten-week journey, informing the director, Chen Liangfu, that he and Sun Yun-suan would depart from Miami by train to Washington, DC.[23] Sun and Zhang stopped by the Library of Congress to gather research materials before reporting to the New York office on May 27.[24]

Convinced of the comprehensiveness of TVA's training program, Sun and Zhang abandoned their original plans and headed out to Knoxville, Tennessee. Sun had applied for a twenty-one month posting to the transmission and relay department at Westinghouse or GE to learn more about long-distance power transmission; Zhang Guangdou, who had supervised the Wanxian hydropower project, was planning to study dam building at the Bureau of Reclamation.[25] Chen Liangfu persuaded them to take on a broader training program at the TVA. Both arrived together at Knoxville, Tennessee, on June 23, 1943. Zhang also became the only trainee for dam building in his cohort, as his fellow hydroelectric engineer Jiang Guiyuan was confined in a sanatorium after being diagnosed with tuberculosis upon his arrival in the United States.

The TVA not only provided hands-on experience in project management to the Chinese trainees but also hosted activities to rally support for China's war effort. Seven out of the thirty-five engineers received advanced training at TVA facilities. Other trainees attended regional conferences held at TVA project sites. The Chinese engineers exhibited their professionalism by attending these events dressed neatly in suit and tie. They also expressed their appreciation for American assistance by delivering carefully scripted statements to the press. The dates of the conferences also coincided with important anniversaries. The second local meeting between May 15 and 18, 1943, at Wilson Dam in Florence, Alabama, was held in conjunction with the tenth anniversary of the TVA. The third regional conference at Chattanooga on August 13, 1943, was scheduled on the same day as the sixth anniversary of the battle for Shanghai.[26] The TVA Engineers Association put up a statement of support for their Chinese counterparts and used these occasions to reaffirm Sino-American friendship.

Diplomatic pleasantries aside, Chinese trainees at the TVA kept themselves busy with a packed training schedule. Zhang, whose training laid the groundwork for a career in dam construction, offered a glowing assessment of the TVA in his first training report submitted to the NRC's New York office, listing "the rapid progress of hydroelectric development with up-to-date construction methods and practices, the unified plan of the development, the close resemblance of our Commission and the Authority in ideas" as key advantages.[27] Zhang started by working on designs and drafts. In his August 1943 report, Zhang, however, complained that this training method was "quite inefficient. Because every single drawing takes about one month, and most of the time is used in making fine drawings according to standards."[28] Zhang's supervisors approved his transfer to a higher level, where he studied design features and key structural elements of hydroelectric projects. Zhang obtained firsthand experience on the division of labor and specialization required for the planning and building a hydroelectric plant.

Sun, alternately, gained expertise in power generation and transmission. He did not stray from his objective of "paying close attention to the flexible application of power grid designs, hereby ensuring the stability of transmission voltage and frequency" spelled out in his application.[29] He learned about cost-effective power grids, emergency electrical power deployment systems, lightning protection mechanisms, and interconnected regional power systems. Every segment of his training covered a lot of ground. His time at the power operation department was divided into six sections. He started out with the system operations division for eleven weeks, where he learned power statistics, load dispatching, substation operation, and relay and protection. He spent six weeks covering hydroelectric and thermal power production, then moved on to electrical testing, power system servicing, substation maintenance, and transmission line maintenance for one week each.[30]

Besides summarizing key lessons, training reports also contained vital intelligence that helped the Chinese government negotiate favorable terms for technical assistance. Within the first month at the TVA, Sun and Zhang learned about Soviet-US cooperation in hydroelectric projects. On July 18, 1943, Zhang Guangdou sent a memorandum to Chen Liangfu informing him that five Soviet engineers arrived in Knoxville to seek TVA's assistance with hydroelectric dams under the Lend-Lease Agreement.[31] Sun, who had studied under Soviet instructors at Harbin Engineering University, spoke with the Russians.[32] He even helped design eleven of the eighteen small-scale hydroelectric plants for the Soviets during his first month of training in the TVA's design department.[33] In his first training report sent to the New York office, Sun expressed his amazement with these stations that were "Equiped [*sic*] with supervisory control and tele-metering equipment, which means that stations can automactically controled [*sic*] from the center control station."[34] These small hydroelectric stations had a power output of less than 5,000 kW and directly connected to high-voltage transmission lines that delivered electricity over longer distances. This resembled the NRC's plan for Sichuan. Zhang thus urged the Resources Commission to request technical assistance on terms similar to the Soviets. Chen Liangfu wrote to Lilienthal after receiving reports from Sun and Zhang but was told to seek authorization from the State Department, Lend-Lease Administration, or the Office of Foreign Economic Administration.[35]

The trainees also saw the TVA as a model for techno-nationalist development. In his August 1943 report, Zhang expressed amazement with the TVA's "unified plan of development . . . to reach the most economical and maximum benefit as a whole."[36] Zhang acknowledged that building power systems solely with government resources can be cumbersome but believed that the benefits outweighed the costs.[37] He recognized that private firms may be more efficient in catering to market demand, as they could carefully evaluate "cost, desirability, necessity, or soundness prior to adoption and authorization."[38] That said, the nine advantages of public developments outweighed the benefits from market efficiencies. The government incurred lower interest rates, improved social welfare by bringing electrical power to underdeveloped regions, and considered other long-term benefits such as national defense and improvements in overall living standards. Zhang concluded, "After a decade of controversy, the trend of power utilities proceeds in the direction of public control, although very slowly. The progressive minds believe that for the public weal [sic] the power utilities should be developed and regulated by the government."[39] Zhang's observations of the TVA provided talking points for the NRC to present a compelling case that a state monopoly over the electrical utilities would yield the greatest good to the greatest number of people.

Zhang saw abuses by private contractors as the main source of wastage in US government projects. After transferring to the Bureau of Reclamation in 1944, he compared the operations of the TVA with those of the Bureau of Reclamation. He observed that the TVA tended to complete projects on its own, while the bureau mostly hired private contractors. He conceded that contractors achieved cost savings by using the same set of equipment for multiple projects and quickly adjusted labor inputs to meet project deadlines. Zhang, however, argued that the disadvantages outweighed the benefits. At the Bureau of Reclamation, Zhang noticed that contractors reaped excessive profits by sacrificing work quality to save on labor and materials. He thus concluded that the TVA's exclusion of contractors allowed it to maintain greater quality control. Granted that there was too much red tape and substantial waste of material and labor, Zhang asserted that most TVA employees "believe that these wastes will be greatly outbalanced by the low quality and over-profit of a contractor system."[40] Taking over as chief

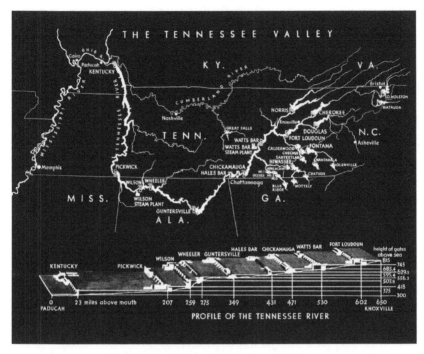

FIGURE 5.2. Profile of the Tennessee River (1944). This map appears in
Sun Yun-suan's report to the NRC office in New York. The integrative
development of the Yangtze River is a scaled-up version of the Tennessee River
hydropower projects. Source: RG82: Tennessee Department of Conservation
Photographs, box 69, courtesy of Tennessee State Library and Archives.

engineer of the National Hydroelectric Commission in 1946, Zhang modeled
its workflow after the TVA by centralizing aspects of planning for hydroelectric
projects. Zhang found this arrangement optimal, as he believed contractors in
China were ill-equipped to construct massive dams.

The TVA's power network inspired the Chinese trainees to imagine a future
when surplus power could be seamlessly transferred across the nation. In his February 1944 report, Sun expressed his fascination with how the TVA maintained
a consistent load of 700,000 kW year round by transferring surplus power with
a 154 kV high-voltage transmission line between its western and eastern regions,
which were 125 miles apart, as shown in the map in figure 5.2:

In the spring, rainfall is more abundant, and the hydro-electric plants
downstream in the western region are able to provide more electrical

power. During this time, the hydroelectric plants upstream on the eastern tributaries begin to store water and generate less electricity, in order to prepare for the dry seasons. During this time, the western regions will transfer its surplus energy to the east.[41]

The process was reversed during the dry season in the fall. The Yangtze Gorges project, which will be discussed in the second half of this chapter, could be thought of as a scaled-up version of the power network in Tennessee described by Sun. Attempts to scale up the TVA's projects on a national level proved to be unfeasible, as existing technology made it unrealistic to build transmission cables that would carry high-voltage current seven hundred miles from the upstream areas to the industrial centers in the Lower Yangtze.

The TVA's cost-competitiveness also caught the attention of the NRC's senior officials. Liu Jinyu, who had built Kunming's wartime power stations, visited the power office in Chattanooga. He described the private operator ConEdison's performance as "rather mediocre," and its electric tariff of 2.41 cents per kWh as "rather high in this country," after learning that the TVA charged a fraction of its cost.[42] Most important, Liu concluded, citing TVA's power policy, "the interest of [the] public in the widest possible use of power is superior to any private power interest."[43]

Yun Zhen, who had fended off Westinghouse's attempt to impose American voltage standards on China, shared Liu's sentiments. Yun led a delegation to the Tennessee Valley between December 3 and 13, 1944, which visited several major dams and heavy industries powered by abundant electricity. His distrust of private enterprise further deepened during this trip. While all other heavy industries in the Tennessee Valley welcomed the NRC, the Aluminum Company of America (Alcoa) declined to host the Chinese delegation.[44] The TVA office warmly welcomed Yun and his colleagues. Its chief engineering office presented the delegates with a detailed profile and map of the Tennessee River System that had appeared in Sun's earlier report (figure 5.2).[45] Yun later drew on his field notes from the visit and trainee reports to strengthen his case for the nationalization of the power sector. Yun Zhen's note of appreciation to Lilienthal stood in stark contrast to his short but cordial thank-you notes sent to private corporations that hosted their visit. After returning to New York, Yun told Lilienthal:

> The services that the Authority rendered, and is rendering, in the interest of the people give us immeasurable encouragement as to how well government-operated enterprise could attend to the betterment of people's

livelihood. As men devoted to the development of national resources in China, we find TVA a source of inspiration and, no less, a big challenge.[46]

The NRC eventually realized the immensity of the challenge when they attempted to apply the lessons learned at the TVA in China with the planning and design of the Yangtze Gorge project between 1944 and 1947.

Replicating the TVA in China

Personal contact between TVA trainees and American dam builders paved the way for Sino-American cooperation in integrative river-basin development. A letter from Zhang Guangdou in September 1943 stoked the imagination of John Lucian Savage, the chief designing engineer for the Bureau of Reclamation. Zhang introduced himself to Savage before his transfer from Knoxville to Denver. His letter came months before the State Department memo to Savage telling him that his work is "expected to be useful to the war effort."[47] Zhang saw in Savage, to quote Sneddon, "key traits of the idealized technical expert: virtuosity in terms of technical knowledge, a capacity to innovate, and an almost metaphysical understanding of how technology might be best applied to promote human welfare."[48] Zhang presented China as a land of opportunity for massive integrative river basin developments by attaching a report of China's postwar electrification plan, which included "14 major hydro-electric power plants in all with an aggregate generating capacity of 400,000 kW, to be constructed within five years after the conclusion of the war."[49]

Zhang explicitly made the connection between industrialization and hydropower development on top of providing a snapshot of the hydrological surveys conducted in Sichuan, Yunnan, Guizhou, and Hunan since the Nationalist retreat to Southwest China. The 40,000 kW Zi River hydropower station, for example, would be built near the antimony production center of Xinhua, whereas the 20,000 kW plant at Guizhou would support the extraction of newly discovered bauxite.[50] In early October 1943, Savage responded with great interest and expressed his hope "that the negotiations between the Chinese representatives in Washington and State Department officials will terminate in a decision to send me to China."[51] Savage informed Zhang of his trip to India and requested follow-up communication through diplomatic mail, which Zhang duly conveyed to his supervisors in New York.

Having opened up direct channels of communication, Chen and chief representative K. Y. Yin (Yin Zhongrong) began securing Savage's passage to China.

They sought clarification from Savage if his trip to India was "sponsored by the American Government or the British government and through what Department of the United States Government were the arrangements effected."[52] The NRC then obtained the necessary clearance to bring Savage to China by December 1943. Sun Yun-suan followed up by introducing Savage to his colleagues at the Electrical Power Bureau in Chongqing.[53] Zhang asked his supervisors to expedite the process in his March 1944 memo to Chen Liangfu. He pointed out that the construction of Fontana Dam and Kentucky Dam was ending in a few months, which would free up the TVA to assist China. Zhang added that TVA had an unparalleled scope of expertise, ranging from hydraulics, power network management, and farmland and forest conservancy.[54] Zhang Guangdou spent eight months working on various phases of the Fontana Dam project in North Carolina and saw firsthand how the TVA completed the project in about two years.[55] Postwar China needed to build dams quickly, and only American federal agencies like the Bureau of Reclamation and TVA were able to deliver such accelerated development.

Savage set aside his work in India and departed for China in June 1944, as he found the Yangtze Gorges project to be more promising. His project along the Sutlej River at Bhakra in Punjab faced opposition from community leaders who feared that water from the reservoir would submerge a local temple. The locals also demanded that Savage scale down his project.[56] The Yangtze Gorges project was the complete opposite of the case in Punjab. Hydropower development in China came under the NRC's centralized command. This afforded Savage the luxury of focusing on site selection and cost-benefit analysis, while his Chinese hosts dealt with the politics of population displacement. Savage arrived in China in June 1944, just as the Japanese captured strongholds in Hunan like Changsha and Hengyang during the Ichigo Offensive. Savage steered clear of military conflict by starting his surveys of the four Yangtze tributaries near the wartime capital Chongqing, namely Min River, Tuo River, Jialing River, and Wu River. Some four hundred miles downstream, Guomindang forces of the sixth war area were conducting a three-month campaign against Japanese forces in Yichang— the proposed site of the Yangtze Gorges Dam. Savage traveled to Xiling Gorge in western Hubei on September 20, 1944, when the fighting ended. Savage was within three miles of Japanese forces but took the risk to venture to the place that he described as "an engineer's dream site."[57]

Accompanied by the commander of the Upper Yangtze Defense Force Wu Qiwei and Hydroelectric Bureau Chief Huang Yuxian, Savage traveled by boat along a two-and-a-half-mile stretch between Shipai and Pingshan, approximately fifteen miles from Yichang (see figure 5.3). During the ten-day

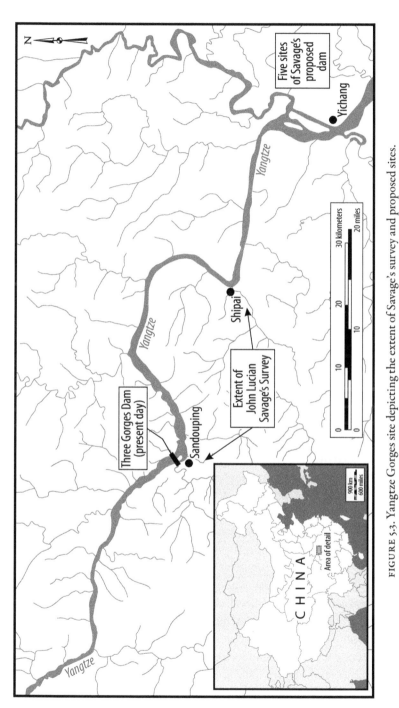

FIGURE 5.3: Yangtze Gorges site depicting the extent of Savage's survey and proposed sites. Savage did not visit the area near Yichang during his first visit to China. Map by Mike Bechthold.

survey, Savage lacked the equipment to conduct a full geological profile, so he resorted to picking up rock samples while hiking along the banks.[58] The local garrison showed Savage a topographical map captured from the Japanese, on which Savage based his calculations for drainage volume. Savage then returned to Changshou and spent the next forty days drafting his "Preliminary Report on the Yangtze Gorge Project."[59]

Savage presented a wildly optimistic plan. He proposed the construction of a "concrete gravity dam" about 738 feet high that would raise the reservoir water by 525 feet. The power plant, with a firm power output of 7.8 million kW and secondary output of 2.7 million kW, contains "twenty combined diversion and power tunnels on one side or both sides of the river."[60] The multipurpose development project would also store enough water to irrigate ten million acres of farmland and be equipped with canal locks that allowed ships of 10,000 tons to sail up the Yangtze to Chongqing. Savage estimated the entire project to cost 935 million US dollars and bring in 154 million US dollars of annual revenue.[61] The TVA's Muscle Shoals Plant inspired the funding mechanism for its construction. George Reed Paschal, an engineer with the Foreign Economic Administration, proposed a "fertilizer for dam" financing scheme in August 1944. Paschal calculated that 400 cubic feet of water flowing through the narrow gorge at Yichang each second generated 10.5 million kW. A fertilizer plant with an annual capacity of five million tons will use five million kW, leaving the Chinese with the balance. At 4 percent interest, the Chinese would have paid off the loan in fifteen years, while the Americans acquired fertilizer at half the market price.[62]

Savage based his extrapolations on fragmentary and flawed data. Savage calculated the flood volume of the Yangtze with three data points—the flood water volume from 1896, 1905, and 1931.[63] When he exhausted the NRC's sources and his direct observations, he turned to travel accounts by foreign observers, such as H. G. W. Woodhead's *The Yangtsze and Its Problems* dating back to 1931.[64] Savage qualified his report stating that much work remains to be done. He noted that a review of preliminary economic studies, hydraulic model tests, stress and stability studies, contract designs, and final specifications had to be completed before construction. A large project like the Coulee Dam required at least 15,000 drawings.[65] Savage also neglected to mention that a geological survey in the prospective sites had not yet been completed.

Savage stood on firmer ground with his detailed proposal for smaller hydropower projects on the upper Yangtze tributaries, which was submitted shortly after completing the Three Gorges report. He completed the drawings for three sites in Sichuan (Dadu River and Mabian River; Upper Minjiang and Guanxian; Longxi River) and one in Yunnan (Tanglang River) between September and November 1944.[66] Savage stayed at a guesthouse near the Qingyuandong hydropower station in Changshou, which had been constructed by the NRC during the war and had supplied electricity to Chongqing since 1942. He thus had the benefit of making on-site observations and drawing on recent engineering reports completed by the NRC and their consultant Hugh L. Cooper and Company. His itemized cost estimates for these smaller project sites were much more precise. The Dadu River–Mabian River (1.2 million kW) and Upper Minjiang (820,000 kW) projects, projected to cost 336.0 million US dollars and 199.7 million US dollars respectively, were not only described as "feasible in respect to their construction" but also "SUPER-PROJECTS of great potential value to China." The Longxi River and Tanglang River projects, which would have been expansions of existing hydropower stations, were 5 to 10 percent the size of the first two projects. The Longxi River project called for the construction of three hydropower stations with a combined total output of 49,500 kW around the existing 3,000 kW Lower Qingyuandong hydropower station at 11.2 million US dollars. The Tanglang River project, which diverted water from Kunming Lake, Anning, and Fuming, required a 16.3 million US dollar budget and added 80,000 kW of generating capacity.[67]

Upon completing the preliminary report, Savage sought final approval for the conclusion of a cooperative agreement from the Department of the Interior and the State Department. He cabled the commissioner of the Bureau of Reclamation on September 26, 1944, to inform him of Chinese interest in the Yangtze Gorges project within days of submitting his report. The commissioner supported Savage's actions and notified the secretary of the interior on October 18, 1944, that the Bureau intended to grant assistance. Ten days later, the commissioner cautioned Savage that "any final plans must be prepared by the State Department," and that the Bureau prioritized domestic work over the demands of the Chinese government and undertook no financial responsibilities for the project.[68] Negotiations of the cooperative agreement stalled, as Sneddon duly noted, because the desire to "put its prodigious technical capabilities to work on a once-in-a-lifetime project" on the part of Savage and the Bureau of Reclamation came into conflict with the State Department's cautious approach.[69] During a meeting on March 10, 1945, First Assistant Secretary of the Interior Benjamin Strauss objected to a clause of confidentiality as the Bureau had a long-standing policy of making engineering techniques and discoveries open to the world.[70] In May 1945, the State Department withheld its approval for the agreement raising concerns about the project's feasibility stating that, "such as vast quantity of power as the project would develop appears to be out of line with the capacity of China's industry to utilize that power in the foreseeable future." Strauss replied by invoking the TVA's experience that the availability of cheap hydropower and irrigation water led to accelerated development of agriculture and industry and that assisting China was in line with the United States's national interest.[71] Savage's retirement from the Bureau on May 9, 1945, released him from bureaucratic encumbrance.[72] But without a formal Sino-American cooperation agreement, Savage was unable to achieve anything.

Savage delegated his China portfolio to John S. Cotton, formerly a senior engineer of the Federal Power Commission's San Francisco regional office. On December 9, 1944, the NRC appointed Cotton as the "Chief Investigating Engineer to work on river-basin and electric investigations."[73] Cotton outlined his plans for a comprehensive survey of the Yangtze River basin from his home in San Francisco before meeting the globe-trotting Savage in Denver on March 14, 1945.[74] He departed for China with Zhang Guangdou in April 1945, not before getting immunized for cholera, typhus, typhoid, small pox, and bubonic plague, while securing thirty-five pounds of excess baggage for his technical notes.[75] Cotton did not even bid farewell to his wife before his departure. He forwarded her a copy of his reimbursement and insurance policy through the NRC's New York office.[76]

In the early months as chief engineer of China's National Hydroelectric Bureau, Cotton took a conservative approach by prioritizing the construction of small hydropower stations around Chongqing over the hydrological surveys for the Yangtze Gorges. Cotton started by presenting his tentative outline for his report on the economies and planning of the Yangtze Gorges project in May 1945.[77] He released a report in August 1945 proposing the construction of low dams near Yichang, "with the view of early and rapid construction of about 1,500,000 firm kW to meet the great and urgent demand for industrial power in the central region of China." The low dams were half the height of that in Savage's proposal and created a smaller catchment area; the powerhouse was equipped with twenty 75,000 kW generator units, rather than ninety-six generators. With no reliable data, Cotton drew on Savage's initial calculations and estimated the project to cost between 197 million and 313 million US dollars.[78]

Cotton also obtained a more precise understanding of the topography near Yichang. The Hydroelectric Engineering Bureau lacked the equipment to conduct detailed surveys. It acquired office supplies such as compasses, slide rules, photo-stat machines, and calculators from the American military forces about to withdraw from Chongqing.[79] The NRC attempted an aerial survey on its own. It borrowed an F-5 reconnaissance plane in August 1945 from the Aviation Affairs Commission to take aerial photographs that would be used to prepare a 1:25,000 map.[80] Cotton reached out to the TVA for help in September 1945, only to be told that "under the authority of the TVA act we cannot undertake work on a reimbursable basis directly from a foreign government."[81] The TVA, however, offered to prepare a map from the photos taken. Avenues for assistance such as Lend-Lease were being closed off as the war ended.

With the lack of accurate topographical data and guidance from their American advisers, the National Hydroelectric Engineering Bureau shelved the Yangtze Gorges project until the cooperative agreement was concluded on October 1, 1945. The commissioner of the Bureau of Reclamation assigned the Yangtze Gorges project to the Branch of Design and Construction, but even then formal studies did not begin until June 3, 1946.[82] Before March 1946, Savage was tied up with projects in India and Palestine.[83] With the Yangtze project in limbo, Cotton directed his energy toward the smallest hydropower project proposed by Savage's November 1944 report—the Longxi River hydropower project. In fact, Cotton only visited the Yangtze Gorges project site for the first time in February 1946, just as Savage was planning to return to China.[84]

The Longxi River hydropower project promised a quick reprieve from Chongqing's power shortage. According to Cotton's planning report published on September 28, 1945, Chongqing had only four power stations with a combined output of 12,000 kW, while its industries supplied their own electricity by installing generators with a total capacity of 17,900 kW. Cotton estimated that Chongqing required an output of 135,000 kW. Based on this projection, there would be 200,000 new residential customers in Chongqing with an average individual annual consumption of 300 kWh. Damming the Longxi River, a Yangtze River tributary ninety kilometers downstream from Chongqing, would create a 2,900-square-kilometer reservoir with enough water to power four hydropower plants totaling 64,000 kW, which would meet half the projected demand. An additional steam power plant added 16,000 kW of firm power. Cotton adopted a cautious approach by dividing the development into two stages and choosing the cheapest construction method. The initial development required the construction of the Lion Rapids Reservoir and Dam and one 15,000 kW power plant at Lion Rapids and one 10,000 kW power plant at Upper Qingyuandong. In the second stage, a 6,000 kW plant would be built in Huilongzhai and 30,000 kWs of additional capacity would be installed at the existing Lower Qingyuandong plant. His cost estimate of 7.494 million US dollars was lower than Savage's earlier proposal. The submerging of 43,500 *mu* of arable land would displace 7,229 inhabitants, but Cotton surmised that increased employment from local industries outweighed the loss of farmland.[85]

Cotton did not appear to know about the resentment arising from hydropower development near Chongqing. The damming of the Longxi River in 1942 had already increased local flood risk. Tensions over the hydropower plant boiled over after a flood on September 30, 1945, just two days after Cotton transmitted his Longxi River planning report. The waters of Peach Blossom Creek surged after a severe thunderstorm resulting in a flood near Ranshi Bridge. The villagers reported that flood waters washed away a flour mill and damaged the sweet potato harvest. The villagers blamed a defective sluice gate that did not open fully to release the flood waters. The chief of Duzhou Village, Zhou Zhilu, demanded compensation from the Longxi River Engineering Office.[86] Huang Yuxian, then director of the Longxi River project and later promoted to the head of the National Hydroelectric Engineering Bureau, insisted that the sluice gate had nothing to do with the flooding. He pointed out that local residents had been receiving a monthly allowance after a 1944 flood to maintain a flood-prevention team and neglected their duties of opening the sluice gate during the September 1945 flood. Huang then asserted that the damages at Duzhou Village were no more serious than those in neighboring villages, then questioned why the residents did not move their food and personal property to higher ground to avoid a few inches of floodwater.[87] After months of investigations, the NRC approved compensation of 700,000 CNC dollars, amounting to about 10 percent of that demanded by the villagers in May 1946, while acceding to an increased monthly flood-management allowance.[88]

Blissfully unaware of local opposition, Cotton went ahead with drafting invitations for bids shortly after filing the report. The National Hydroelectric Bureau imposed strict deadlines on prospective contractors to safeguard against default. For example, the contractor had to deliver two 9,000 kW turbines along with the governor, butterfly valve for the Lion Rapids plant to Shanghai, with the first complete unit arriving within 250 days and a second one within 350 days. The delivery of one 10,000 kW power plant for Upper Qingyuandong had to be completed within 150 days. Once awarded the contract, the contractor had to post a performance bond of "not less than 50 percent of the estimated aggregate payments to be made under the contract." They were also subject to fines of 20 US dollars per day for each item delivered late.[89] As no suppliers in China were capable of fulfilling these orders, the NRC's New York office conducted the invitation to tender in the United States. The NRC only had a budget of 4 million US dollars, which was lower than the prevailing price of the equipment. It even had to remove concrete from the purchase list to stay within budget—a move that ultimately hampered dam construction at Lion Rapids.[90]

Savage's return to China in March 1946 forced Cotton to redirect his focus to the Yangtze Gorges survey and set aside the work on the Longxi River project. After arriving in Chongqing, Savage headed to Changshou to convene with Cotton and his colleagues at the Hydroelectric Engineering Bureau.[91] Savage assembled a stellar engineering team. W. C. Beatty, retired chief mechanical engineer of the Bureau of Reclamation, supervised and coordinated all aspects of the Yangtze Gorges study. Fred O. Jones, a thirty-four-year-old geologist who had worked on the Columbia River Basin project, directed the testing of the limestone bedrock near Yichang. Jones recognized the enormity of the mission, describing the gorge as one "deeper and steeper than our Columbia River canyon." Morrison-Knudsen sent a high-ranking drilling superintendent to complete the geological drilling after company representatives met with Savage in Shanghai. Savage also set up contract negotiations for an aerial topographical survey of the Yangtze Basin between the vice president of Fairchild Aerial Survey and the Chinese government.[92] Before his departure, Savage left general instructions for further studies of five possible project sites and more detailed topographical studies. He had little to say about the Longxi River project after a routine inspection, besides stating that he found the design and plans to carry out construction by manual labor to be satisfactory.[93]

News of Savage's efforts to jumpstart the next phase of the Yangtze Gorges project generated great excitement. The project came to symbolize hopes for a prosperous future under a strong central government. A Guomindang party cadre Sun Danchen went as far as to compare Savage to Li Bing, the Qin statesman credited with building the Dujiangyan—an irrigation system that transformed Sichuan into a productive tax base, which ultimately gave the Qin an advantage over their Warring States rivals. Besides relieving flooding in western Sichuan and irrigating 5.2 million mu of land in eleven counties, the Dujiangyan opened a transportation channel for timber to be shipped from southern Sichuan.[94] News bulletins referred to the Hydroelectric Bureau as the Yangtze Valley Administration (YVA). The author went further to suggest that the YVA had greater potential than the TVA by virtue of its wider geographical coverage and larger projects.[95] If, in the words of David Ekbladh, "the TVA's mission was to turn the ornery Tennessee River from an unpredictable force into a docile servant for regional development," then the YVA aimed to transform the Yangtze into the driving force for national development.[96]

Further feasibility studies conducted after June 3, 1946, began exposing problems with Savage's preliminary proposal. The Bureau of Reclamation's chief engineer delayed the assignment of engineers to the Yangtze project until May 30, 1946.[97] Within two weeks, Chinese engineer C. Y. Pan alerted Savage to inaccuracies in the topographical maps that served as the basis for his 1944 report. Savage's team had determined the contour intervals on the map to be thirty meters by looking at the numbers marked out on a series of contours. An accurate translation revealed that these were not elevation figures in the first place. Furthermore, the map was made with aerial photographs taken in July 1940 when the water level at Yichang was fifty meters higher than normal. These threw off all the initial estimates. With no further information, the engineers assumed that the contour intervals were ten meters, as per mapping conventions and calculated that a dam at proposed number 4 site needed to be 2,400 feet longer than the original.[98] Later reports revealed elevation figures to be "inaccurate as much as 600 to 1,000 feet." The survey team had to wait until January 28, 1947, before reliable data from aerial surveys became available.[99]

The scale of the Yangtze Gorges project expanded uncontrollably with the completion of further electrical equipment studies. Savage's 1944 report proposed the installation of sixty power units of 175,000 kW for a total capacity of 10.5 million kW. In his revised plan in March 1945, Savage conceived something even bigger. He planned on building twenty diversion tunnels each thirty feet in diameter and adding another four after the completion of the dam. Each of these twenty-four tunnels would be installed with four 110,000 kW power units, bringing the total installed capacity to 10.56 million kW. When Beatty took over from Savage, he factored in the flow rate of the river and increased the power rating and number of generators required. Beatty's August 1946 plan estimated configurations of 96, 114, or 126 units of 175,000 kW each.[100]

The technology that existed at that time made it difficult if not impossible to generate such a vast amount of electricity and transmit it over a long distance. A further study about the suitability of 200,000 kW turbines proposed for the Yangtze by Champe Lu in January 1947 noted that "the maximum head of the Yangtze turbine of 535 feet as compared to 355 feet on the Grand Coulee Unit introduces greater stress in such parts as scroll cases, speed ring, wicket gates and runners." Lu added that the parts were made of cast steel and should have ample strength to deal with the conditions on the Yangtze.[101] Plans to construct 330 kV transmission line from the Yangtze Gorges Dam would, however, prove to be less feasible. General Cable Corporation pointed out that the highest operating voltage available was 230 kV. To reach the desired voltage, the pressure rating for the oil-filled transmission cables had to be doubled from 250 psi (pounds per square inch) to at least 500 psi, which was outside the operating range of existing oil pressure control equipment.[102]

Geological studies also called into question the suitability of four out of five sites. Between October and December 1946, Hou Defeng and four other geologists from the Central Geological Institute conducted a preliminary survey in Yichang. The geological team from Morrison Knudsen did not arrive until November 1946, as a dock strike in the United States delayed the shipping of drills. On their own, Chinese geologists constructed a complete geological column of the Yichang area, covering an area of one square kilometer. Fred Jones did a follow-up study in March and June 1947. The results were disappointing. The bedrock for sites number 4 and 5 was tertiary Shimen conglomerate, susceptible to leaking. The bedrock at site 1A was crushed limestone, which was not an ideal foundation for dam construction. The geological team found site number 1 to be even less suitable due to a crumbly layer of shale. Only the bedrock at sites number 2 and 3 was firm enough. Even then, the large numbers of caves in site 3 rendered it unsuitable, as the dam would have to be extended by about 2,000 meters.[103]

Scrutiny during the final months of the study raised concerns on the durability and benefits of the proposed megadam. The severity of sedimentation only came to light two months before the termination of the Yangtze Gorges survey. E. W. Lane, with the assistance of Chinese engineer W. Y. Yu, calculated the sediment volume by using fragmentary historic data and gathering samples with sediment traps from nine gaging stations along the Yangtze and two tributaries (Jialing and Wu Rivers). The data did not yield an accurate estimate but raised enough cause for concern. Lane noted that the wide variation of sediment discharge from the Jialing River placed the annual discharge volume between 150,000 to 600,000 acre-feet. He calculated the total suspended load to be between 350,000 acre-feet to 775,000 acre-feet, yielding an approximate lifespan for the dam of between 127 and 207 years. The report also suggested that deposition near Chongqing during the dam construction would affect the navigability of the Yangtze by reducing the river's depth and limiting the draft of the vessels sailing up to Chongqing.[104]

Cost estimates skyrocketed as these detailed studies emerged. The cost of river diversion had to be revised upward from Savage's preliminary estimate of 154 million US dollars to 392 million US dollars following detailed topographical surveys that offered a full calculation of the drainage area.[105] As planners increased the generating capacity to reduce average cost of electricity, they not only called for the installation of larger and more expensive turbines but also increased the number of diversion tunnels that needed to be built. The financial study released in May 1947 placed the cost of the power generating facilities alone between 1.683 and 1.8165 billion US dollars, almost double that of Savage's original estimate. The lower-end estimate was achieved by deferring part of the structural work after the completion of the upstream construction.[106] The average construction cost of 67.76 US dollars per kilowatt power generated was one third that of the Dajia River Integrative Development undertaken by the Japanese in Taiwan during World War II.[107] Three days after the release of the financial study, the Nationalist government implemented the emergency economic program and suspended the Yangtze Gorges study. Two months earlier, the Nationalist government had withdrawn forces from Hubei and curtailed funding to the project. Despite pleas from Cotton for the "existing contract [to] be kept alive or small scale by use of existing staff," the Executive Yuan terminated the services of the geological surveyors Morrison Kundsen Company.[108] At Yun Zhen's request, the Bureau of Reclamation issued a status report of all work completed by August 15, 1947, and preserved the studies and working files. Beatty expressed regret about the demise of the study and noted that it was "difficult to leave the work in proper shape for completion in the future."[109]

Despite these troubles, Savage and Cotton remained hopeful for the revival of the Yangtze Gorges Dam project. In a 1948 interview, Savage believed that the project was "still alive," as he "sees it as a great, strong heart, pumping the life-blood of electrical power, new industry and renewed agriculture into the veins of a revitalized China."[110] Cotton moderated his expectations. Earlier projections that the Yangtze Gorges Dam could be completed in eight to ten years proved to be unrealistic. In a speech to his Chinese colleagues, Cotton reminded them that the Three Gorges Dam was five times the size of Grand Coulee Dam, which was then the largest hydropower station. It took thirty-four years to complete the Grand Coulee.[111] All the work was not in vain, however. On April 12, 1948, the Chinese representatives in Denver sent back two big boxes of materials to Nanjing, which included 193 sheets of topographical maps of South China from the US Army, thirty-eight sheets of navigational maps of the Yangtze River from Yichang to Chongqing, and several copies of climatological data.[112] Their work laid the foundation for future hydropower development in the Yangtze.

Faced with this setback, Cotton redirected his focus to the Longxi River project in Sichuan. Excavation and damming at Longxi River had barely gotten any attention thus far, as the Hydroelectric Bureau diverted its attention to the Yangtze Gorges survey in March 1946. Half of the Hydroelectric Bureau's 305 staff members were tied up with the Yangtze Gorges, while the remaining staff was sent to

Northwest China, western Sichuan, and the Qiantang River to conduct surveys.[113] As early as July 1946, Cotton alerted Huang Yuxian of the material shortage at the Longxi River project, stating that, "The only good construction equipment at Upper Tsing Yuan Tung [Qingyuandong] is one compressor, one wagon drill and four jackhammers." Only one out of four trucks obtained from the NRC was working.[114] By September 1946, only the design work has been completed; funds for the purchase of generators never came through. With no bids received from the earlier call for tender, Cotton fell back on an earlier batch of components that had been purchased in 1941 through the Lend-Lease Act. Hundreds of parts for four 1,000 hp turbines, designated for the Lower Qingyuandong Power Plant, finally arrived in Shanghai in November 1946 after sitting in storage in India for years. They remained in hundreds of crates scattered across four warehouses in Shanghai, as the Longxi River project remained at a standstill.[115]

Problems with the building materials caused the construction of the dam to grind to a halt. In September 1947, Cotton issued a concrete testing report, which noted that the sandstone from a nearby quarry was so brittle that it could "be scratched with a pocket-knife" and "much more porous than ordinary sedimentary rock." Bleeding, where water comes out to the surface, occurred in all different grading of sand. The durability and strength of concrete mixes were also questionable. Furthermore, due to the poor conditions of the laboratory, engineers were unable to complete permeability testing.[116] Zhang Guangdou had projected the Upper Qingyuandong hydropower station to enter service in the middle of 1948. This did not happen. Labor shortage further slowed down the progress. Engineers at the Upper Qingyuandong reported that workers continued working on the project in August 1948 even though they did not receive their wages. Construction activity stopped one month later when workers returned to their fields to tend to their harvests.[117] The National Hydroelectric Engineering Bureau, which aspired to be China's TVA, lacked the capability to complete the smallest hydropower plant in its project pipeline.

SETBACKS IN THE Yangtze Gorges and Longxi River projects did not deter visionaries like Savage and Cotton. Just as further studies revealed major flaws in Savage's ambitious 1944 proposal for the Yangtze Gorges Dam at Yichang, Cotton drafted similar plans for integrative developments elsewhere in China. He conveyed the wild optimism in Savage's visions for his plans in Guangdong, Fujian, Yunnan, and Hunan. In his April 1948 report, Cotton envisioned the construction of a large synthetic nitrogen fertilizer plant capable of producing 240 tons of ammonium sulphate per hour, which would consume 20 percent of the newly available electricity from the Weng River hydropower project.[118] The attempt to replicate the Muscle Shoals fertilizer plant simply shifted from the upper Yangtze to the Pearl River Delta.

Contrary to existing studies that painted Savage as a visionary who was ahead of his time, this chapter brings into focus his failure to consider the cost of accelerated development promised by integrative river basin projects. Efforts to replicate the TVA's achievements in China also exposed the limitations of this developmental model. The engineers had worked on the assumption that the war had wiped the slate clean and paved the way for China's electrical industries to cast aside troubled beginnings and start afresh. Development plans for the Yangtze Gorges promised the deliverance of exponential economic growth but glossed over negative effects like farmland loss and population displacement. Even as they embarked on small-scale hydropower projects, the National Hydroelectric Bureau of the Guomindang regime struggled to deal with resistance from the local population and confronted shortages in funds and construction equipment.

The colossal scale of Savage's proposed dam required a massive transformation of the natural environment that not only impacted the local ecosystems but would also result in a gestalt shift in the geosphere and biosphere of the Earth system. While the term *Anthropocene* had not come into existence during Savage's time, the dam builders understood their projects as attempts to overcome geological and hydrographic constraints shaped by deep time. Projections of the dam's lifespan measured in tens of decades suggest the acceleration of changes in topography and hydrography that would have otherwise taken millennia. Interventions meant to extend the lifespan of the dam required humans to act as geological agents. Using the newly harnessed electricity to power the Haber-Bosch process would also speed up the nitrogen cycle and lay the groundwork for explosive population growth. The movement of vast amounts of electricity over long distances with high voltage networks called for the redesign of power transmission systems. Savage's earlier developmental vision evoked a future where mankind would create the stratigraphic and functional change for perpetual economic growth.

Savage's plan should not be dismissed as a footnote in hydropower development on both sides of the Taiwan Strait. Sun Yun-suan, who moved to Taiwan and rose through the ranks in the Taiwan Power Company, oversaw the development of major hydropower projects in Taiwan with American aid.[119] In mainland China, the proliferation of small hydropower stations in the 1950s followed in the wake of the initial setbacks in grandiose plans for integrative river basin developments. Zhang Guangdou, who had worked closely with Savage and Cotton, picked up where the two visionaries left off and advocated for its construction in the 1970s and 1980s. Construction for the Three Gorges Dam began in 1994. It entered service in 2003. The belief that man could conquer nature that emerged in the dreamscapes of Savage's Yangtze Gorges Dam is still alive and well. During the massive summer floods in 2020, Chinese state media acknowledged that the Three Gorges Dam was slightly deformed as it held back the stormwater but readily refuted claims that the megadam was structurally unstable as unscientific.

The National Hydroelectric Bureau had tried to turn the tide of China's declining fortunes by leveraging its connections with the TVA. It ultimately spread itself too thin by juggling small hydropower projects and the Yangtze Gorges survey. Cotton had a nagging concern throughout his tenure as chief engineer. The Little Fengman Dam along the Sungari River in Northeast China remained in a state of disrepair. The Communists soon took advantage of the destabilization of local society resulting from the power shortage and established a foothold in the Northeast. They would soon learn how to use electricity as weapon in urban warfare and turn the tide of the civil war in their favor.

CHAPTER 6

Waging Electrical Warfare

I N AUGUST 1946, CIA officer Carleton B. Swift wandered into an industrial
wasteland in Northeast China. In a report submitted to President Harry
Truman by the US representative to the Allied Reparations Committee
Edwin Pauley, Swift estimated that the Soviets removed 56 percent of 1.786
million kW of installed capacity, leading to a reduction in peak energy load of
71 percent.[1] The Pauley mission had surveyed eight out of the fourteen largest
electric power plants but was unable to access the remaining ones either because
they were too far away or were located in areas under Communist control. The
Soviets dismantled steam boilers and generators from power plants across the
northeast, leaving power stations littered with rubble (figure 6.1). High-voltage
power cables disconnected from transformers dangled from the transmission
towers. The industrial base in Northeast China, which had grown rapidly on
the back of massive Japanese investments in the electrical power sector, came
crashing down with the Japanese defeat. Founded in Changchun in 1934 with
start-up capital of 90 million yen, the Manchuria Electric Company grew from
strength to strength with industrial expansion. By the time it was amalgamated
with the Bureau of Hydropower Construction in 1940, its paid-up capital had
grown to 640 million yen.[2] World War II's victorious powers would jostle with
each other to pillage these assets.

Swift and his colleagues at the Pauley Commission were witnessing the first
stage of a war to control China's electrical industries, which played a pivotal
role in tilting the power balance from the Nationalists under Chiang Kai-shek
toward the Communists under Mao Zedong. The takeover of the electrical infrastructure enabled the Communists to seize control of the economic, social,
and intellectual capital in urban areas and paved the way for the state-building
enterprise of the Communist regime. An account of the Chinese Civil War
(1946–1949) that places electricity at its center challenges the conventional
narrative that the Communists defeated the Nationalists simply by encircling
the cities from the rural areas. Power not only emerged from the barrel of the

FIGURE 6.1. Fushun Power Plant after the Soviet takeover, circa 1945. Soviet forces removed one of the 25,000 kW generators and its steam turbine. The power plant is littered with insulating material from the steam pipes. Source: Pauley Commission Report on Japanese Assets in Manchuria, July 1946.

gun but was amplified by taking hold of the energy lifeline of millions of China's wealthiest urban residents.

There were three distinct phases in electrical warfare during the Chinese Civil War. Phase one began with the scramble for reparations after the Japanese surrender in August 1945. The Soviets claimed large amounts of power-generating and power-using equipment before the Nationalist forces arrived on the scene. The Communists capitalized on the power vacuum resulting from the collapse of the industrial economy to establish a foothold in Northeast China. The Nationalist government meanwhile faced mounting losses in their takeover of the Beiping-Tianjin-Tangshan network in North China left behind by the Japanese, while they handed the asset-rich Shanghai Power Company to its American owners. It only succeeded in taking over Japanese power assets in Taiwan.

Phase two played out on the battlefield. Both the Nationalist and Communist military forces devised strategies through trial and error to cripple their adversaries' electrical infrastructure. After taking over the cities, the belligerents saw the restoration of electricity as a crucial step in reinstating civil and economic order. The third and final phase arose from the demand for technical expertise in the maintenance of power networks. The Communists secured the defection of the engineer-bureaucrats and electrical workers who had been serving the Nationalist regime. Caught in the crossfire of the Chinese Civil War, the engineer-bureaucrats threw their lot in with the Communists, upon seeing that an alliance with workers and soldiers offered the best chances of fulfilling their long-time vision of nationalizing the electrical sector. Besides working with the Communists to impose a power blockade, they also protected vital assets of the Guomindang regime's state-owned industries and allowed the Communists to inherit the power infrastructure largely intact amid a bloody civil war.

Chen Yun's slogan "Wherever the People's Liberation Army goes, the lights will come on," sums up how the Communists came to understand the centrality of electricity to urban governance and subsequently national unification.[3] After making this statement in August 1948 at the Sixth All-China Labor Congress held in Harbin, Chen Yun and the People's Liberation Army (PLA) went about fulfilling this promise, as they captured one city after another from the Nationalists. Stung by the defection of many of its most experienced engineers, the Nationalists in Taiwan guarded against Communist subversion of the electrical industries. Liu Jinyu, who had built the power stations in Kunming during the War of Anti-Japanese Resistance and then coordinated the rehabilitation of Taiwan's power grid from 1945 to 1950, became a casualty in the war for control over China's electrical industries. Accused of conspiring with the Communists based on highly circumstantial evidence, one of the Guomindang regime's most experienced system builders was sentenced to death by firing squad in July 1950.[4]

Scrambling for Reparations

The scramble to claim Japanese industrial assets ended in a stalemate between the Nationalists, the Communists, and the foreign powers. The Soviets had seized the power equipment of the Manchuria Electric Company before the Nationalists arrived on the scene, and the North China power grid left behind by the Japanese turned out to be a liability that bogged down the Nationalist regime. In the Lower Yangtze, the Nationalists facilitated the restitution of the electrical industries and handed the assets seized from the occupation forces back to their rightful owners.[5] Private operators in the largest electrical markets struggled to stay afloat and faced severe losses much like their state-run counterparts. The

struggles of the Guomindang regime to restore order to China's electrical industries foretold its collapse in 1949.

The career of Sun Yun-suan, the TVA-trained engineer who later became the
premier of Taiwan, tracked closely with the failures of the Guomindang regime
in mainland China and its success in Taiwan. Sun was appointed to the Ministry
of Economic Affairs committee for the takeover of industrial assets in Northeast
China in October 1945 after completing his training at the Bonneville Power
Authority in Portland, Oregon. Having studied under Russian instructors at the
Harbin Institute of Technology, Sun was conversant in Russian and could possibly communicate with the Soviets based in Northeast China. Sun could not
accomplish his objectives. After entering Manchuria on August 9, 1945, Soviet occupation forces started hollowing out the industries in Northeast China to claim
their share of Japanese reparations. American intelligence sources were only able
to assess the impact of the Soviet occupation by August 1946. They found out that
the Soviets not only targeted the power equipment and hydraulic turbines owned
by the Manchuria Electric Company but also crippled industrial production by
seizing generators from coal mines, cement plants, and steel factories.[6]

Sun, however, never left for Northeast China, as he remained behind in
Chongqing to assist an American engineering team. Sun would not have been
able to apply the lessons learned at the TVA. The Japanese legacy system in Manchukuo was also a far cry from the highly integrated networks administered
by the TVA. It was a patchwork of disjointed hubs and spokes with three to
four medium-voltage transmission lines stretching out from each power station.
The Japanese in Northeast China faced the same problems as the Guomindang
regime in Southwest China. During dry seasons, there was too little water to
generate hydropower, so the Japanese depended on coal-fired power stations to
make up for the shortfall. Sun observed, "Despite advocating 'hydropower first,
thermal power second,' they ended up building more coal-fired power stations."[7]

Sun was later transferred to Taiwan, where he made important contributions
to the Nationalist takeover of the Japanese-owned Taiwan Power Company. The
NRC had sent a skeletal crew of twenty-three engineers to take over a massive
power conglomerate with a total staff strength of 6,400. As Lin Lanfang noted,
with about 3,000 of its staff slated for repatriation (Jp: *hiki-age*), the power sector in Taiwan faced a massive shortfall in manpower. The Nanjing government
responded by transferring an additional twenty-six engineers from state-owned
power stations in Chongqing and Kunming to Taiwan.[8] Liu Jinyu, who had
spearheaded wartime Kunming's electrical infrastructure development, had
arrived in Taiwan in October 1945 to take the lead. Sun left for Taiwan in

December 1945 as part of the second team of engineers to support the takeover. Sun took up the posting to Taiwan for both professional and family reasons. Sun was also an ideal candidate to coordinate the NRC's rehabilitation efforts in Taiwan, having spent most of his time in the power planning divisions at the TVA, Pacific Power and Light, and Bonneville Power Authority.[9] The power grid in Taiwan, which was centered on the hydroelectric dam at Sun-Moon Lake (Ch.: *Riyuetan*) in Nantou County, resembled the TVA's Main River projects. Sun's mother also fled to Taiwan after the Communists occupied their hometown in Penglai, Shandong Province.[10]

Liu also saw Taiwan as the perfect place to replicate the TVA's model of building integrative hydroelectric project management through federal authority. Writing to the NRC's director of the Electrical Bureau Chen Zhongxi, Liu expressed his wish "to transform Taiwan into an area with surplus electricity, establish an organization like the TVA's Promotion Department to encourage the use of electricity and provide electrical power at the lowest cost."[11] Despite its compact size, Taiwan had 3.5 million kW of potential hydroelectric power and its hydropower output per unit of land area was comparable to Switzerland's.[12] Furthermore, the Japanese left behind a well-integrated transmission system. The thirty-four power stations in Taiwan were connected either to the western or eastern power transmission network. A high-voltage transmission line running 231 miles from Songshan in the north to Kaohsiung in the south was connected to the largest hydroelectric facility, Sun-Moon Lake Power Station, at the center. Its transmission frequency at sixty cycles was the same as that of the United States.[13]

On May 1, 1946, Taiwan Power Company (hereafter abbreviated as Taipower) was formally incorporated. The nationalization of electrical power in Taiwan occurred by chance. As no private businesses in Taiwan had enough capital to buy the shares previously held by Japanese banks and corporations, all the assets and liabilities of the Taiwan Power Company were "automatically transferred to the Chinese Government at the disposal of National Resources Commission and the Taiwan Provincial Government."[14]

The rehabilitation proved to be straightforward. In a report addressed to Taipower, the American consultants J.G. White & Company noted, "The power installation was not subjected to indiscriminate bombing but was instead hit at key positions," and as a result, "generating stations were comparatively unharmed but the transformer and switching stations were completely wiped out."[15] They thus recommended that Taipower deploy undamaged transformers and circuit breakers from other substations on the island to Sun Moon Lake. This redeployment increased power output from 40,000 kW to 100,000 kW.[16] Chiang Kai-shek was pleased with the rapid rehabilitation of the Sun-Moon Lake hydropower station. When he visited Sun-Moon Lake on October 23, 1946, he personally thanked a Taiwanese engineer, Zheng Kaichuan, and a Japanese engineer named Suzuki and offered a token of appreciation of 100,000 Taiwan dollars.[17] Within eighteen months of the American consultants' reports, Taipower repaired the damage caused by the typhoon of 1944 and two rounds of bombing in 1944 and 1945.[18] Riots in the aftermath of the February 28 incident in 1947 barely disrupted the rehabilitation efforts. Rioters surrounded the company headquarters, while aboriginal tribes in the highlands threatened to kidnap the directors of key hydropower installations. Taipower staff continued working through the social unrest. Speaking on a radio broadcast in the official Chinese language and Taiwanese dialect on July 15, 1947, Taipower's second-in-command, Huang Hui, assured the public that there had been no power outages since the outbreak of violence on February 28, 1947. The daily average electric supply of 1.65 million kWh remained at 83 to 90 percent of the preconflict level.[19]

The smooth takeover of Taiwan stood in stark contrast to the chaotic situation in Northeast China, the Lower Yangtze, and North China. Based on site visits in Northeast China between May and July 1946, the Pauley Reparation Mission report states that "of a total electric generating capacity of 1,786,253 kW, the Soviets removed 1,008,300 kW, plus an additional 385,000 kW of equipment most of which was in the process of being installed, amounting to 200 to 300 million US dollars."[20] The Soviets also targeted the incomplete Sungari hydropower project. They removed four generating units as well as construction equipment from the site. Though 89 percent complete, much concrete remained to be placed in the haphazardly constructed dam to prevent it from breaching and flooding an area with millions of inhabitants.[21]

The Chinese Communists condoned Soviet pillaging despite public criticism. Suzanne Pepper pointed out that the press blamed both the Nationalists and Communists for not standing up against the Soviet Union.[22] A March 16, 1946, letter by Fan Yuanzhen, the wife of future deputy minister of Water Resources, Li Rui, revealed the strategic purpose of the ensuing chaos. Writing to her husband, who was then serving as Chen Yun's secretary in Northeast China, Fan described Soviet plunder as a necessary evil: "It is important to prevent American economic forces from making inroads into the Northeast. . . . This ensures that the Americans would not use the reconstruction of the Japanese industries to transform the Northeast into an anti-Soviet base."[23] By keeping the Americans at bay, the Chinese Communists made inroads into key cities in Northeast

China. In April 1946, the PLA occupied the city of Harbin in present-day Hei-
longjiang Province, which would remain the largest city under Communist control until the middle of 1948.

Besides ceding parts of Northeast China to the Communists, the Nationalist regime handed China's largest power market in Shanghai to foreign capitalists. Between 1946 and 1948, *Dagongbao* journalist Xu Ying traveled around
China to interview many of the nation's leading industrialists for a biographical
compilation.[24] Summing up his interviews with leading electrical engineers, Xu
Ying criticized the NRC for wasting its time and energy on the restitution of
foreign-owned power assets. Shanghai Power Company, which was handed back
to its American owners, generated about 99 percent of Shanghai's electricity.
The Japanese had hollowed out Chinese-owned power companies after the fall
of Shanghai in November 1937, leaving them to resell electricity acquired from
Shanghai's main power station on Yangshupu Road. The NRC also returned the
power stations in Suzhou, Hangzhou, Jiaxing, and Zhenjiang to their owners
and held a minority stake in the power industries in the capital Nanjing.[25] The
NRC reaped no financial or political gain for all its work.

The power assets in North China left behind by the Japanese turned out to be
a liability. The general managers of the state-owned North Hebei Power Company, Guo Keti, Zhang Jiazhi, and Bao Guobao, detailed the weaknesses of the
Japanese legacy system in their interviews with Xu Ying. On paper, its predecessor *Kahoku Dengyō* employed 4,000 workers, operated about 2,000 miles
of power lines, and provided maximum instantaneous power of 100,000 kW
to customers in Beiping, Tianjin, and Tangshan. Operationally, its peak output was closer to 86,000 kW. NRC chairman Weng Wenhao admitted that the
"demand is being kept within the margin by cutting down power and lighting
users."[26] These TVA-trained engineers realized that the North China grid was
a far cry from the integrated networks they saw in the United States. Xu wrote,
"The Power Company only relies on a trunk line to transmit electrical power. It
is an exaggeration to call the Beiping-Tianjin-Tangshan power grid a network. It
should be called a power snake. If one electrical pole breaks down, or if a ceramic
insulator is smashed, the entire line will go out of service."[27]

Guo Keti, who had joined the team of engineers building the power station in
wartime Kunming after graduating from the electrical engineering department
at Purdue University, faced the unenviable task of cleaning up the mess left behind by the Japanese. When Guo stepped into Shijingshan Power Station in the
western suburbs of Beiping in November 1945, the installation work for the largest 25,000 kW turbine was only 77 percent complete. Manufactured jointly by

Mitsubishi and Hitachi, the turbine was not designed to combust higher-grade anthracite coal readily available in the nearby Mentougou coal mine. Instead, it used lower-grade pulverized bituminous coal that had to be transported from the Kailuan coal mines in Tangshan about one hundred miles away. This was a relic of the Japanese strategy of centralizing the distribution of coal by funneling coal supplies through Kailuan. In addition, as the Japanese turbines were imitations based on Swiss products, there were no readily available replacement parts.[28] Beiping's industries clamored for more electricity, leaving the Chinese engineer-bureaucrats with no choice but to pick up where the Japanese left off. It was too late to tear everything down and start from scratch.

Political instability in North China led to fraud and profiteering. In March 1946, the NRC uncovered a plot by a former Japanese collaborator Wang Bao-hua, who posed as a deputy manager of the electric company to collect fines from its customers after falsely accusing them of misusing electricity. Wang allegedly pocketed several hundreds of thousands in fines before the military police caught up with him.[29] With coal pilfering rampant in the winter months, the North China power grid often ran out of fuel. During the cold months from October to February, the power company hired fifteen armed escorts to protect the rail convoy of coal from Tangshan to Tianjin and Beijing. Despite added security, the power company failed to obtain the 18,000 tons of coal required to maintain its operations. Company records showed coal mysteriously leaking out from the slits of the worn-out coal carriages. Coal raids also broke out at Fengtai Railway Station in the southern suburbs of Beijing and at the Shijingshan Power Station.[30]

Within a few months of the Nationalist takeover, North Hebei Power Company faced staggering losses. Between March 1 and October 31, 1946, the company took in 10.891 million CNC dollars but incurred expenses of 3.919 billion CNC dollars (5.6 million US dollars in November 1945 exchange rates in Tianjin).[31] It attempted to balance the books by raising tariffs and pushed many of its paying customers to the brink of collapse. As one of the largest customers in Beiping, the water works bore the brunt of the rate increase. In August 1946, Beiping mayor Xiong Bin implored the NRC to reduce electric tariffs for the city's water works. He wrote:

The North Hebei Electric Company of your esteemed commission is continuously increasing electrical tariffs. Right now, the electric tariff owed to the Power Company is equivalent to one third of the water work's revenue. With higher labor expenditure and rising cost, our revenues cannot cover our expenses. In the long run, this will affect the water supply to our city.[32]

With spotty power supply, state-owned industries delayed their production and fell further behind on their payments. The NRC's industries in the North China industrial hub of Tianjin owed 413 million CNC dollars in electrical tariffs. Full repayment of these debts would have increased its deficit by 13 percent.[33]

The power company imposed rolling outages to stabilize the situation. After the number 1 generator's rotor at the Tianjin power station broke, the Power Company divided Tianjin's industrial users into six groups and cut off power for one day to each group starting on November 15, 1946.[34] The power curtailment hampered the operations of state-owned industries. The general manager of the Central Machinery Works, Du Dianying, noted that his factory had rescheduled the workers' off days by making them rest during scheduled outages. Unscheduled outages between December 5 and 11, 1946, ranging from thirty-five minutes to seven hours, caused further disruptions.[35] The sudden halt resulting from the abrupt power cut damaged delicate components in the machines. The Nationalist regime incurred losses from the operations of the power company and state-owned industries. The power company recovered after conducting timely repairs on the generators. Tianjin's power output in 1947 increased by 5.76 percent over the previous year.[36] With industrial activity picking up, its debtors began paying their dues. The Central Machinery Works, for example, cleared 54 percent of its CNC 85.53 million dollars in arrears as early as February 1947.[37]

Struggling with the economic uncertainty arising from hyperinflation, private power companies did not do as well as state-owned ones. The rapid devaluation of the Chinese national currency *fabi* made it difficult for the Shanghai Power Company to determine a reasonable public utility rate. Imported oil accounted for four-fifths of its fuel cost. The Chinese government mitigated the impact of hyperinflation by granting a fixed exchange rate of one US dollar for 12,100 CNC dollars solely for oil purchases, while the open-market exchange rate was one US dollar for 50,100 CNC dollars. The subsidized rate ended in October 1947, forcing the power company to shift the price increase onto consumers. Between July and October 1947, the nominal operating expenses increased by 56.537 billion CNC dollars based on an exchange rate of one US dollar to 50,100 CNC dollars. By the time the report was completed, the open-market rate shot up to one US dollar for 56,000 CNC dollars. While the company's costs tripled and quadrupled due to hyperinflation, Shanghai momentarily recorded the lowest electrical tariffs in the world at 6.82 US cents per kilowatt-hour, about two-thirds that of major American cities.[38] Low prices did not translate into better affordability, however. A low-ranking government worker who took home about 200,000 CNC dollars could barely afford his rent and food, let alone pay for electric lighting priced at 3,800 CNC dollars per kWh.[39]

The American owners of the Shanghai Power Company proposed to address the problems by merging all of Shanghai's power companies into a single entity called the United Power Company. The new entity formalized the arrangement that had been in place since 1938, in which the Shanghai Power Company generated almost all of Shanghai's power and distributed it to the smaller power companies for sale within their franchise areas. The Shanghai Municipal Assembly objected to the proposal, accusing the Americans of capitalizing on the crisis to monopolize Shanghai's electrical power market.[40]

While the Guomindang regime and foreign capitalists were struggling to restore order to the electrical power sector in major cities, the Communists took their first shots at running small power stations in areas under their control. Between 1945 and 1946, the Communists administered the city of Zhangjiakou (Kalgan) as the "second red capital" after Yan'an and took over a Japanese-built power plant that had been part of the Mengjiang Electrical Corporation.[41] They appointed Ren Yiyu, an engineering graduate from the National Beiping University, who had spent most of his time in the Communist bases as both a public health school instructor and a military mechanic. Anecdotal accounts list Ren Yiyu as part of a team that built a makeshift sulfuric acid manufacturing facility in Communist areas.[42] As these engineering graduates lacked the necessary experience, electricians with only a year or two of vocational training took charge of most of the power plant's daily operations. With the loss of Kalgan in September 1946, the largest power station under Communist control fell into the hands of Nationalist forces. Ren retreated with the defeated forces to small cities in the Shanxi-Chahar-Hebei base area, where he continued to administer small power stations.

Nationalist forces reported that the Communists inflicted minor damage on power generation and transmission without causing catastrophic destruction. In February 1947, Communist guerrillas entered Tongzhou County, approximately fifteen miles east of Tiananmen Square. They fired at transmission lines and transformers and burned the electric meters at the railway station, county government office, police station, and hospital, but did not attack the main power station, apparently because it was "too remote."[43] The Communists did not capture or control Tongzhou after the raid but did enough to cripple the local government's daily operations.

Communist operatives faced major setbacks in their attempts to mobilize electrical workers. In October 1945, the underground Communist Party members in the Shanghai Electric Company formed the Democratic Unions Planning Committee. These labor movements organized a number of strikes that forced foreign capitalists to accept some of the workers' demands for better working conditions. Guomindang agents targeted these instigators of labor unrest. By April 1948, the Nationalists crushed the Communist network within Shanghai Power. The underground Communist Party leader of Shanghai Power, Wang Xiaohe, and fourteen of his colleagues were accused of throwing iron filings into the turbines to trigger massive power outages in Shanghai. The Nationalists arrested and executed Wang and many Communist Party members, dealing a heavy blow to the Communist grassroots organization in Shanghai's power industries.[44] After the purge, there were only forty-eight Communist Party members out of 4,000 employees at the Shanghai Power Company.[45]

By April 1948, the Nationalists, foreign capitalists, and the Communists had failed to bring the electrical industries under control. Saddled with debt and hit with hyperinflation, the electrical industries in North China and the Lower Yangtze struggled to remain solvent. The Communists were in a weaker position. By mid-1948, the Communists controlled 586 cities, most of which were in Northeast China and none with a population higher than one million.[46] They not only lacked experience in administering large-scale power systems of major cities but also had yet to build the mobilization structures within the electrical industries of key urban centers.

The Communists ultimately achieved decisive victories in urban warfare by exploiting the weaknesses of the electrical infrastructure that had emerged during the chaotic takeover. In his 1967 documentary *China: The Roots of Madness*, American journalist Theodore White recalled Mao saying, "It is true that the Japanese and Chiang Kai-shek had electricity, airplanes, tanks, and we have nothing." In that same conversation, Mao imagined himself as a modern-day George Washington, who was about to pull off a victory against a formidable adversary and added, "But then . . . the British had all those things, and George Washington didn't have electricity, and yet George Washington won."[47] Mao understated the importance of electricity in his interview with White. His generals were hatching a grand plan to take over the electrical industries even before marching into the cities.

Seizing the City's Lifeblood

In October 1948, the Communists signaled a major tactical shift in the battle to control China's electrical industries. Prior to this, the Communist and

Guomindang military forces focused their effort on preventing power assets from falling into enemy hands. Retreating forces blew up the power station and crippled the urban economy. Having seen how cities without electricity became death traps and having dealt with the aftermath of the destruction of the electrical infrastructure, the Communists realized that it was unwise to march into cities and seize the electrical infrastructure at gunpoint. They called on a small group of underground operatives to work stealthily within the electrical industries to secure the support of electrical workers and engineers. These agents persuaded engineer-bureaucrats serving the Guomindang regime to hand over the power assets to the Communists.

The North Hebei Power Company, which had developed into the largest state-owned electric utility under the Guomindang, became the key battlefield in the struggle to control the nation's electrical industry. The PLA under the command of generals Peng Zhen, Nie Rongzhen, Lin Biao, and Luo Ronghuan executed the first successful power blockade in Beiping (Beijing) between December 1948 and January 1949, which trapped Nationalist Fu Zuoyi's troops. Having gained the support of the engineering elite of the power station, the Communists secured the power station supplying most of Beiping's city before encircling the old capital in December 1948. The "power blockade," which had worked as a blunt tool prior to October 1948, was transformed into a precision strike weapon. The PLA was able to briefly restore power supply for a few hours to relieve the inconveniences of Beiping's residents. Building on the alliance between military, workers, and engineering elite, the Communists were able to use the blockade to effect Beiping's "peaceful liberation" in January 1949. They replicated the strategy elsewhere, allowing them to inherit the electrical infrastructure intact following a bloody civil war.

The Communists were the first to suffer the effects of a power blockade. After the Communists established the first people's government in Harbin in May 1946, the Nationalists launched a counteroffensive in the fall by occupying the Little Fengman Hydroelectric Plant in neighboring Jilin Province and cutting off power to the city.[48] The Communists mobilized rank-and-file workers to restore power to Harbin. The Xinhua News Agency and *Renmin ribao* (People's Daily) reported that power plant workers salvaged a few old and disused generators and reactivated them.[49] The Communist propaganda machine heaped praise on a veteran electrician named Xiao Chaogui and when asked why he was only an ordinary worker in the old days, Xiao wiped the coal ash off his face and reiterated the party's slogan, "Who cares about the past? As you can see, the factories belong to the workers now!"[50] The report followed a set script: The self-taught worker of humble origins, energized by revolutionary spirit, achieved a technological breakthrough to save the people. This narrative of "worker's innovation" would resurface during the mass mobilization campaigns in the 1950s and the Cultural Revolution.[51]

The power outage in Harbin alerted the Communists to the importance of electricity in urban warfare. During military campaigns in Shandong and Shanxi, Communist forces reported that Nationalist forces blew up power stations before retreating. About a month after surviving the power blockade in Harbin, the PLA entered Shuo County in Shanxi Province only to learn that the former warlord Yan Xishan had ordered the destruction of the county seat's power station. According to the Xinhua News Agency, it took twenty-seven days to repair the damage, and after which "the bright lights of 15,500 lamps celebrated the return of brightness."[52] The same thing happened after the Communist capture of Xinjiang County in southern Shanxi and Cang County in southern Hebei in July 1946. These power stations had a generating capacity of 80 kW, and the Communists were able to execute the repairs within a week.

The Communists publicized successful electrification projects to allay fears among those who doubted their competency as urban administrators. In November 1947, the Xinhua News Agency declared that the Mudanjiang-liberated region had tripled the length of its power distribution lines and fully electrified suburbs within the ten-kilometer radius of Harbin. They claimed to achieve a basic level of rural electrification, as the peasant cooperative in the Zhixin district in Yanji city had invested in a small power plant, which allowed the peasants to mill grain with electrical machinery.[53] Despite these purported successes, the Communists were capable of managing small power systems, which constituted a few medium-voltage power lines connected to a single power station operating a low-capacity turbine.

The inability to manage complex electrical power networks hindered the PLA's advancement into larger cities. One example was their struggle to restore electrical power at Linfen in Shanxi Province. In March 1948, Communist and Guomindang forces clashed at Linfen, a strategic transportation hub between the provincial capitals of Shanxi, Henan, and Shaanxi. The Linfen power plant straddled the northern and eastern city walls. Guomindang forces defended the city by barricading the power plant with three trenches, two lines of barbed wire, and one electric fence and garrisoned one brigade of soldiers in the power plant. Although the Communists managed to breach the defense after five days of intense bombardment and relentless assaults, the turbines and piping were completely destroyed.[54] As dead bodies and shrapnel were cleared out of the power station, technicians painstakingly welded short pieces of wire recovered from the warehouses and salvaged old power-generating equipment left behind by the Japanese. Repairing this medium-sized power station with a generating capacity of about 800 kW, ten times the size of power systems in small county seats, took about five months. The PLA needed to avoid a repeat of the destruction that occurred in Linfen as it moved on to other major cities.

The Communists tried to gain the cooperation of the workers and engineers, which would allow them to secure the electrical infrastructure before marching into cities. The deputy commissar of the Liaodong Military Zone Chen Yun, who rebuilt the economies of major urban centers in the northeast, recognized the importance of electricity. Elected chair of the All-China Labor Congress in Harbin in June 1948, Chen called on the workers to join the revolutionary struggle by proclaiming, "Wherever the People's Liberation Army goes, the lights will come on."[55] Chen's understanding of the importance of electricity comes through most clearly in his essay about the takeover of Shenyang.

Summarizing five key lessons from the Communist takeover of Shenyang in November 1948, Chen Yun wrote:

> First and foremost, we must restore electrical power. Without electricity, electric lights will not shine, telephones will be disconnected, tap water cut off, trains and trams stalled, the city would become a ghost town, and order cannot be restored. Shenyang depends on electricity from the outside. With Fushun liberated, we were able to send power here, and earned the praise of the people. There is one key condition to doing our work well. We need enough technicians. We brought hundreds of technical cadres from Harbin. They are all brave, loyal, and well-trained. . . . The local people were surprised and do not feel that "the Communists are country bumpkins with no technological knowledge.[56]

Shenyang's fate stood in stark contrast to another city in the northeast. Between May and October 1948, the Communists besieged Changchun, the capital of Jilin Province. The city held out for months but finally surrendered on October 17, 1948—a day after the Communists cut off electricity to the city.[57]

Chen Yun's essay expressed a desire to avoid a repeat of the tragedy in Changchun, as the PLA advanced into heavily populated cities farther south. Having accumulated considerable experience from running Harbin's power grid, the Communists handled the takeover of Shenyang's electrical infrastructure with relative ease. Chen was also aware that it was not enough to rely on the support of rank-and-file electrical workers. The Communists needed to secure the defection of the engineer-bureaucrats who had played an instrumental role in building the power infrastructure and nationalizing the power sector.

The Communists started by forging alliances with the workers. Even before the siege on Beiping, Peng Zhen, one of the PLA's commanders in the Beiping-Tianjin campaign who later became party secretary and mayor of Beijing, spoke to workers in the Beiyue district in Beiping in July 1948. He urged workers to gather information about the political views of managerial staff in key industries and also called for the organization of party branches within the power station.[58] These party branches were known as Factory Protection Committees or "safety committees." Workers participated in these committees, as they sought to protect their livelihoods. Wen Tingkuan, a Communist underground activist in the Shijingshan Power Plant, persuaded workers to look out for saboteurs during secret patrols, reminding them that they would lose their jobs if the power-generating equipment was destroyed.[59] The small size of these committees allowed them to evade surveillance. The safety committee of the Central Electrical Manufacturing Works at Shanghai, for example, had only fifteen active members. They met once every few weeks to coordinate patrols and emergency relief for the workers. Workers who participated in these patrols received a small stipend and an extra meal of gruel from the communal kitchen.[60]

Protection committee members served as liaisons between the military and engineers. While electrical workers did the manual work to keep the turbines running, it was the engineers who knew the power generation, transmission, and consumption patterns of the power grid. Reaching out to the Guomindang's engineering elite was a risky venture. Many of these engineer-bureaucrats were Guomindang members, some of them were genuinely loyal to Chiang's regime while others had been forced to take up membership in order to qualify for higher office.[61]

Oral histories offer some clues on the workings of Communist operatives. In an interview with the Chinese Society for Electrical Engineers in 2008, the children of the chief engineer of the Shijingshan Power Plant Cai Changnian recounted secret visits from underground Communist party members. According to Cai's third daughter, Cai Xiangfen, her father became acquainted with underground Communist member Wu Zuguang a few months before December 1948 and was subsequently introduced to another underground activist Jiang Hongbin. These two "special guests" called on their heavily guarded residence. Engineer Cai sent his chauffer to pick up these two men, who sat in the rear passenger seats concealed by black curtains drawn over the windows and sneaked past the sentry. Cai then met the agents in his locked bedroom. He sent his children out in the courtyard to keep lookout, while playing shuttlecocks. The children were told to sit on the living room couch and not move when other guests called on the Cai residence. His daughter realized many years later that her father concealed Communist printed material in a secret compartment of that couch. Once assured of engineer Cai's support, the operatives asked him to introduce them to Bao Guobao—the general manager of the North Hebei Power Company.[62] T. V. Soong was planning to appoint Bao Guobao as the director of the Guangzhou Power Bureau. The Communist activists, however, persuaded Bao to turn down the offer.[63] During the siege on Beiping, Bao Guobao remained inside the city and liaised with his colleagues at the power station, which had fallen under Communist control.

Certain of the support from the engineers and a loyal core of electrical workers, the PLA ordered the capture of the Shijingshan Power Station. On December 11, 1948, the Central Military Commission of the Communist Party ordered troops to surround enemy forces in North China without attacking them.[64] PLA forces captured the suburbs and the power lines running into the old capital. Two days later, the North China Bureau of the Communist Party relayed an order to its underground cells in Beiping and Tianjin to prepare for the takeover. The order specifically tasked the party committees of Beiping's neighboring counties to protect the assets of the Shijingshan Power Station and Steel Factory, Mentougou coal mine, and Tongzhou Power Station.[65]

On December 15, 1948, the PLA launched a surprise attack on the Shijingshan Power Station. According to battlefield dispatches dated December 23, 1948, and published in the *Renmin ribao* on January 12, 1949, an eight-man machine gun team slipped into the Shijingshan Power Station when sentries from the Guomindang forces were changing shift. With assistance from the Communist underground, they occupied the coal tower—the power station's highest point. Communist forces fired at an enemy battalion that tried to retake the power plant, while operatives ran up and down the coaling tower to replenish supplies of food, water, and rifle-cleaning cloth. Communist reinforcements arrived the next day, forcing the Nationalists to retreat.[66] Recognizing the significance of this episode, party histories and official gazetteers retell the heroic feats of the battle for Shijingshan Power Station.[67]

Beiping plunged into darkness for at least twelve days. Xu Ying, the *Dagongbao* reporter who had been working on the collective biography of Chinese industrialists, chronicled the ensuing chaos in his diaries.[68] Cut off from the main source of electrical power, the winter became exceptionally harsh for residents of the old capital. Streetcars ground to a halt. Two thousand tram drivers remained idle. Electric pumps that supplied tap water to one third of Beiping's residents stopped running. Residents ran around the city to collect water from wells. The price of food increased exponentially within days.[69]

Bao Guobao, the general manager of the North Hebei Electric Company, remained elusive during the early days of the blockade. A few days into the power blockade, Xu Ying contacted Bao, who appeared to be trapped inside Beiping with Guomindang forces. According to Xu's December 19, 1948, diary entry, Bao reportedly said that the company had activated back-up power by repairing an old 500 kW power generator tucked away in storage. He also authorized the transfer of electricity from Tongzhou County to keep some of the trams running. After telling Xu Ying that he could do nothing more, Bao then left his office to skate on the frozen ice at Beihai Park. In the days without power, Beiping residents hung red oil-lamp lanterns outside their doors for safety, but the cold winter wind always extinguished the flames before sunrise. A folk song circulating around Beiping went: "Every household that hangs a red lantern joyously welcomes Mao Zedong" (Ch. *jiajia gua hongdeng, yingjie Mao Zedong*).[70] The Communist takeover was imminent.

Beiping was no stranger to armed conflict, but the hardship imposed by the power blockade was unprecedented. On Christmas Day 1948, a light snow blanketed Beiping and covered 1,700 tons of garbage that could not be removed. Beiping was still without power. Reflecting on the events of the past two weeks, Xu Ying wrote:

> The red kerosene lamps that light up the streets of Beiping show that we have regressed to a living standard from thirty years ago. Within two weeks, Beiping underwent a drastic transformation in clothing, food, accommodation, and transportation. The siege of Beijing by the Japanese after the Marco Polo Bridge incident and the Boxer Rebellion cannot be compared to this.[71]

Two hundred thousand troops were garrisoned in a city of 1.2 million people. Armed sentries guarded every tall building and intersection. With no light and power, Fu Zuoyi's troops were cut off from the outside world.

Beiping's residents began to express resentment about the inconvenience caused by the power blockade, prompting the Communists to adjust their strategy. The Communists ordered the partial restoration of electricity to Beiping. The engineers in Beiping had to reconnect the 33 kV transmission line between Shijingshan and Beiping. The secretary of the Beiping electric company used his personal connections to persuade Fu Zuoyi into accepting the Communists' offer. It so happened that the secretary's father was not just Fu's teacher at the Baoding Military Academy but was also the godfather of Fu's chief of staff. With Fu's tacit approval, the TVA-trained engineer Wang Pingyang and two of his colleagues ventured out of Beiping's city walls with a shortwave radio set and established communications with their colleagues in Shijingshan. The military representative of the Shijingshan Power Station, Ren Yiyu, who had served as the general manager of the power station at Zhangjiakou, relayed Peng Zhen's order that Shijingshan Power Station would supply electricity to the 1.2 million people in Beiping but not to the Guomindang military. To achieve this objective, the power station generated 20,000 kW of electricity, just enough to power the water pumps and electric lights, and some semblance of normalcy returned on December 27, 1948.[72]

Xu's account mapped onto that of a foreign missionary who was then teaching at Yenching University. Ralph Lapwood, who had worked closely with Rewi Alley at the Chinese Industries Cooperatives during the second Sino-Japanese war, also detailed the power blockade and expressed amazement at the Communists' ability to bring back power so quickly. The blackout at Yenching University lasted from December 19 to 21. According to Lapwood, a political commissar of the PLA addressed the students at Yenching University. Besides sketching a rosy picture of the future, "emphasizing democracy, production, education, rising living standards" among other things, the political commissar promised the students and faculty that the Communists would resume electrical supply to the city within days. When the lights came on as promised, Lapwood remarked, "Having become inured to delay and prevarication under the Nationalist officials, we were astonished that the promise was promptly honored. As time went on, we found this unexpected efficiency a regular feature of the new regime."[73]

The Communists curtailed the power supply when they learned that Fu Zuoyi broke his promise of limiting electricity to civilian use. Writing to Marshal Ye Jianying on January 5, 1949, Peng Zhen noted:

> There has not been a single interruption of power since we started sending electricity into Beiping in December last year. However, the enemy prevents the people from listening to the Xinhua Radio Broadcast Station and diverts power for military use. Besides the availability of tap water, the residents of Beiping have little to gain. We decided to change the way we supply electrical power to the city. From now on, we will supply electrical power for two hours in the morning and afternoon, so that the civilians can collect tap water and cut off power at night to weaken the enemy.[74]

With the city's lifeline under Communist control, it was only a matter of time before the Communists formally took over Beiping from Fu Zuoyi. Demoralized and incapacitated by the energy shortage, Fu Zuoyi's forces agreed to a ceasefire on January 22, 1949.[75] The orchestrated blackout and intermittent power restoration allowed the PLA to capture Beiping with minimum casualties, while sparing the electrical infrastructure from destruction. This strategy, inspired by Chen Yun's slogan of the PLA bringing light and power to the people, enabled the Communists to capture major cities and maintain social order shortly after the takeover.

Misled by earlier conspiracies and counterintelligence, Nationalist forces in the south failed to notice that the Communists were preserving the electrical power assets in anticipation of power transition. On January 27, 1949, the Shanghai Public Works Department received a warning from the US Army International and Research Unit. American secret agents, who had infiltrated the Far East Intelligence Bureau of the Comintern, claimed that Communist Party members within the Shanghai Power Company were conspiring with a bomb squad to blow up Shanghai's main power station.[76] The report appeared consistent with earlier accounts of guerrilla operations, in which Communist-backed forces damaged power-generating and transmission facilities to stir unrest. The case of Wang Xiaohe, the Communist electric worker executed in April 1948 after he was accused of sabotage, was still fresh in the minds of members of the Shanghai Public Works Department. The Nationalists remained on the lookout for similar conspiracies.

Contact between underground Communist operatives and engineer-bureaucrats in state-run industries caught the Nationalists off-guard. The Nationalists thought that they had already crushed the underground cell at the Capital Power Station in Nanjing by arresting its mastermind Zhang Guobao in April 1948. Chen Shenyan replaced Zhang as the leading operative in Nanjing. Chen got in touch with the power station manager through the manager's female cousin, who was already a Communist Party member. The manager agreed to establish secret patrols and preempt any attempts by the Guomindang regime to dismantle the power-generating equipment. The operations of the Capital Power Station remained undisrupted throughout the PLA's siege on Nanjing on April 23, 1949. Workers locked the gates and prevented the military police from entering the premises.[77] Similar accounts of "factory protection" unfolded across China.

Before marching into Shanghai, the PLA secured the defection of Yun Zhen—the NRC's engineer in charge of electrical equipment manufacturing. After the Japanese surrendered in August 1945, Yun moved to Shanghai to oversee the implementation of the Westinghouse technology transfer agreement. Communist activists capitalized on labor disputes and gained the support of rank-and-file workers. Ge Helin, who was factory manager of the state-owned wire manufacturing plant in Shanghai and the chair of the Communist "factory protection committee," served as an intermediary between the PLA and the Guomindang's engineering elite in Shanghai. Ge enjoyed a personal friendship with Yun Zhen. In the 1930s, the president of Shanghai Jiaotong University expelled Ge Helin for participating in leftist labor organizations. Yun Zhen employed Ge Helin as a personal assistant and arranged for Ge to complete his studies at the University of Nanking. In 1948, Yun Zhen appointed Ge as the manager of the state-owned wire production facility in Shanghai, as Ge was highly experienced in settling labor disputes.

Based on Yun Zhen's recollections, Chen Yi, commander of the Third Field Army and future mayor of Shanghai, contacted him before the PLA's capture of Shanghai. Chen Yi asked Yun Zhen for the addresses of all the electrical equipment warehouses and the phone numbers of the warehouse supervisors. Chen relayed this message to Yun Zhen through Ge Helin, who received it from another Communist underground activist.[78] On May 26, 1949, the PLA dispatched troops to the electrical equipment plants by using the information provided by Yun Zhen. Two days later, military representatives marched into these factories and took over from the factory protection committees.[79]

The battle for electrical resources revealed how the Communists secured their ultimate victory by paying close attention to practical economic concerns. The Communists mobilized allies across different social classes by aligning their interests with the party's tactical objectives. The engineers who had worked tirelessly to nationalize the electrical power sector believed that the Communists would fulfill the dreams of strengthening the nation through state-driven industrialization. Huang Yuxian, the hydroelectric engineer who surveyed the Three Gorges Dam with American engineer John L. Savage, defected to the Communists. According to a self-introduction written at the Central Training Corps in 1944, Huang Yuxian graduated from Tsinghua University and Cornell University and worked in the United States for about six years. Upon hearing about the Japanese occupation of Manchuria in September 1931, Huang gave up his job that paid three hundred US dollars a month and returned to China to work for the National Defense Design Committee, pledging that "from now on, I will establish close ties with the [Nationalist] party in order to strengthen my work."[80] Huang became disillusioned with the Nationalists and came to see the Communists as a force of national reconstruction. Huang remained critical of Chiang's regime. In a May 1963 report by the *Dagongbao* newspaper in Hong Kong, Huang accused T. V. Soong of siphoning funds from the construction of hydropower dams. The rapid completion of the Lion Rapids Dam under the Communists stood in stark contrast to the perpetual shortage of funds and manpower during his one decade of service under the Nationalists.[81]

Plagued by war fatigue, the engineer-bureaucrats of the Nationalist regime were receptive to the Communists' promises of stability and efficiency. Having been on the run since the outbreak of Sino-Japanese hostilities in 1937, the engineers were reluctant to take apart the electrical infrastructure they had painstakingly rebuilt and withdraw to Taiwan with the Nationalists. Back in 1938, Bao Guobao was ordered to blow up the Guangzhou Power Station just before the Japanese captured the city. One decade later, Bao Guobao, in his role as the highest decision maker of the North Hebei Electric Company, chose to facilitate a peaceful transition by working with the Communists. Yun Zhen also remained on the mainland instead of relocating power equipment to Taiwan. Chen Yi's promise to distribute food and money to the electrical workers and stabilize Shanghai's economy was enough to persuade Yun to work with the Communist forces.[82]

Grappling with White Terror

The battle for electrical power raged on even after the Guomindang retreated to Taiwan. In the early months of 1950, secret agents of the Nationalist regime claimed to uncover a conspiracy to subvert Taiwan's electrical power system. Liu Jinyu, who had built the electrical power industries in wartime Kunming and had taken over as the general manager of Taipower in 1946, was accused of agreeing to turn over the Sun-Moon Lake hydropower plant—the heart of Taiwan's electrical power system—to the Communists. Spooked by the loss of the power infrastructure in mainland China, the Nationalists executed their most experienced engineer after completing a hasty investigation and summary trial.[83]

Terrified by the defeat on the mainland, the State Secrets Bureau suppressed leftist social movements in Taiwan under a campaign of "White Terror." Secret service agents, who had once served under spymaster Dai Li, began hunting down leftist sympathizers from all walks of life. Liu came under suspicion, not only because he was contacted by his former colleagues on the mainland but also because his sons had escaped to mainland China. The charges laid out by the military tribunal state that three of Liu's five sons had fled Taiwan for Shanghai in the spring of 1949 after being implicated in leftist campus movements. His eldest son, Dengfeng, "told his father to preserve Taiwan's electrical industries and hand them to the Communists" in the event of a Communist invasion. Dengfeng also informed his father that a spy named Wang Yanqiu would get in touch with him. The State Secrets Bureau learned that Taipower purchasing department head Yan Huixian, who had arrived in Taiwan from Shanghai via Hong Kong in July 1949, carried a message from Chen Zhongxi, the former chief of the NRC's Electrical Department Bureau who had defected to the Communists. Chen made the same request as Dengfeng. The meeting between Liu Jinyu and Wang Yanqiu allegedly took place in November 1949.[84]

The investigation against Liu was haphazard from the start. Gu Zhengwen, the main investigator, joined the secret service in 1935 when he was a student at Peking University. In oral interviews published in 1995, Gu claimed that the State Secrets Bureau had investigated the Defense Ministry's deputy chief of staff Wu Shi and branded him a Communist collaborator. Drawing on Wu's confessions, the secret agents detained Cai Xiaoqian—the alleged leader of the Communist movement in Taiwan. During his interrogation, Cai mentioned that two of Liu Jinyu's sons, who fled to mainland China, sent a letter to their father. The sons told their father that the Communists were preparing to attack Taiwan and asked Liu to persuade his colleagues at the NRC to support the Communists.[85] Mao Renfeng, who had succeeded Dai Li as the spy chief, doubted Gu Zhengwen's story and asked for concrete evidence. Gu replied that there was no evidence, but he would "find a way to make him confess."[86]

Gu's recollections contained inconsistencies. He claimed that he visited Liu Jinyu at the Taipower headquarters on Heping East Road in January 1950. Liu was said to have asked Gu to hand a pile of cash to a Mr. "Zheng Xianghui," which happened to be the alias of the alleged Communist leader Cai Xiaoqian. Gu maintained that the secret service detained Liu three days later.[87] These events, even if true, could not have occurred before April 1950. Liu made his last public speech on March 19, 1950, when he visited Fengshan Training Camp to lend support to thirty-six Taipower employees who volunteered for political training.[88] Subsequent issues of the Taipower bulletin from May 1950 onward scrubbed all mentions of Liu Jinyu, suggesting that the arrest took place in late April rather than in January.

Liu's deputy Huang Hui was appointed acting general manager on May 3, 1950, while Liu's detention was kept secret. The Taipower high brass were simply told that Liu was on extended leave.[89] Gong Debo, a newspaper editor detained for criticizing Chiang Kai-shek, shared the same cell as Liu. Between 9 pm on May 2, 1950, and 4 am the next day, interrogators tied Liu to the "tiger bench," forcing him to sit upright with his knees pinned down and his ankles elevated on several bricks.[90] Two days after torturing Liu, the NRC informed Taipower that Liu was being investigated and relieved of his duties without giving further details. On May 15, 1950, Sun Yun-suan, praised as "forward-looking, hardworking and law-abiding," took over Huang's former position of chief engineer.[91] The Nationalists in Taiwan now carefully vetted every engineer placed in a leadership position within the power company.

Interrogators did not extract any clues from Liu about undercover Communist agents. Liu mentioned a section chief Chen from the Finance Ministry during his questioning. The Secret Service ended up arresting the wrong person. Gu Zhengwen even admitted that the investigation of Liu was too hasty. Nonetheless, the secret service extracted a written confession under duress from Liu. When the chairman of the provincial government of Taiwan Wu Guozhen (K. C. Wu) vouched for Liu Jinyu's innocence, Chiang showed him Liu's statement. Wu would later resign from office and spend the rest of his life in the United States.[92]

The state-controlled media in Taiwan started a smear campaign against Liu once his arrest was publicized in mid-May. Liu defended himself against accusations of conspiracy to no avail. According to a *Xinshengbao* report published after his execution, Liu maintained that he was a scientist with no interest in politics. He tried to convince his interrogators that he was a staunch anti-Communist by pointing to his friendship with Yu Bin, the archbishop of Nanjing who was later elevated to cardinal in the Catholic Church under Pope Paul VI. Liu also argued that he was the leading coordinator of American aid and would not have spoken to the Americans if he was a Communist spy.[93] News reports brushed aside his pleas of innocence. A commentator for the *Niusi zhoukan* under the pseudonym Zhuge Ming called Liu a cowardly and deranged ingrate, who "sold his soul to the Communist bandits and betrayed the trust of his country." He criticized Liu's lavish lifestyle. According to him, Liu was chauffeured everywhere in a posh Buick sedan and insisted on having his shoes polished before stepping out of his car.[94] The *Xinshengbao* reporter chaffed at Liu's statement that "a poor country should cherish talent like myself." He retorted, "I do not see how our nation is regarded as a poor country, and why such a 'talent' is worth 'cherishing.'"[95] The heroic engineer who had built the power system for Kunming's defense industries during the war was now branded a self-entitled traitor.

The military tribunal revised the charges to bolster its case against Liu. In the original charges on June 21, inquisitor Wang Youliang of the Taiwan Peace Preservation Corps proclaimed Liu and Yan guilty of spying for the Communists. Chief of Staff Zhou Zhirou noted in the amended charges that it was much more appropriate to charge Liu with obeying seditious orders from the enemy and failing to abide by discipline, both of which were still punishable by death. Zhou pleaded for a reduced sentence for Yan, but Chiang Kai-shek refused to accede and ordered the execution of both accused men.

Zhuge Ming gave a blow-by-blow account of Liu's final moments. On July 17, 1950, at 4 am, Liu and his alleged co-conspirator were hauled to a brightly lit military courthouse in Taipei. The judge ordered the military police to untie the accused so that they could write their last words. Liu protested, "I am innocent, why am I sentenced to death?" The judge retorted by saying that Liu committed an unforgivable crime. Liu lowered his head and sobbed as he wrote his farewell note.[96] Liu raised his trembling hands and begged the courts to bring him a Catholic priest to say his final prayers. The judge turned down his request, saying that it was no longer possible to delay the execution. When he was done writing, Liu pulled out a comb from his coat pocket and asked it to be handed over to his family. The escorts tied up the two men on death row. They then forced a warm loaf of bread and low-grade red cordial liquor down Liu's throat. Like all other political prisoners, Liu Jinyu and Yan Huixian were executed by firing squad at Machangding. Shortly after the execution, the Military News Agency put out a news release announcing the execution of a traitor for "attempting to make use of an important national asset to preserve his personal status and secure promotion and prosperity."[97]

Was Liu Jinyu guilty as charged? One thing is for sure: Liu's sons had fled Taiwan for mainland China. One other fact is well established: Yan Huixian, who fled Shanghai for Taiwan via Hong Kong, carried a letter from Liu's former supervisor Chen Zhongxi addressed to Liu. The inconsistencies in the case records point to a miscarriage of justice.

Gu Zhengwen barely mentioned Yan Huixian in his later accounts. The charges also made no mention of Liu's alleged contacts with underground Communist activists such as Wu Shi and Zhu Feng, calling into question the credibility of Gu's investigations. National security agents acted on the slightest suspicion that Liu had agreed to turn over the island's electrical power industries to the Communists and most likely extracted a confession under duress. In the frenzied struggle for survival, the Guomindang regime in Taiwan did everything in its power to cling to the only electrical power network that it successfully took over during the scramble for reparations.

THE STRUGGLE FOR control over China's electrical industries played a significant role in shaping the outcome of the Chinese Civil War. The scramble for reparations illustrates the perils of compressed economic development. The turmoil arising from the rapid collapse of Northeast China's industries after the Japanese surrendered in 1945 led to a power vacuum that was filled by the Communists. The Nationalist government shouldered the immense cost of returning private power companies to their rightful owners and running a sprawling network with many inherent flaws left behind by the Japanese. Not only did assets easily become liabilities, but the rapid reversal of fortunes starting around July 1948 also suggests that positions of strength concealed weaknesses and vice versa. The Communists, who did not control many of China's electrical resources in the beginning, were not bogged down by the heavy responsibilities that came with administering legacy systems that came into being over decades of haphazard development. Starting with small-scale rehabilitation projects and learning from tactical mistakes, the Communists adapted to conditions of material scarcity and devised strategies to subvert the electrical industries.

After grappling with a world of invisible threats in a catastrophic civil war, neither the Nationalists nor the Communists would ever relinquish control over the power grid. The power blockades executed by the PLA exemplified how the loss of electrical power could result in the loss of political power. The Nationalists never fully understood why their American-trained and party-indoctrinated engineering elite went over to the Communists. The climate of fear was no less intense on the mainland. The centralized power networks, jointly managed by the engineering elite, military, and party cadres, would transform China's workforce into a militarized industrial army. The civil war had never ended and looming fears of catastrophic power loss continue to shape the outlook on energy security on both sides of the Taiwan Strait.

CHAPTER 7

Manufacturing Technocracy

C LASHES BETWEEN NATIONALIST and Communist forces continued even after the founding of the People's Republic in October 1949. On February 6, 1950, seventeen bombers from the Nationalist Air Force dropped around seventy bombs on Shanghai. The city's four major power stations, Yangshupu, Zhabei, Chinese Merchants' Electric Company, and the French Power Company, were the primary targets. Between April 1949 and May 1950, Shanghai experienced seventy-one air raids that resulted in 4,500 deaths and injuries.[1] The Communists were well aware that if the Nationalists were to stage a comeback, they would start by recapturing Shanghai through military force or economic subversion. The February 6 bombardment struck at the heart of Chen Yun's promise that the lights would come on wherever the People's Liberation Army goes.

The Communists bounced back from the energy crisis following the February 6 bombardment to create a technocratic energy regime based on a shared consensus between engineering elite, military, and party leadership, ultimately contributing to the centralization of state power. This chapter follows Miguel Centeno's definition of technocracy, which he takes to mean "the administrative and political domination of a society by a state elite and allied institutions that seek to impose a single, exclusive policy paradigm based on the application of instrumentally rational techniques."[2] Put simply, the result in Shanghai was not a dictatorship of engineers but rather a situation in which the "reds" and "experts" leveraged their strengths.[3] Party cadres delegated the daily operations to electrical engineers, while they exercised their authority to coordinate power consumption across the industrial sector. Engineers tacitly supported these mass campaigns, as they believed chronic power shortages could only be addressed with political solutions. The engineers, military, and party grassroots, who had forged an alliance during the civil war, united themselves around two common objectives: fortify the security of power stations and maximize production capacity by maintaining electrical industries near peak load, while scheduling industrial production around the clock to fully tap additional electrical energy generated.

In the quest to secure electrical power in Shanghai, soldiers blended into the labor force, while workers were transformed into conscripts of an industrial army.[4] Even before the seizure of American assets in December 1950 following the Korean War, military representatives entered the power station to restore order and functioned as intermediaries between the technical staff and municipal authorities. The Communist regime increased the electric supply without major investments in electrical equipment through the copper conservation campaign of September 1951 and the implementation of "peak-load management" to coordinate power demand. As party cadres with military credentials took control of the Shanghai Power Bureau and used their organizational skills to micromanage the power consumption of 450 of Shanghai's largest factories, Shanghai's workers were forced to comply with highly regimented work schedules. The rationing of electrical power not only inconvenienced many workers but also led to the militarization of civilian life. The Shanghai Power Bureau and later the East China Power Bureau also made use of public-private partnerships to seize control of the long-distance transmission lines under the ownership of private power companies founded by Chinese entrepreneurs. It integrated a patchwork of foreign-controlled power conglomerates (e.g., American-owned Shanghai Power Company) and power stations owned by Chinese businessmen (e.g., Chinese Merchants' Power Company, Pudong Electric, and Zhabei Electric) into a coherent regional network.

Regional networks across the nation adopted measures introduced in Shanghai after the February 6 bombardment. Power systems in North China also attempted to optimize energy use by flattening the power consumption curve. The Chinese Communists went beyond emulating Lenin's idea that "Communism is Soviet Power plus the electrification of the whole country." They transformed electricity into a technology of mass mobilization, which allowed the new regime to control the urban population and catalyze economic development from the mid-1950s onward. By remaining on a wartime footing, China's electrical industries generated the economies of speed, through which the expansion of centralized electrical power systems extended the reach of centralized state power.

Militarizing Civilian Life

As the PLA secured the electrical infrastructure during the closing stages of armed conflict during the civil war, Mao urged soldiers to take on new responsibilities for urban governance. In a February 1949 telegram addressed to the second field army in Shanxi-Chahar-Hebei and third field army in East China, Mao wrote, "The military is not simply a fighting force, it is primarily a workforce. The cadres of the military must learn to take over and manage cities, understand how to deal with imperialism, Guomindang reactionaries, and the capitalist class." This required the military to "address problems related to food, fuel, and other necessities, and appropriately handle financial and fiscal issues."[5] Military personnel soon became part of the labor force in China's power industries, as 2.1 million soldiers from the PLA's four field armies were deployed to the nine southern provinces along with 53,000 supporting party cadres, following the end of the three major battles of the Chinese Civil War in late January 1949.

Mao also rightly reminded the PLA not to forget its primary role as warriors. Stung by the loss of Shanghai in May 1949, the Nationalist Air Force retaliated by bombing the largest power stations in the Lower Yangtze on February 6, 1950. The American-owned Shanghai Power Station on Yangshupu Road was not spared. An internal report to the Communist Party Shanghai Municipal Committee expressed surprise, "Before the air raid, it was commonly believed that Shanghai Power would not be bombed since it is owned by the American imperialists. Enemy jets flew overhead several times before and did not cause any alarm."[6] The bombing proved what the Communists knew all along: The Nationalists were not going to surrender their power assets without a fight, and the Communists' control over the electrical industries was far from complete.

The air raid resulted in a brief power outage in Shanghai. The Municipal Committee reported that the Shanghai Power Station resumed operations on February 8, 1950, at 7:50 am, nearly forty-three hours after the bombardment. Instantaneous output reached 2,022 kW, but increased to 16,618 kW by the end of the day. The crisis offered an opportunity to rally the workers against a common threat. Sixteen days later, that figure went up to 74,000 kW, more than half of the pre–air raid levels. On February 8, 1950, the Party Committee established a command center with five subdivisions—air defense, propaganda, organization, monitoring, and general affairs. The Shanghai Power Station became a fortress. Air defense command installed twelve concrete flak towers around the buildings housing the power turbines. Workers manned anti-aircraft guns ready to shoot down enemy planes. Party cadres ordered workers and volunteers to pile sandbags and dig air raid shelters in the open spaces of the power station.[7]

Deep-seated tensions threatened to derail recovery efforts. The Power Company could have achieved pre–air raid levels of output, had the workers removed a thirty-four-ton component from a damaged generator more expeditiously.

Four days after the bombing, the owners approved the military representative's request to dismantle the damaged generator but warned them about the immensity of the task. Party cadres accused the owners of enlisting old workers to delay progress. After much persuasion, these old workers cooperated and devised a novel strategy to transport the massive component out the door and onto the truck within seven days. This was quite an accomplishment, considering that it had taken two months to install this piece of equipment.[8] This minor confrontation was a precursor to further disputes to come.

The Communist Party recognized the urgency of drumming up support among rank-and-file workers. Lou Xichen, a party propagandist, noted that many of these rank-and-file workers had either little knowledge of the Communist Party or lacked the courage and determination to organize other workers.[9] Cadres began identifying model workers as possible Communist Party recruits. They publicly commended the actions of party member Zhang Junsheng, who remained at his post at boiler unit number 3 even during the bombardment. They quoted him as saying, "I am a party member. . . . If there is a power outage, factories will stop working, there will be no tap water, air raid sirens will not sound. I am willing to sacrifice myself for my work."[10] Propagandists also blasted messages over loudspeakers to drive home the message that Chiang Kai-shek was responsible for the workers' suffering and deaths. The work report offered an anecdote of Zhang Laifa, who died from injuries during the air raid, who told his son Zhang Dinghai (also an electrical worker) to avenge his death. The elder Zhang told his son, "You must know that I am killed by the American imperialists and Chiang Kai-shek. You must convert all my work points into public bonds, so that we can attack [and liberate] Taiwan."[11] This foreshadowed large-scale mobilization campaigns that would take place a few years later, aimed at extracting more donations from power station workers.

Beneath the revolutionary and patriotic fervor, party cadres, military representatives, and rank-and-file workers distrusted each other. Party cadres reported that "the military representatives did not understand the conditions of the power station and lacked technical competence. They also did not understand the masses, so the branch party secretary took the lead for the inspection and devised repair plans, which were later authorized by Mayor Chen Yi."[12] Military representatives, who had been stationed in Shanghai's industries after the Communist takeover, appeared to do little besides maintaining order and escorting Soviet experts to survey the damage. When the threat of aerial bombardment abated, workers refused to contribute further to air defense operations. Students and workers from outside the power station were left to fill and pile up sandbags.

Party members also shirked responsibility. The military representative flew into a rage when no one responded to the alarm for an air raid drill. The party member on duty at the central command had left for a drink and was not at his post. One month after the air raid, the municipal committee acknowledged that the air defense command was merely an empty shell.[13] The Communist Party attributed its failure to rally the workers during the crisis to a "disconnect with the masses." With no security threats on the horizon, the Communist Party began scaling back their air raid precautions.

China's entry into the Korean War revived the urgency to secure Shanghai's power network. After assessing the vulnerability of Shanghai's transmission network, the Public Utilities Party Committee noted in a report dated October 28, 1950, "There were three hundred substations under the control of Shanghai Power Company and West Shanghai Power Company, of which eleven are primary substations," and that these substations located in remote areas were once guarded by sentries during the Japanese occupation and Civil War, but were now left totally unprotected.[14] Within a month, the Public Utilities Party Committee established a dedicated security unit to protect public and private power stations.[15]

The Communist regime subsequently used its emergency powers to seize the Shanghai Power Company from its American owners. On December 18, 1950, Prime Minister Zhou Enlai ordered the seizure of all American assets in China.[16] Within two weeks, the Communist Party appointed Cheng Wanli, a long-time veteran of the Communist Party branch in Shanghai Power, as Shanghai Power's military administrator. The military representative was no longer a glamorized security guard. He took command of the power company. In addition to the Shanghai Power Company, the military established direct control over 115 other American companies in Shanghai, including Texaco and Standard Oil.[17] Party officials also mobilized 3,000 power plant workers to participate in war rallies and resource conservation campaigns. They reported that workers in Shanghai donated RMB (old) 250 million to buy bullets for the volunteer army and aid North Korean refugees during the Korean War. This was a paltry donation, considering Shanghai's wealth and the size of the Shanghai Power Station's workforce.[18] However, this marked an improvement over the situation after the February 1950 bombardment. Party recruitment also picked up. Before 1949, no more than one hundred employees from the Nanjing Road office of the Shanghai Electric Company showed up at strikes and demonstrations. On March 4, 1951, more than four hundred workers from that office participated in rallies to protest American plans to rearm Japan. Eighty truck drivers from the Shanghai Power Company, who had long stayed away from political movements, participated in political discussions about the Korean War and donated RMB (old) 1.8 million to support the war effort.[19]

Material scarcity resulting from the mobilization for the Korean War also resulted in future conservation campaigns in Shanghai's power sector. Power companies across Shanghai, most notably Shanghai Electric, Zhabei Electric, and Pudong Electric, competed with each other to reduce wastage and improve performance standards. Pudong Electric reported reducing coal consumption by 10 percent without affecting power output, saving RMB (old) 100 million that could be diverted to the front line.[20]

The copper conservation campaign launched in September 1951 by the Shanghai Federation of Industry and Commerce was also part of a series of austerity measures. Yun Zhen, who had taken charge of developing China's electrical equipment manufacturing capabilities during the War of Anti-Japanese Resistance, fully supported efforts to find copper substitutes and divert high-grade copper to the production of electrical equipment. By then, Yun was director of the electrical equipment manufacturing division in Shanghai. To make 1,000 yards of copper wire, one needed at least forty-nine pounds of copper, while a 3,000 kW generator required two-and-a-half tons of pure copper.[21] Due to the exhaustion of known domestic copper deposits and the international blockade, China's electrical industries scoured the country for scrap copper to support its expansion. Unable to estimate how much copper was in the hands of its people, removing copper coins from circulation became the primary means to secure the much-needed raw material. Instead of using copper, private manufacturers made flashlight casings with cardboard.[22] Such measures, however, only reduced copper usage by 47,000 pounds every month.[23]

Metals with poorer conductivity had to be used to make power transmission lines due to the copper shortage. The problem was not unique to China. Yun Zhen pointed out that the Germans used aluminum as wiring material to get around the lack of copper. As aluminum had poorer conductivity, the wiring had to be very thick to ensure that electrical power could be transmitted safely. This was not a viable option in China due to the lack of aluminum deposits. Ultimately, the engineers increased the steel to copper ratio in transmission cables. Although the resistance of steel was five times that of copper, it was much cheaper and readily available.[24] The system builders traded efficiency for lower cost in the face of material shortage.

Besides conserving copper, Shanghai's industries also had to curtail power usage while increasing production output. Subordinated to the Ministry of Fuel Resources, the Shanghai Power Administration had to comply with the central government's orders of "making best use of the old machines" and patiently awaited budgetary approvals for investments in new power-generating equipment.[25] The lack of electrical power was no excuse for failing to meet ever-increasing production quotas, and Shanghai's factories had to raise the output of paper by 20 percent, matches by 50 percent, cigarettes by 20 percent, while textile mills had to double or triple production quantity. New food processing plants, such as flour mills and oil presses, also competed with preexisting industries for the limited electrical supply.

The military administrator of the Shanghai Power Administration objected to these targets. Citing figures by Soviet experts, Cheng Wanli noted that improvements in labor efficiency without further investments in new power generators would only generate a 10 to 13 percent increase in power output. Cheng noted that the capacity of Shanghai's power industry was close to that of 1926 Leningrad. To achieve enhancements in power efficiency, the work that took the Soviets several decades to complete had to be finished within five years.[26] During a September 1951 meeting, Cheng admitted that the maximum instantaneous output of the city could only be maintained at 162,000 kW, which was only about 77 percent of the installed capacity.[27] Cheng's conservative approach incurred the wrath of Communist Party cadres.

Dissatisfied with Cheng Wanli's unwillingness to cooperate and frustrated with the lack of support among rank-and-file workers, the Shanghai Power Communist Party Committee plotted to remove Cheng Wanli. The 1952 work report blamed the poor performance of Shanghai Power on Cheng's incompetence. It claimed that in the second half of 1951, there were "sixty-six power reductions, which led to reduced power output of 5.5 million kWh, four cases of accidental death, eight major power outages that lasted as long as 353 minutes. Inspection and the time taken to repair and inspect boilers and turbines was more than double the standard stipulated by the Ministry of Fuel Resources. . . . [Shanghai] was now lagging behind the advanced power industries of the rest of the nation."[28] Apart from poor performance, party cadres accused Cheng of failing to garner support from rank-and-file workers. Despite its best efforts, the Communist Party branch added only 110 new members, amounting to 3 percent of the power company's workforce, way below its 10 percent target. The Communists complained about the "complicated political situation" at Shanghai Power, which "concentrated all the scum of old Chinese society." Party members conducted background checks on 675 employees and noticed that "many senior technicians maintained connections to the American imperialists, while they assumed positions of power and controlled the most important production process." The control room director, Lin Huawan, for example, was a major-general of Wang Jingwei's collaborationist army.[29] In January 1952, the Communist Party Committee began a massive purge within the Shanghai Power Company by sacking Cheng Wanli on accusations of mismanagement and corruption. One after another, Cheng Wanli's associates were forced from office on accusations of bureaucratism and individualism. The economist Gu Zhun, who had urged the Industrial Bureau to be more explicit about power usage standards, was accused of being a counterrevolutionary and was removed from office.[30] A party cadre with military connections would replace Cheng Wanli to deliver the results.

Li Daigeng, the new chief military representative, brought with him experience in military command, labor organization, and indoctrination of party ideology. A native of Henan, Li joined the Communist Party in February 1938 and was assigned to the New Fourth Army in October 1938. He was deployed to northern Jiangsu, where he worked to raise funds from peasants to sustain anti-Japanese guerrilla warfare and mobilized peasants to advocate for rent reductions. In August 1942, Li took over the organization of the Communist underground near Yangzhou, which allowed the Communists to secure the movement of operatives in and out of Shanghai and Nanjing. At the end of the War of Anti-Japanese Resistance, Li was appointed chair of the Hangzhou Labor Union and vice chairman of the Zhejiang Labor Union. In December 1951, he assumed the role of military administrator of the Shanghai Power Administration.[31] His experience in military logistical support came in handy, as he eventually took on the responsibility of coordinating power consumption across hundreds of factories.

Unlike his predecessor who half-heartedly carried out the Three-anti Campaign aimed at eradicating corruption, waste, and bureaucratism, Li Daigeng went all out with his "tiger hunts." Between February and May 1952, almost all 3,000 workers were caught up in the anticorruption drive. Li reported that the party "captured 72 'big and small tigers.' 240 corrupt people who pocketed more than RMB (old) 1 million. 1,972 workers confessed to stealing from the power plant." Private contractors pocketed illicit gains amounting to RMB (old) 1.01 billion. All in all, party cadres purged 118 people, forty-eight of whom were sacked, eleven arrested, thirty-five sent to the Public Security Bureau for training, and sixteen sent to East China Revolutionary University for reform. Party cadres busted a gambling and prostitution ring in the power line repair crew, which involved around forty workers from a team of fifty-four repairmen. Another hundred workers were allegedly lending money to their colleagues at exorbitant interest rates.[32] The electrical workers took the Communist's claims of being the champions of workers' rights more seriously after the anticorruption drive. Li went on to reorganize the party committee by appointing loyal party members. Of the twenty-six men and one woman who assumed leadership positions, nineteen of them had joined the Communist Party before 1945 and stayed with the party during trying times.[33] As military administrator, Li brought his experience with centralized command and demands for loyalty to transform Shanghai's labor force into conscripts of a new industrial army. The enforcement of discipline percolated outward from the power station to the hundreds of factories in Shanghai. For Shanghai to overcome power shortages, workers adopted a regimental lifestyle by complying with work schedules centrally coordinated by the Power Administration and Industrial Bureau.

Flattening the Curve

Shanghai's electrical industry remained on a perpetual war footing, even after the Korean War ended in 1953. The Communist regime remained reluctant to increase Shanghai's generating capacity, as they predicted that Shanghai would bear the brunt of a Nationalist counteroffensive.[34] Even after Chiang's defeat at Dachen Islands in February 1955, Su Yang, a former militia leader who was elected chairman of the Shanghai Power Company, called for the curtailment of Shanghai's industrial growth.[35] To meet production quotas, the Shanghai Power Administration devised a two-prong strategy for the "rational use of electrical power." First, it implemented peak-load management, which forced Shanghai's industries to cut back production during daytime peak hours and schedule more shifts at night to fully utilize off-peak power. Under Li Daigeng's leadership, the Shanghai Power Administration micromanaged power usage across Shanghai's industries, causing workers to reorganize their lives around the peaks and troughs of Shanghai's power distribution system. Second, it spearheaded initiatives to modify manufacturing processes for improved energy efficiency. Engineers, who had once served under the Guomindang regime, handed the authority to allocate electrical power to these administrators. Administrators, who had no engineering background, appropriated the ideas of scientific rationality and efficiency and refashioned themselves as technocrats.

Peak-load management addressed the uneven distribution of power consumption between day and night and increased total energy input without the installation of new generators. The Soviets had devised the strategy during World War II to cope with power shortages. To accurately predict hourly power demand, the power administration had to chart the load curve daily and obtain the average daily load for each season by synthesizing these records every quarter.

Soviet experts invited by the Ministry of Fuel Resources formally introduced the concept of "peak-load management" in a series of lectures delivered in August 1951.[36] Peak-load management as practiced in Shanghai deviated from the recommended practices of the Soviets. The Shanghai Power Administration did not have the luxury of time to study power consumption patterns and scheduled power consumption by decree. Due to its limited generating capacity, power stations in Shanghai constantly ran their generators near full capacity and did not follow the Soviet recommendation of scheduling lull periods to take the generators offline for maintenance.

Shanghai Power Administration had already introduced measures to spread out power demand by June 1951. The Ministry of Fuel Resources imposed mandatory rest days across industrial sectors to relieve the power load. Steel, nonferrous industries, and chemical plants ceased production on Mondays, whereas textile mills, construction material, and fuel processing plants stopped work on Thursdays.[37] By September 1951, the Shanghai Power Administration stepped up efforts to balance supply and demand. Cotton mills, which accounted for 44 percent of overall power consumption, worked three day shifts and five night shifts and consumed more electricity at night than during the day. Even after these adjustments, Shanghai consumed 40,000 kW less electricity at night than during the day.[38] This gap represented "excess output" that had to be utilized. The administrators saw the night as a void to be filled.

By the fourth quarter of 1952, the Shanghai Power Administration implemented peak-load management more aggressively. In his report to the Communist Party Committee on February 1953, Li Daigeng noted that the maximum instantaneous output of Shanghai's power grid was 235,000 kW for October, 246,000 kW for November, and 248,000 kW for December 1952, all of which exceeded the maximum electrical load of 226,000 kW. Li Daigeng had to work within the parameters established by the Ministry of Fuel Resources and opted to "maximize the potential of existing systems." With power wastage at 8 percent, power conservation efforts did little to relieve the pressure of the power grid. Li then concluded, "Our experience shows that adjusting the load factor is the most economical solution. Based on preliminary results, adjusting the power usage patterns can allow us to reduce peak output by 10,000 kW."[39]

Coordinating a labor force with great precision tested the management capabilities of a centrally planned economy still in its infancy. Shanghai Power Administration targeted cotton mills, not only because they were the largest power consumers but also because they were able to adjust production schedules most flexibly. It worked together with the East China Textile Management

Board to devise a master schedule for all the textile factories in Shanghai. In August 1953, the Shanghai Power Administration ordered all private textile mills to work five day shifts and seven night shifts. State-owned textile mills adopted a three-shift system, with workers working five day, six afternoon, and seven night shifts.[40] Textile mills were divided into seven clusters according to their designated rest day. To prevent them from overloading the power grid, the start times of each cluster were spaced five minutes apart. Within each factory, workers were divided into three groups, with each group taking turns for meal breaks. Staggering start times and breaks helped smooth out the power demand and avoid underutilization or power surges. This scheduling change inconvenienced the workers. By starting morning shifts in the early hours and ending afternoon shifts at night, cotton mills shifted more of their production into off-peak hours. Out of twenty-one shift times, ten of them started or ended between 10 pm and 5 am when buses and most trams had stopped running. Morning shifts started as early as 5:45 am, about forty-five minutes after the first buses and trams started running. Even workers assigned to the 2:05 pm afternoon shift ended work five minutes after the last bus. They also endured erratic meal times. With the rest day rotated once a month and the assigned rest group changing weekly, a morning shift worker might head out for lunch at 9:35 am or 11:25 am at different points within the year.[41]

Flattening the power demand curve allowed the Shanghai Power Administration to supply more electricity without installing new power-generating equipment. Shanghai's power output increased by 19.56 percent and 8.06 percent in 1953 and 1954 respectively (see table 7.1). These were the years when the Power Bureau aggressively pushed for peak-load management. The total installed capacity of the city remained at 303,140 kW up until 1955. In order to keep its industries running, the Shanghai Power Administration had to achieve a load factor of 84.2 percent. This was significantly higher than the average load factors in power systems across the country. In neighboring Nanjing and Wuxi, the load factor ranged from 70 to 77.79 percent, while that of power systems in Shandong Province of North China were another 4 to 10 percentage points lower.[42]

Peak-load management generated much discontent. The Municipal Committee warned the Shanghai Power Administration that draconian measures could be counterproductive. Workers complained about health problems resulting from the constant adjustments to their daily schedules within months of implementation. Shanghai Power Administration also punished factories that consumed too much electricity in the daytime to shift more production into the night. It penalized the largest flour mill in Shanghai by ordering it to move all its production to

TABLE 7.1. Power output figures for Shanghai, 1949–1956

Year	Generating capacity (10,000 kW)	(% gain/ loss)	Power output (million kWh)	(% gain/ loss)	Utilization time (hours)	(% gain/ loss)
1949	25.96		1,009		3,886	
1950	23.56	−9.24	881	−12.69	3,699	−4.81
1951	26.59	12.86	1,192	35.30	4,662	26.03
1952	29.79	12.03	1,319	10.65	4,707	0.97
1953	30.11	1.07	1,517	19.56	5,266	11.88
1954	30.16	0.17	1,711	8.50	5,643	7.16
1955	31.17	3.35	1,526	−10.81	5,054	−10.44
1956	34.44	10.49	1,891	23.92	5,779	14.35

NOTE: Area shaded in gray denotes years of significant increase in power output despite marginal increase in generating capacity. Source: Data adapted from *Shanghai dianli gongye zhi* (Gazetteer of Shanghai's electrical industries), 44.

nighttime. As the flour mill was designed for continuous operation, the cessation of daytime operations lowered output. Military industries also came up with excuses to back out of the citywide effort to coordinate power usage.[43]

Not only were workers pushed to breaking point but machines broke down more often from overuse. In the first quarter of 1954, the East China Textile Administration was too focused on increasing production and wore out many of its electrical machines. This was also the case at the privately owned Guangqin Cotton Mill, which put off repairs until the last minute and ended up burning out forty-nine motors in 1954.[44] Faced with ever-increasing production quotas, the Power Administration had no choice but to push the electrical power grid from 84.2 percent load factor to 89 percent. The city's power supply hinged on a 25,000 kW generator, and Li Daigeng acknowledged that "should the main generator break down, we will face a massive power shortage."[45]

The Power Administration tried a different tack in September 1953 to implement peak-load management in tandem with power conservation. The electrical industries took the lead in the "increase production, practice economy" (*zengchan jieyue*) campaign. Li Daigeng made clear that "power conservation calls for improvement in production processes on the basis of safe use of electrical power. It is attained through improvements in product quality, not through the passive reduction of power consumption."[46] When the Power Administration launched its pilot program in January 1953 directed at 450 large factories, it aimed for an 8 percent reduction in power wastage across Shanghai's major industries (textiles, dyes, petrochemicals, paper, machinery, flour, and rubber).[47] It curtailed power usage by making factories install machines with lower power ratings and removing excess transformers. Factories were required to disconnect transformers during their off-days. The Power Administration also worked with factories to adjust manufacturing processes for improved energy efficiency. The pilot program reduced power consumption by 5.55 percent, falling short of the original target. Even then, the 4.75 million kWh reduction, amounting to the energy consumption of six 50,000-spindle cotton mills, bolstered the administrators' confidence and led them to postpone the construction of a new power station for three years.

The Shanghai Industrial Production Committee promoted power conservation more forcefully in 1954. Shanghai Power inspected factories more than 1,000 times. It organized a five-day conference in June 1954 for five hundred participants from Shanghai's textile and electrical industries to exchange ideas on power conservation. Administrators pushed for the implementation of power-conservation initiatives introduced in 1953. Dongfang Steel (forerunner of Bao Steel) replaced copper bearings in its steel rollers with resin parts and reported using 21 percent less electricity, while increasing output by 40 percent.[48] The 1954 report reiterated the efficacy of this method and added that the steel mills introduced a nonoxidizing process from the Soviet Union to reduce the time taken for smelting. Flour mills also competed with each other to cut down power input. In 1953, Shangwan Flour Mill reported increasing daily output from 15,000 to 17,000 bags with an 8 percent energy saving. In 1954, Hongfeng Flour Mill reported a 26 percent decrease in per-unit-output power consumption. Textile mills claimed that the reduction in per-unit power consumption went up from 17 percent in 1953 to 25 percent in the following year. The Power Administration went further in the 1954 report to say that conservation efforts amounted to "constructing a 10,000 kW power station," which would have cost the state 10 billion (old) RMB.[49]

In the midst of the conservation campaigns, the Shanghai Power Administration formed public-private partnerships with small privately owned power stations as part of a broader strategy to seize control of Shanghai's power transmission lines. While these electric companies contributed to a small share of the city's power output, they owned sections of the 33 kV transmission lines, which served as the building blocks for the East China regional power grid. As early as

December 1952, Zhabei Electric volunteered to enter a public-private partnership with the Shanghai Power Administration to secure capital injection from the state.[50] Li Daigeng chaired the board of directors of the new public-private entity. Zhabei Electric effectively became a subsidiary of the Shanghai Power Bureau. In December 1953, Pudong Electric followed suit.[51] In Zhabei and Pudong, shareholders retained ownership of the company's shares and received dividends, while government and party officials controlled operational matters by virtue of their majority representation on the boards of directors and supervisors. The government appointed seven out of eleven board members who were allocated "public shares," while private owners retained four seats on the board.

The Shanghai Power Administration plotted to acquire Chinese Merchants' Electric Company by igniting a factional struggle within the board of directors around December 1953. Sun Zhifei, a protégé of Green Gang boss Du Yuesheng, balked at the idea of surrendering ownership to the government. His rival Huang Bingquan, however, argued that the capital injection would allow the company to keep paying out dividends and retain the support of shareholders.[52] As Huang's faction emerged victorious, the Chinese Merchants' Company was reincorporated as a public-private partnership in August 1954 and renamed Nanshi Power Company. Su Yang, a former militia leader, was appointed the leading public director. Su placated shareholders by assuring them that the state would respect private ownership.[53] Three months later, the newly incorporated company began a shareholder registration exercise, during which 1,977 shareholders laid claim to 1.765 billion shares. About 227 million unclaimed shares were converted to public shares, allowing the government to further increase their stake.[54]

The Communists then used the Five-anti Campaign to further weaken private control in the newly incorporated Nanshi Power Company. Huang Bingquan, who had supported the public-private partnership, ran through the company's accounts to identify more than RMB (old) 31 billion that the Chinese Merchants' Electric Company had taken from the state. Huang reported in the meeting that "the Chinese Merchants' Electric Company has stolen a lot of the state's wealth," which included RMB (old) 9 billion of enemy reparations, RMB (old) 8.9 billion in loans from the Guomindang government, RMB (old) 2.3 billion in foreign exchange gains from the purchase of a 4,000 kW generator, RMB (old) 9.6 billion in subsidies from the Guomindang government, and RMB (old) 1 billion in unpaid taxes.[55] The Communists then converted debt and owed taxes into government-owned shares and secured a majority stake for the government.

By 1955, the Communists established total control over Shanghai's power generation and transmission. The first Taiwan crisis in 1954/1955, which saw

Chiang's evacuation from the Dachen Islands, diminished the likelihood of a Nationalist counterattack on the mainland. Disasters of the previous six years exposed the vulnerabilities of the electrical industries in eastern China. The Power Administration recognized the need to integrate power stations across the Lower Yangtze to reduce the likelihood of power outages during disasters and divert excess power capacity from neighboring cities into Shanghai. The Shanghai Power Administration established public-private partnerships with private power companies that controlled the 33 kV power distribution lines around Shanghai. The centralization of electrical industries occurred on January 1, 1955. Following the abolition of the East China Electric Power Administration, Shanghai's power industry was placed under the direct control of the central government.[56]

Two events in 1955 reduced the need for peak-load management in Shanghai. First, the flood in February 1955 damaged the cotton harvest in East China, which sharply reduced the power demand from Shanghai's cotton mills. This resulted in a 10.81 percent decrease in overall power consumption in 1955 compared to the previous year.[57] Second, China began domestic production of its first boiler-turbine systems, which were installed at newly constructed power stations. Before his transfer to Beijing on January 1, 1953, Yun Zhen incorporated the Shanghai Boiler Factory in the Minhang district of Southwest Shanghai by merging the old boiler repair plant of American-owned Anderson Meyers & Co. Ltd. with the turbine factory of the now-defunct National Resources Commission.[58] Back in 1953, the Shanghai Electric Motor Factory acquired designs for 6,000 kW turbines from Skoda, after Wang Daohan, who was then deputy minister of the First Ministry of Machine Building, concluded a technology transfer agreement with Czechoslovakia. The first set of 6,000 kW generators were installed at Tianjia'an Power Station in Anhui Province in 1956.[59] The Shanghai Electric Motor Factory later acquired designs for 22,000 kW generators from Skoda. The Wangting Power Station in Suzhou completed in 1958 was equipped with four of these generators.[60]

The year 1955 also marked a turning point for the electrical industries nationwide. The Ministry of Fuel Resources broke up into the Ministries of Electric Industry, Coal Industry, and Petroleum Industry after restructuring. Liu Lanbo, appointed as minister of Electric Industries, had a career trajectory similar to Li Daigeng's. He joined the anti-Japanese resistance in Northeast China, served as political commissar of anti-Japanese militias in Manchukuo, and was later promoted as party secretary and military committee chair of Andong Province, an old administrative area that stretched along China's side of the Yalu River.

Liu summed up the achievements of the past six years before pointing the way forward. In his September 1955 speech at the National Congress of Model Workers in the Electrical Power Sector, Liu listed productivity improvements achieved nationwide. He noted that apart from 1952, the electrical power sector met or exceeded production targets and developmental goals. Between 1949 and 1954, fuel consumption generated fell from 0.89 kilogram per kWh to 0.614 kilogram per kWh; line loss reduced from 22.33 to 10.36 percent; and energy consumed by the power station during generation fell from 8.28 to 5.28 percent. Conservation efforts reduced electrical energy consumption by 181 million kWh in 1953 and 218 million kWh in 1954.[61] Having unleashed the full potential of existing generating equipment through modifications and balancing power demand, the Ministry of Fuel Resources had put off infrastructure expansion. The Electric Industries Ministry that took over from it then had to complete the heavy lifting within the last two years of the first Five-Year Plan. Liu noted that 56 percent of infrastructure development had to be completed in 1956 and 1957. The Ministry of Fuel Resources had struggled to execute forty projects in 1955. To complete forty-six new power stations as planned called for a 15 to 20 percent reduction in construction time. Quality would have to be sacrificed for speed. Breakdowns due to installation errors were already happening. Liu cited the example of Fuxin Power Station, which reported slippages in the machine base of its number 2 generator shortly after installation.[62] Inefficiencies from compressed development began creeping into the electrical industries at the national level.

Projecting Centralized State Power

Power conservation campaigns and demand-management operations similar to those in Shanghai took place across the nation. Between 1950 and 1955, initiatives introduced in Shanghai were transferred to other parts of the country. Shanghai's electrical industry, which had developed independently as an outlier, became fully integrated as a national economy. Across the People's Republic, the electrical industry followed the same organizational logic of prioritizing industrial use over residential use. Throughout this period, Soviet experts mostly provided general guidance, while administrators appointed by the Chinese Communist Party made the key decisions. The electrical industries in the early years of the People's Republic thus deviated from the Leninist GOELRO (State Commission for Electrification of Russia) model, as rank-and-file workers and party cadres strengthened their grip over the means of production. The engineer-bureaucrats, who had charted the development of the electrical industries in the early years, became sidelined just before the end of the first five-year plan.

Throughout the early years of the People's Republic, the Soviets played a limited role in the electrical industries. Soviet experts, who arrived shortly after October 1949, mostly provided prescriptive advice. In November 1949, Stalin's former minister of railways Ivan Kovalyov inspected the power stations in Fushun and Fengman in Northeast China and Shijingshan. The Soviet delegation identified eight common problems with these power facilities. They advised the Communist regime to quickly appoint engineers to manage the operations of important power plants, rectify flaws with power-generating equipment to reduce wear-and-tear, strengthen the will of the workers to protect the power-generating assets, and absorb engineers into the Communist Party.[63] Soviet experts spelled out eight key principles; the first one being that all national plans for electrification must be devised "in accordance with a unified national economic plan, while taking into account the geographical location of users, water resources, fuel, railroads, navigation routes and construction materials."[64] The engineer-bureaucrats who had defected to the Communists were already making recommendations similar to those of the Soviets'.

The development of China's electrical industries between 1950 and 1955 deviated significantly from the Soviet Union's. Lenin's formula "Communism is Soviet Power plus the electrification of the whole country" first appeared in Chinese translation on the inside cover of a special issue of the electrical power journal *Renmin dianye*, which commemorated the third anniversary of Sino-Soviet cooperation. The editorial board made general statements to express gratitude to the Soviet advisers but only listed one substantial achievement from the cooperation. By following the recommendations of Soviet advisers, China's electrical power sector increased power output by 15 percent without installing new equipment.[65]

Accepting the advice of the Soviet experts came with some risks. The increased output of the Tianjin Power Bureau was the result of assurances from Soviet advisers that equipment once thought to be faulty was functioning normally. The Tianjin Power Bureau had curtailed power output after both its generators broke down in 1946 and 1947 to avoid further wear and tear. Soviet advisers inspected the turbine shroud ring and stators and found no damage and recommended that operators gradually increase power output.[66]

The developmental pattern of North China's electrical power sector mirrored that of Shanghai's. Following the Shanghai air raid on February 6, 1950, delegates of the Second National Electrical Industries Conference held in Beijing in March 1950 reaffirmed the importance of "safety," which not only called for the elimination of accidents but also the "establishment of defensive mass organizations and the eradication of counterrevolutionary saboteurs." Adjustments aimed at "making full use of existing inventory" also unleashed the full potential of North China's power industry. The Tianjin Power Bureau in North China reported increasing instantaneous output by 3,500 kW and achieving a 100 percent load factor in June 1950 after three repair operations.[67] In 1950, North China's regional network supplied 18.06 percent more electrical units than the previous year.[68]

Just like Shanghai, the North China Power Bureau attempted to predict power demand but appeared to be less successful. Instead of peak-load management, the Beijing-Tianjin-Tangshan power grid practiced "economic dispatching," which required the power stations to dynamically adjust its output based on projected demand. The North China Power Bureau shut down some of its generators during off-peak hours to conserve fuel but supplied additional power only on occasions when they were able to anticipate demand. For example, on the second day of Chinese New Year, urban residents switched on their lights between 5 and 7 am to welcome the god of fortune, which increased instantaneous load from 29,000 kW at 4:30 am to 37,000 kW at 6:30 am. The Power Bureau activated two more generators to cope with additional demand. Predicting industrial power consumption, however, proved to be much more difficult. Some factories failed to plan production, while some refused to provide power consumption data on the grounds of secrecy.[69]

Similarly, urban residents cut back on personal power consumption for lighting to free up electricity for factories. Between 1950 and 1955, industries accounted for approximately 55 percent of Beijing's power consumption. The remaining 45 percent was classified as "daily usage" (shenghuo yongdian), which included illumination of public places such as streets, schools, and hospitals.[70] The brightness level of elementary and middle school classrooms was ten lux, which is about one-tenth the minimal light level required for transitory areas like lifts and hallways. Some schools installed light bulbs with lower power ratings to meet power conservation targets. The Beijing Municipal Committee claimed that long-term exposure to dim lighting would lead to increasing occurrence of myopia among school-age children.[71] It cautioned against sacrificing the health of the urban residents to pursue industrial growth.

Ultimately, pushing a power system with limited capacity to the brink came with greater risks of accidents and system failure. The Shijingshan Power Station, which was captured by the PLA just before the siege on Beiping, prided itself as a trailblazer of New China's electrical power sector. It later suffered from frequent breakdowns due to overuse. As was the case in Tianjin, Soviet advisers recommended that the Shijingshan Power Station operate its 25,000 kW generator at full capacity.[72] The Shijingshan Power Station, however, skipped a major repair in November 1950, as the generator had to keep running to make up the shortfall due to the system overhaul at Tianjin Power Plant. The operators did not remove the stator for cleaning during the minor repair in January 1951 and did not detect any abnormalities over the next three months of operations. On April 21, 1951, a fire broke out in one of the generators when the stator burned out. Investigators discovered that large amounts of dirt had clogged the generators' vents, and the insulation on the wiring was badly worn.[73]

Despite taking remedial actions, the Shijingshan Power Station was unable to avert an embarrassing catastrophic failure three years later. On October 21, 1954, around 6 pm, during Jawaharlal Nehru's visit to China, the number 6 generator at Shijingshan, which supplied 31,000 kW of power to Beijing, broke down. Street lights went off for nineteen minutes; water pressure for the city's tap water system fell sharply; seventy-nine factories were forced to stop work; and hospitals carried out surgeries with flashlights.[74] It was no coincidence that the Ministry of Electric Industries directed its attention to infrastructure development after 1955. The policy of "optimizing power output with existing equipment" had simply caused power bureaus across the country to run their aging equipment into the ground. It was now no longer viable to postpone capital investment.

In the early years under Communist rule, the Chinese electrical industries turned Lenin's GOELRO model on its head. Chinese engineers in the early years of the People's Republic were content to let political appointees handle the political problem of power allocation. With party-appointed administrators gathering data on power usage, engineers no longer groped in the dark to project power demand. Such an arrangement, however, ran counter to Lenin's ideal that discussions on the question of electrification "marks the beginning of that very happy time when politics will recede into the background . . . and engineers will do most of the talking."[75]

Engineers were also not sheltered from politics. While they benefited from earlier mass campaigns to secure support for electrical infrastructure development projects, some of the engineers who defected to the Communists suffered during the massive purges of the 1950s and 1960s. Yun Zhen, who had actively participated in the copper conservation campaign in 1951, convinced Chen Yun to import several tons of silicon steel for the new turbine manufacturing facility in Shanghai, as both Anshan Steel Works and Taiyuan Steel Works were unable to manufacture the materials.[76] Yun Zhen was transferred to Beijing in 1953, where he was appointed as the chief engineer of the First Ministry of Machine Industry. There, he joined the Jiusan Society (September 3rd Society), a democratic party legally recognized by the Communists. His political connections, however, did little to shelter him from persecution. As early as October 1955, investigators began gathering evidence to accuse Yun of being a Guomindang spy during the height of the counterrevolutionary campaign. Eleven men, who had either participated in Guomindang political indoctrination activities with him or worked under him, came forth with evidence. His investigators cited Yun Zhen's suppression of the labor unrest in Kunming in 1945 as an example of his counterrevolutionary activities. In addition to accusing Yun of harboring Guomindang spies, they also listed ten infractions that Yun committed, ranging from criticizing the Communists to selling raw materials from state-owned enterprises to private businesses.[77] Yun was labeled a rightist in 1957 and sidelined. He was assigned to work as an interpreter at an industrial intelligence office and was later sent to work at a machine plant in Guizhou and to teach electrical engineering at a local technical school. In his oral interviews with Zhang Baichun in the late 1980s, Yun Zhen reiterated that he had long renounced his Guomindang membership. Despite being labeled a rightist for two decades, Yun expressed no regret about his decision to remain on the mainland. Compared to Liu Jinyu, who had been executed on charges of colluding with the Communists, and many of his former colleagues who were sidelined in Taiwan, Yun had the fortune of living a long and fruitful life and witnessing the phenomenal growth of China's electrical industries.

Bao Guobao, Yun's former colleague at the NRC, survived the counterrevolutionary campaign of 1955. Having collaborated with the Communists during the siege of Beiping when he served as the general manager of the North Hebei Electric Company, Bao was later appointed director of the Power Management division in the Ministry of Electrical Industries. Speaking in the presence of Premier Zhou Enlai in April 1956, Bao echoed the argument that Yun Zhen made just before the Communist victory in 1949. He said, "In the old era, technical personnel were often isolated and unable to approach the problems holistically. We focused on the electrical industries and did not see the entire plan for national economic development."[78] Bao made the point that the Communists succeeded where the Nationalists failed. The Communists not only understood how to integrate the work of the engineers in their plans for national reconstruction but had also transformed electricity into a weapon of mass mobilization.

THE ACCELERATED DEVELOPMENT of China's electrical industries came about as the People's Republic remained on a war footing. The aerial bombardment of February 6, 1950, meant to cripple the power industries, triggered a series of events that catalyzed its expansion. Party cadres with military experience assumed leadership positions in power bureaus. The state exercised emergency powers to seize Shanghai's largest power station from the Americans. The climate of fear after the civil war compelled Power Bureau administrators into adopting draconian measures to manage power demand. The militarization of China's electrical power industries not only led to heightened security but also provided the organizational structure for state administrators to impose regimented work schedules on the workers. Initiatives to flatten the power demand curve required workers to make radical readjustments to their lives.

The Communists maintained a fragile alliance among the state, industrialists, and engineering elite by appealing to the interests of the stakeholders. Well aware of its tenuous control over the cities, the Communists avoided confronting the capitalists head-on during the early years of the People's Republic. Instead of forcibly taking over the private power industry, the Communists exploited rifts between shareholders by satisfying the demands of some of them, thereby gaining support for the public-private joint management of the power sector. Engineer-bureaucrats, who had served the Guomindang regime, facilitated the consolidation of a highly fragmented electrical power sector and the creation of regional power networks under state coordination.

In 1954, the Ministry of Fuel Resources published a translation of S. F. Shershov's *Leninist-Stalinist Electrification of USSR*, which had been released three years previous in the Soviet Union to commemorate the thirtieth anniversary of Lenin's GOELRO.[79] The book begins with this quote from Wilhelm Liebknecht's biography of Marx: "That King Steam who had revolutionised the world in the last century had ceased to rule, and that into his place a far greater revolutionist would step, the electric spark. . . . In the wake of the economic revolution the political must necessarily follow, for the latter is only the expression of the former."[80] Marx's prediction rang true in 1950s mainland China. The capitalists, who had once accumulated massive wealth with their steam-powered mills, had voluntarily entered public-private joint management agreements and surrendered control of their assets to a state that controlled the flow of electrical power.[81] State-owned factories fueled by centrally planned power networks soon dominated China's industries. In a way, the Chinese even outshone their Soviet elder brother by achieving the dream of workers seizing the means of production. While the Soviet Union relied on highly educated technocrats to manage its electrical industries, many of China's power stations had appointed soldiers and workers as superintendents.[82]

Accelerated development under conditions of material scarcity came at a price, however. Power bureaus tolerated high levels of line loss and fuel consumption largely due to a lack of electrical equipment. Even as China alleviated the shortage of electrical equipment through domestic production, systematic weaknesses had already crept into the hastily constructed power systems in the early stages. Several small hydropower stations in the Lower Yangtze completed in the first few years of the Great Leap Forward (1958–1962) fell into disrepair after only operating for a year or so. The energy intensity of the Chinese economy increased, and the growth rate of electrical supply outpaced that of industrial production, which pushed China into a transition toward a carbon-intensive economy.

The phenomenal growth of China's electrical industry up until 1957 overlapped with the beginnings of the second stage of the Anthropocene—the Great Acceleration. As Will Steffen, Paul Crutzen, and John McNeill observed, "the lessons absorbed about the disasters of world wars and depression inspired a new regime of international institutions after 1945 that helped create conditions for resumed economic growth."[83] Despite being marginalized from the global economic order dominated by the United States and its allies, the People's Republic of China strived to catch up with the industrialized world by boosting its industrial output by all means necessary. The electrical power sector, which had to increase output by fully utilizing its existing inventory of obsolete generating equipment, served as such an engine for the pursuit for exponential economic growth. Excessive resource use and environmental deterioration resulted from this relentless pursuit for wealth and power. When Beijing was blanketed in gray snow in 1951, its administrators complained about the wastage of unburnt coal ash that was escaping through the chimneys instead of lamenting air pollution. In the early stages of China's shift toward a carbon economy, productivity was valued above the pristineness of its natural environment and its workers' health. China today continues to grapple with the human cost of the compressed development of its electrical industries, as it pivots quickly toward renewables.

Conclusion

Hauntings from Past Energy Transitions

S IXTY-SEVEN YEARS AFTER Taiwan Power Company's first general manager Liu Jinyu was executed by firing squad on charges of conspiring with the Communists, politicians in Taiwan are still grappling with the fear that the loss of electrical power will lead to the collapse of the regime. The latent fear surrounding the vulnerabilities of Taiwan's electrical infrastructure emerged during an island-wide power outage on August 15, 2017. The power went out across Taiwan after two engineers made a serious mistake during a regular maintenance run at Taiwan's largest natural gas power station in Ta-tan. At 4:45 pm that fateful afternoon, one of them disconnected power from the redundant system. Two and a half minutes later, the controller system rebooted the system upon detecting a power failure, causing two electrical valves to shut down and cut off the natural gas supply for two minutes. All six generators tripped at 4:50 pm, cutting off the power supply to 6.68 million residential and commercial units.[1] After recounting the incident report, Dai Ligang, the host of CTI Television's "News Tornado" exasperatingly exclaimed, "If anyone were to invade Taiwan, they could simply bribe a power plant worker to cut off [our] natural gas supply, and the whole island would be paralyzed! Wouldn't it be so easy to invade Taiwan?"[2] He was not alone in his criticism.

Echoing these sentiments a week after the blackout, the founder of the grassroots network Nuclear MythBusters, Huang Shih-hsiu, wrote, "Only the Presidential Palace was lit up when the entire Bo-ai district [in Taipei] plunged into darkness. Do you know what I was thinking? This is the best time for the People's Liberation Army to attack. We'll be doomed if they send a missile flying over."[3] Their anxieties echoed the fears shared by the prosecutors who relentlessly established Liu's guilt sixty-seven years ago.

My family was in Taiwan during the August 2017 blackout. The possibility of a PLA invasion was the last thing on our minds that evening. We had spent the afternoon at Keelung's National Museum of Marine Science and Technology, which was housed in a building converted from an old steam power station constructed by the Japanese. The museum's lights and air-conditioning were still on, so we were blissfully unaware of the massive power outage until we embarked on our return journey to Taipei. The lights were out at the train station. We walked right past the electronic fare gates that had stopped functioning, then headed through a pitch-dark tunnel to get to the platform. Thankfully, the trains that ran on electricity still arrived on schedule. Peering out of the train windows, we caught glimpses of candlelight flickering in the darkened houses along the railway track. Not everyone was impacted. People within the same vicinity experienced the effects of the blackout differently. My family grabbed dinner at the Taipei Main Station with no trouble, but diners at the Q-square Food Court behind the train station had to eat in the dark.

The outage was mostly over by 9:40 that night, but the political fallout went on for months. For a start, the minister of economic affairs Chih-kung Lee, a Cornell-trained civil engineer and theoretical physicist, assumed responsibility and resigned that day. The Guomindang opposition quickly seized on the vulnerabilities exposed during the blackout to mobilize a pronuclear coalition in 2018 to force the Democratic Progressive Party (DPP) government into abandoning its "nuclear-free homeland" platform. Energy security briefly emerged as a campaign issue during the 2020 presidential elections but faded away in the wake of protests against the extradition law in Hong Kong, which transformed the election into a repudiation of the "one country, two systems" policy. The blackout did not lead to regime change, but it would have far-reaching consequences.

The 2017 blackout offered a unique focal point to draw one's attention to political challenges behind the transition to renewable energy not just in Taiwan but also in mainland China. Apart from raising concerns that Taiwan's power infrastructure might not survive the onslaught of all-out warfare, the massive outage also laid bare Taiwan's overdependence on fossil fuels, exposing the inadequacies of its plans to diversify the fuel mix. Their counterparts in mainland China also grapple with the same issues. The history of electrification from the 1880s to the 1950s highlights three key issues that must be considered as economic planners and engineers on both sides of the Taiwan Strait chart their energy future. First, the tumultuous beginnings of the electrical industries from decades of war and revolution shaped the siege mentality of the power sector on both sides of the Taiwan Strait, which resulted in a tendency to favor short-term economic growth over long-term sustainability. Second, wartime mobilization led the Nationalist and Communist regimes to lock into carbon-intensive modes of power generation, as fossil fuel–based power generation offered the fastest and most cost effective way to relieve power shortages. Finally, political leaders in mainland China and Taiwan had to overcome the vast institutional inertia from earlier stages of development.

These three key issues are at the center of what energy economist J. G. Pearson called the "energy policy trilemma," in which the "centre of gravity moves between three policy objectives (energy security; affordability and international competitiveness; and environmental quality)." By examining the shift to renewables from 2006 to 2020 in mainland China and Taiwan in this conclusion, I hope to address what Pearson considered to be "three areas of energy transition that can be enriched by historical analysis," namely (1) duration and speed of transition; (2) path dependence, lock-in, and the role of incumbents; and (3) sustainability transitions and innovation theory.[4]

The centralization that has facilitated the accelerated growth of electrical industries in China and Taiwan thus far has turned out to impede the institutional innovations needed to promote the adoption of renewables. This is yet another example of how strengths can become weaknesses. Envisioned during decades of war and revolution from the 1880s to the 1950s, the centralization of power distribution at the national level has already been achieved on both sides of the Taiwan Strait. As mentioned in chapter 6, the Nationalist government assumed full control over the electrical power sector by default after taking over Taiwan in 1945. In mainland China, the State Grid Corporation of China was incorporated in 2002. The interconnection of regional power networks through ultra-high-voltage networks not only allowed the State Grid to deploy surplus power across different macroregions but also helped to avert massive regional blackouts like those that happened in New York on July 13, 1977, or in India in July 2012.[5] The State Grid reflects the strong administrative capacity and the limitations of the central government in Beijing. Power-generating units at the local level sometimes defy the central government's demands for higher electrical output, as they struggle to supply electricity at low cost in the face of rising fuel costs. In 2011, some small cities experienced rolling blackouts, as the regional power networks tried to safeguard the power supply of large urban populations.[6] As will be discussed, centralization allowed the State Grid to amass resources to bring about an exponential growth of generating capacity in the wind power sector, but it also resulted in diseconomies of scale and poor coordination. Economic planners in Taiwan are also walking the same tightrope of balancing short-term economic growth and long-term environmental sustainability. Pressure from the ballot box forced policy makers in a multiparty democracy to rely on tried and tested modes of energy production to minimize potential economic disruptions.

The Perils of Rapid Response

Just as the electrical power infrastructure expanded hastily during an age of armed conflict in the 1930s and 1940s, the exponential growth of the renewable sector within half a decade resulted from a hurried response to climate change. The formative years of the power networks on both sides of the Taiwan Strait overlapped with the age of total warfare. The construction of the power infrastructure on the ashes of war started the transition from an organic to a carbon economy for mainland China and Taiwan. The Japanese had expanded the generating capacity of Taiwan's power sector to fuel the industrial growth that was central to its southward expansion. Liu Jinyu, the engineer who rehabilitated Taiwan's power network discussed in chapter 6, had built wartime Kunming's power network from the ground up under the threat of Japanese air raids. At the end of that chapter, we saw how the Nationalists altered the charges to justify executing Liu on the slightest suspicion of contact with Communist agents. The paranoia of enemy infiltration was no less intense on the mainland.

The climate of fear emerging from this frantic quest for survival fostered the "development-at-all-cost" mindset. Economic planners in Communist China saw the electrical power sector as the vanguard of all industries and made it expand faster than the overall economy. Even exogenous shocks to fuel supply did not derail these plans. In 1956, the year before the Great Leap Forward, floods and a shortage of wood to prop up coal mine tunnels caused a production loss of 2.38 million tons of coal. Despite this, China's electrical power output increased by about 25 percent.[7] The power demand management strategies described in chapter 7 and improvements in combustion efficiency brought about such increases. The militaristic approach to infrastructure development also carried on well into the 1970s, long after the end of the war. The expansion of the Wangting Power Station in 1973, for example, followed the *dahuizhan* (grand battle) model. This all-hands-on-deck strategy made it possible for the construction of the exterior building to be completed within three months and for 300,000 kW of hydropower capacity to come online within forty months. A series of "campaigns" over the next three years would address fifty-nine major defects that had emerged from this hasty construction.[8]

Fast forward to the early twenty-first century, the Chinese government recognizes air pollution and high levels of carbon dioxide emissions as an existential threat to economic growth and the well-being of its people. Much like how the Guomindang regime transformed the frontiers into a resource base in the War of Anti-Japanese Resistance between 1937 and 1945, the People's Republic has now transformed the northern frontier regions into production bases for renewable energy in this war against carbon emissions. Following the enactment of the Renewable Energy Law in 2006, the wind energy industry posted a 100 percent growth rate for four consecutive years.[9] China announced the establishment of seven wind power bases of at least 10 GW each in Gansu, Xinjiang, Hebei, Jilin, eastern and western Inner Mongolia, and coastal Jiangsu in 2008.[10] As early as September 2007, China's government announced that it sought to nearly double the proportion of renewables from 8 percent in 2006 to 15 percent in 2020. In fact, they exceeded even those ambitious goals, tripling total renewable energy output between 2005 and 2013.[11] State-owned industries also dominated the market for wind turbines.[12] The underlying logic of the state-driven approach to renewables is reminiscent of the NRC's strategy to dictate industrial standards back in the 1940s, which sought to prevent China's electrical sector from being beholden to the interests of American electrical equipment makers.

The rallying cry for China to catch up with advanced industrialized nations that had featured prominently in military campaigns and mass mobilizations from the 1930s to the 1960s has simply reemerged in a different form. In 2014, the National Development and Reform Commission (NDRC) declared that green, low-carbon development was not only important for sustainable development, "but will also demonstrate to the world that China is a responsible country committed to making an active contribution to protecting the global environment."[13] This echoes the anxieties voiced by the proposers of the Anthropocene, who have been alerting the public about the dangers of anthropogenic environmental change in this new geological epoch.

The wind power sector ran into strong headwinds, as the residual inefficiency from the massive carbon lock-in within the northern, northeastern, and northwestern regions prevented the added renewable capacity from being transmitted in the grid. After four years of phenomenal growth from 2006 to 2010, China's wind farms suffered a 30 percent decline in annual income between 2010 and 2013, as around 10 to 20 percent of the wind power was not put on the grid because the power grid in these frontier regions did not have the transmission and distribution capacity to carry the vast amounts of electricity into the centers of power consumption. Known as wind curtailment, this phenomenon resulted in RMB 10 billion of direct economic loss in 2012. Wind power could only be used locally. In the winter heating months, the thermal supply units that ran on coal also generated electricity, and it became necessary to cut off the electricity from the wind bases to prevent the saturation of transmission channels.[14] To utilize the wind energy locally, users would have to replace existing heaters with electric heaters. This would come at an immense cost—both financially and environmentally. The combined bursts of "high tide" mobilization with organized and ongoing micromanaging of the workforce observed in chapter 7 exacerbated the mismatch between production, distribution, and the consumption of clean energy.

Overcoming Carbon Lock-in

Developments in the formative years of China's electrical industries determined the extent of carbon lock-in. Coming of age during decades of armed conflict and revolutionary upheaval, China's power producers and consumers continually grappled with the persistence of obsolescence. Capital shortage led power producers and consumers to favor incumbent technologies over newer and more energy-efficient modes of production. Such practices occurred as far back as the late Qing and Republican era and continued right up until the 1970s. The electrification of textile manufacturing discussed in chapter 1 showed that the arrival of new energy technologies did not bring about Schumpeterian "creative destruction," in which the introduction of novel technologies forced firms to either exit or catch up. Cotton mills did not immediately switch to the more energy-efficient practice of buying electricity from the power company but went with low-cost incumbent technologies to generate electricity in-house. The unreliability of new energy sources vindicated their choice. In the same vein, in chapter 2, large foreign-owned power systems did not displace small Chinese-owned power plants, as the local elite extracted legal protections from the state and prevented encroachment into their franchise areas.

Developments in the War of Anti-Japanese Resistance locked China into a path of carbon-intensive development. As seen in chapter 3, both the Chinese and Japanese wartime regimes centralized coal distribution to more effectively budget the energy needs of their defense industries. The developmental pattern persisted beyond 1945. As an island on the front line of the Cold War, Taiwan increased its dependence on fossil fuels, as it pursued rapid economic growth to ensure its survival. Sun Yun-suan, the NRC engineer who trained at the TVA (chapter 5) and played an instrumental role in rehabilitating Taiwan's power grid after Japanese surrender (chapter 6), took over as chief engineer of Taipower in the 1950s. Under his leadership, Taiwan moved further from the "hydro first, thermal second" of the Japanese colonial era and went full steam ahead with fossil-fuel power generation. The demand for electricity during Taiwan's economic takeoff in the 1960s required Taipower to rapidly add generating capacity.

Joe Moore, an engineer with the US Aid program, revealed that Sun had pushed through the purchase of gas turbines as "an emergency project" during recurring power shortages in the 1960s. Taipower was constructing major hydropower projects in central and eastern Taiwan, but it would take almost ten years before the new hydropower plants came online. With power consumption doubling every five to seven years, Taipower needed to expand its generating capacity. As Moore pointed out, the gas turbines were "inefficient and expensive to operate but could be delivered and installed much sooner than conventional units."[15] Natural gas, which was adopted as an emergency measure, has evolved into the largest fuel source for power generation in Taiwan in the twenty-first century.

The People's Republic of China followed a similar carbon-intensive developmental trajectory. After inheriting the patchwork of idiosyncratic legacy systems, the Communists had to grapple with all the systematic flaws baked into this hastily built wartime electrical infrastructure. As outlined in chapter 7, the Power Bureau exercised peak-load management and maintained high levels of power output to keep manufacturing running around the clock. Energy intensity, defined as a ratio of energy inputs to production value, increased. The Great Leap Forward further accelerated the transition toward fossil fuels. Small hydropower plants frantically built between 1958 and 1962 failed at a high rate. The Third Front policy, which saw the shift of industrialization toward inland China, also saw the postponement of power plant expansion in coastal cities. Installing oil generators became the fastest way to make up the shortfall in generating capacity during the fourth five-year plan (1971–1975). Between 1973 and 1975, the East China regional grid expanded its generating capacity by close to 80 percent, with construction and installation taking three years and four months. Of the three million kW of additional capacity, 2.68 million kW came from oil generators.[16]

For every three steps made toward the transition to green energy, China seems to slip two steps back onto its pathways of carbon-intensive development shaped seventy years ago. The frontier regions, which had been at the forefront of solar and wind power generation, have also become the sites of highly pollutive coal conversion projects. Such projects capture the policy dilemma faced by the subnational government and coal industry, as they vacillated between curbing coal consumption to meet carbon emission targets and betting on new fossil fuel technologies to propel economic growth. Central policy makers were cognizant that coal conversion was a risky gamble taken to diversify sources of raw materials for petrochemical products "in the face of uncertain global oil supply."[17] The incorporation of coal conversion into oil and natural gas as part of the thirteenth five-year plan offers yet another example of how the rhetoric for national survival overrode broader environmental concerns. The water-guzzling coal conversion projects are mostly concentrated in arid regions of West China. Well aware of potential environmental damage, the Ministry of Environmental Protection released rules limiting water usage and implemented a water rights trade system, but researchers have estimated that the water footprint of coal conversion "is still projected to remain disproportionately large."[18] The frontier not only opened up a new horizon for limitless energy expansion, but it also served as a sacrifice zone.

The Limits of Carbon-Fueled Growth

Judging from its earliest transition toward renewables, authoritarian eco-modernism has become the predominant framework for energy policies.[19] As China struggles to break free from massive carbon locked-in, it appears to be in a "trapped transition," to borrow a term coined by the political scientist Minxin Pei, who argues that "the power of the state is used to defend the privileges of the ruling elites and to suppress societal challenges to those privileges, instead of advancing broad developmental goals."[20] The reluctance of China's state utilities to relinquish control over the centralized grid arose from a desire to maintain status quo, which in turn led to the single-minded approach of growing the generating capacity of renewables without taking into account the limitations of existing power distribution systems. This is consistent with the experiences in other Asian countries and affirms Elizabeth Chatterjee's observation that, "Recent moves to decarbonize electricity sectors have so far more reinforced than disrupted the political logic of developmentalism."[21]

Do the governments in China and Taiwan have the political will to build a sustainable power infrastructure in the face of a global environmental crisis that threatens to overwhelm the Earth's system? The conclusion is not as simplistic as the one suggested by Naomi Oreskes and Erik Conway in their fictional account of the "Great Collapse of 2093," in which an authoritarian China survives the climate crisis of the twenty-first century by relocating more than 250 million people to higher ground, while democratic governments dither in the face of doom.[22] In the case of China, there is no doubt that the one-party authoritarian state is capable of mobilizing vast amounts of human resources and capital to ramp up solar, wind, and hydropower. The curtailment of renewable energy sources since around 2012 suggests that this centralized top-down approach has resulted in overcapacity and a lack of proper coordination. In the case of Taiwan, the democratically elected government has shaped its energy policies in response to the demands of the electorate. Discussions about Taiwan's energy future have been narrowly focused around two issues: air pollution caused by coal-fired power stations and the decommissioning of nuclear power. Following the 2017 blackout, the incumbent and opposition offered green energy development plans with vaguely defined numerical targets and few specifics. Be it authoritarian or democratic, the political leadership on both sides of the Taiwan Strait is struggling to confront the precarity of economies that are running up against the limits of carbon-fueled growth.

Engineers and social scientists studying China's energy transition in the twenty-first century seem to share the views of engineers who had left power demand management to the party leadership in the 1950s: power distribution is a political rather than a technological problem. They largely agree that the central government should continue to play an important role in promoting renewable energy, but they recognize the need for greater coordination between central and local governments. In an article published in 2014, Guo-liang Luo's team at the North China Electric University listed eleven reasons for wind curtailment, all of which resulted from systematic planning failures and poor implementation of the renewable energy laws. His team noted specifically that the establishment of seven 10,000 MW wind farms was out of line with power consumption patterns. China's government leveraged its control of the power sector to promote large-scale wind farm development by disbursing subsidies from the state public budget—an estimated RMB 300 for each kilowatt installed between 2006 and 2011. At the local level, developers complained that they never received the promised subsidies and found it a hassle to apply for such funding. The failure to collect sufficient fees from end users exacerbated the funding shortage.[23] In the end, private wind farm operators were at the mercy of state-controlled power grid operators, who could decide to purchase wind power or order curtailment based on their interpretation of the "guaranteed purchase" clause under the Renewable Energy Law.[24] The problems persisted. Three years later, Dahai Zhang's research team at Ocean College, Zhejiang University, raised identical concerns. They lamented, "Management, strategies, programs and policies are separated by too many departments of the Chinese government, which means there are no plans as a whole."[25]

Han Lin, a political scientist with the Climate and Sustainability Policy Research Group at Flinders University, concurs with these scientists, as she concludes that the major problems with China's energy policy "lay neither in speed nor in scale, but in coordination efficiency, integrity, and quality." She sees the integration of the Department of Climate Change into the renamed Ministry of Ecology and Environment in 2018 as a step in the right direction by the central government but also argues that the problematic relationship between the central, provincial, and municipal governments in climate change mitigation can only be fully resolved with a whole-of-government approach that eliminates fragmentation of policy responsibilities.[26]

Turning to Taiwan, its democratically elected government struggled to contain the political fallout from the massive power outage in August 2017. Right before the accidental power trip at Ta-tan, Taiwan was experiencing one of its hottest summers on record. Generators at Taipower were running close to maximum capacity to keep up with power demand for air-conditioning. Government offices voluntarily switched off air-conditioning during the hottest hours of the afternoon to relieve pressure on the grid, a measure similar to the peak-load management detailed in chapter 7. Back in July 2017, commentators had been ridiculing the government for their quixotic efforts to save electricity, saying that the DPP government would not be in this predicament had it not painted itself in a corner by committing itself to its promise of a nuclear-free homeland. Taiwan's government quickly recovered from this debacle. Days after the August 15 blackout, the Summer World University Games opened in Taipei to much fanfare. The euphoria appeared to have wiped out the fear and confusion from those hours of darkness.

Lingering fears of power shortages came back to haunt the DPP government during the local elections of 2018. Huang Shih-hsiu's Nuclear MythBusters tabled a "Go Green with Nuclear" referendum that forced the DPP government to repeal Article 95, Section 1 of the Electricity Law and halt the shutting down of all nuclear reactors by 2025. During the 2018 local elections, when the DPP government suffered stinging defeat, 59.49 percent of voters who voted on the "Go Green with Nuclear" motion supported it. The Ministry of Economic Affairs amended the laws as required but still proceeded to halt the construction of the fourth nuclear plant in New Taipei City.

In a bid to address the concerns of 5.89 million voters who voted in support of "Go Green with Nuclear," the DPP government and the Guomindang opposition offered competing programs for energy transitions during the run-up to the 2020 presidential election. Days after the second anniversary of the August 2017 blackout, the Industrial Technology Research Institute chaired by the ousted Economic Affairs minister Chih-kung Lee published a report to allay fears of power shortages arising from the shutdown of nuclear power, curtailment of coal, and transition toward solar and wind. For more than four hundred days after the blackout, Taipower's operating reserve did not fall below 6 percent, even as two coal-fired power stations in central Taiwan cut back power production by 10.6 to 16.1 percent. It was also noted that the share of renewables (hydro, solar, and wind) accounted for 6 percent of the energy mix, and went as high as 13 percent in June 2019.[27] The figures tell the story that the DPP government was within striking distance of achieving its goal of 20 percent renewables.

Skeptics remained unconvinced. Huang's Nuclear MythBusters criticized the DPP government for failing to make a much-needed shift away from fossil fuels. Coal and natural gas accounted for 38.8 percent and 38.6 percent of fuel used for power generation in 2018 respectively. The expansion of the Ta-tan Power Plant, the facility at the center of the 2017 blackout, would further increase Taiwan's dependence on natural gas. With Ta-tan taking on a larger share of the island's power supply, critics feared that the next power outage could be much more catastrophic than that of August 2017. Huang later became the energy policy adviser to Han Kuo-yu, Guomindang's candidate for the 2020 presidential campaign, who proposed a target of 50 percent "clean energy" by 2035, which was far more ambitious than that of the DPP's plan. He would later clarify that his plan classified nuclear as "clean energy," and that the share of nuclear energy would increase from 10 to 20 percent.[28]

Energy security and sustainability faded from public attention, as the issues about Taiwan's political identity and the future of its electoral democracy came to the forefront. Before voters had a chance to scrutinize the competing plans, demonstrations against the extradition bill in Hong Kong broke out in June 2019. Han's "green energy plan" would never see the light of day, and he would later be ousted from his post as mayor of Kaohsiung in a recall election. Either way, the plans by both the incumbent and opposition would require massive investments to ramp up the renewable generating capacity. According to the DPP government's Forward-Looking Infrastructure Plan proposed in 2017, more than 80 percent of the 58.49 billion US dollars required to meet the 20 percent renewable target would have to come from private investments.[29] Given the persistent threat of cross-strait conflict and economic uncertainty, it remains to be seen if the private sector will provide the capital to support the electrical power sector that has always been the purview of the government.

If the past offers any indication of the future, one might look toward the energy future of China and Taiwan with trepidation. The governments on both sides of the Taiwan Strait often traded short-term economic gains for long-term sustainability by looking toward fossil fuels to boost economic output. Lessons learned from energy crises are also quickly forgotten. Plans for environmental sustainability are put in the backburner when other pressing issues emerge. There might still be some cause for optimism, however. China's first generation of engineer-bureaucrats showed themselves to be capable of leveraging strong and weak developmental forces to overcome the constraints of economic growth. They engineered many remarkable comebacks in the face of adversity. Their modern-day successors can draw inspiration from the boldness and decisiveness of their predecessors in the face of crisis and find a way to prevent carbon-combustion complexes from overwhelming the biosphere, geosphere, and anthro-technosphere of this integrated Earth system.

Bao Guobao	鮑國寶
bianxiang lueduo	變相掠奪
Cai Changnian	蔡昌年
Cai Xiaoqian	蔡孝乾
Changshou	長壽
changsi	廠絲
Changzhou	常州
Cheng Linsun	程林蓀
Cheng Wanli	程萬里
Chen Liangfu	陳良輔
Chen Shenyan	陳慎言
Chen Yi	陳毅
Chen Yun	陳雲
Chen Zhongxi	陳中熙
Chiang Kai-shek (Jiang Jieshi)	蔣介石
Chu Yinghuang	褚應璜
Chūgoku tsūshinsha	中国通信社
Dachang	達昌
Dadu River (Dadu he)	大渡河
dahuizhan	大會戰
Dai Li	戴笠
Dai Ligang	戴立綱
daotai	道台
Dasheng	大生
Dayouli	大友利
Dazhao	大照
Denki shinpō sha	電気新報社
diandengchang	電燈廠
dianqi gongsi	電氣公司
Dianye jikan	電業季刊
Dong Shu	董樞
Donghua	東華
Du Yuesheng	杜月笙
Erliu hongzha	二六轟炸

fabi	法幣
Fang Gang	方剛
Fu Zuoyi	傅作義
Gao Ming	高明
Ge Helin	葛和林
Ge Zuhui	葛祖輝
Gong Debo	龔德柏
gongsi heying	公私合營
Gu Zhengwen	顧正文
Gu Zhun	顧準
Guan County (Guanxian)	灌縣
Guan Zhiqing	管趾卿
Guangqin	廣勤
Guangyi	廣益
Gui Naihuang	桂迺黃
Guo Keti	郭克悌
Guo Zhicheng	郭志成
Guodian	國電
guomin jingji huifu qi	國民經濟恢復期
Han Kuo-yu (Han Guoyu)	韓國瑜
Hanxue yuekan	汗血月刊
Hengda	恆大
Hengfeng	恆豐
heping jiefang (Peaceful liberation)	和平解放
hiki-age	引揚
Hoku-shi denryoku kōgyō kabushiki kaisha	北支電力工業株式会社
Hou Defeng	侯德封
Huadong dianye guanli ju (East China Power Bureau)	華東電業管理局
Huang Bingquan	黃炳權
Huang Hui	黃煇
Huang Shih-hsiu (Huang Shixiu)	黃士修
Huang Yuxian	黃育賢
huasige	華絲葛
Huaxin	華新
Hubei wenshi ziliao	湖北文史資料
Hui Lianjia	惠聯甲
Huzhou	湖州
Ide Taijiro	出弟次郎
Ishikawa Yoshijiro	石川芳次郎
Jiajia gua hongdeng, yingjie Mao Zedong	家家挂紅燈，迎接毛澤東

Jialing River (Jialing jiang)	嘉陵江
Jiang Guiyuan	蔣貴元
Jianshe weiyuanhui	
(National Construction Commission)	建設委員會
Jibei dianli gongsi	
(North Hebei Power Company)	冀北電力公司
Kachū suiden kabushiki kaisha	華中水電株式會社
Kahoku Dengyō	華北電業
Kajima Jun	加島潤
Kanemaru Yūichi	金丸裕一
Kōain	興亜院
Kōchū Kōshi	興中公司
Kong Xiangxi (H.H. Kung)	孔祥熙
laogui	老規/老鬼
Lee Chih-kuang (Li Shiguang)	李世光
Li Bing	李冰
Li Daigeng	李代耕
Li Hongzhang	李鴻章
Liang Mu	良穆
Li Pingshu	李平書
Li Rui	李銳
Li Yanshi	李彥士
Lin Biao	林彪
Lin Jin	林津
Lin Lanfang	林蘭芳
Lisheng	麗生
Liu Dajun (D.K. Lieu)	劉大鈞
Liu Lanbo	劉瀾波
Liu Jinyu	劉晉鈺
Liu Taotian	劉濤天
Longmao	隆茂
Longxihe (Longxi River)	龍溪河
Lu Yuezhang	盧鉞章
Lu Zuofu	盧作孚
Luo Ronghuan	羅榮桓
Mabian River (Mabian he)	馬邊河
Machangding	馬場町
Majiezi	馬街子
Mao Dun	茅盾
Min River (Minjiang)	岷江
Mingliang	明良
Minhang	閔行
Mitsui bussan	三井物産

Mudanjiang	牡丹江
Nantong	南通
Naigai	內外
Naitō Kumaki	內籐熊喜
Nakashima Kesago	中島今朝吾
Nanshi Power Plant (Nanshi dianchang)	南市電廠
Nie Jigui	聶緝槼
Nie Rongzhen	聶榮臻
Nie Yuntai	聶雲台
Niu Jiechen	鈕介臣
Niusi zhoukan	鈕司週刊
Okabe Eiichi	岡部栄一
Oshikawa Ichiro	押川一郎
Peng Zhen	彭真
Penshuidong	噴水洞
Pingshan	平善
Pudong	浦東
Qian Changzhao	錢昌照
Qian Xuesen	錢學森
Qingyuandong	清淵硐
Qishuyan	戚墅堰
Quanguo dianchang tongji zongbiao	全國電廠統計總表
Quanguo minying dianye lianhui hui	全國民營電業聯合會
Ranliao gongye bu (Ministry of Fuel Resources)	燃料工業部
Ren Yiyu	任一宇
Renmin dianye	人民電業
Riyuetan (Sun-Moon Lake; *jitusu getsu tan*)	日月潭
shangban	商辦
Shen Bao	申報
Shen Sifang	沈嗣芳
Sheng Xuanhuai	盛宣懷
shengguan facai	升官發財
shenghuo yongdian	生活用電
Shenxin	申新
Shijingshan	石景山
Shipai	石牌
Shizui Village (Shizuicun)	石嘴村
Shuo County (Shuoxian)	朔縣
siying	私營
Songming	嵩明
Su Yang	蘇陽
Sugimoto Aki	杉本秋

Sun Danchen	孫丹忱
Sun Yat-sen (Sun Zhongshan)	孫中山
Sun Yun-suan (Sun Yunxuan)	孫運璿
Sun Zhifei	孫志飛
Suzuki Jun	鈴木淳
Suzuki Teiichi	鈴木貞一
Ta Kung Pao (Dagongbao)	大公報
Taipower Company (Taidian gongsi)	台電公司
Tajima Toshio	田嶋俊雄
Tang Mingqi	湯明奇
Tang Wenzhi	唐文治
Tanglang River	
(Tanglang Chuan; Praying Mantis River)	螳螂川
Tanomogi Keikichi	賴母木桂吉
Tao Lizhong	陶立中
Ta-tan (Datan)	大潭
Tian Han	田漢
Tianshenggang	天生港
Tōa denryoku kōgyō	
kabushiki kaisha	東亜電力工業株式会社
Tongming	通明
toufei	投匪
Tsuji Hideo	辻秀男
Tuo River (Tuojiang)	沱江
T. V. Soong (Song Ziwen)	宋子文
tusi	土絲
Wang Daohan	汪道涵
Wang Kemin	王克敏
Wang Jingchun	王景春
Wang Pingyang	王平洋
Wang Shoujing	王守敬
Wang Shoutai	王守泰
Wang Xiaohe	王孝和
Wang Yangming	王陽明
Wang Yanqiu	王雁秋
Wang Yimei	王亦梅
Wang Zihui	王子惠
Wangting	望亭
Weng River (Weng jiang)	滃江
Weng Wenhao	翁文灝
Wu Daogen	吳道艮
Wu Guozhen (K.C. Wu)	吳國楨
Wu River (Wujiang)	烏江

Wu Shi	吳石
Wu Xingzhou	吳興周
Wu Zuguang	吳祖光
Wuhu	蕪湖
Wuxi	無錫
Wuxing	吳興
Xiao Chaogui	蕭朝貴
Xiaofengman (Little Fengman)	小豐滿
Xiashesi	下攝司
Xiling Gorge (Xiling xia)	西陵峽
Xinshengbao	新生報
Xiyuan	西苑
Xu Ying	徐盈
Xu Yingqi	許應期
Xue Yueshun	薛月順
Yan Huixian	嚴惠先
Yan Xishan	閻錫山
Yang Yan	楊琰
Yangshupu	楊樹浦
yangzhuangsu	洋裝素
Yanji	延吉
Yaolong	耀隆
Ye Jianying	葉劍英
Ye Kongjia (K.C. Yeh)	葉孔嘉
Yichang	宜昌
Yiliang	宜良
Yin Zhongrong (K.Y. Yin)	尹仲容
Yu Bin	余彬
Yu En-ying	俞恩瀛
Yun Zhen	惲震
Zeng Guofan	曾國藩
Zeng Zhaojian	曾照鑑
zengchan jieyue	增產節約
Zhabei	閘北
Zhang Baichun	張柏春
Zhang Guangdou	張光斗
Zhang Guobao	張國寶
Zhang Jia'ao	張嘉璈
Zhang Jian	張謇
Zhang Jiazhi	張家祉
Zhang Renjie/Zhang Jingjiang	張人傑/ 張靜江
Zhang Wangliang	張望良
Zhangjiakou	張家口

Zheng Youkui (Y.K. Cheng)	鄭友揆
Zhenhua	震華
Zhenxin	振新
Zhongnanhai	中南海
Zhongyang diangong qicai chang (Central Electrical Manufacturing Works)	中央電工器材廠
Zhou Zhilu	周至祿
Zhou Zhirou	周至柔
Zhu Dajing	朱大經
Zhu Feng	朱楓
Zhu Qiqing	朱其清
Zhu Yongpeng	朱永芃
Zhuge Ming	諸葛明
Zi River (Zishui)	資水
Ziliujing	自流井
Ziyuan weiyuan hui (National Resources Commission)	資源委員會
Ziyuan weiyuan hui gongbao	資源委員會公報
Ziyuan weiyuan hui jishu renyuan fumei shixi shiliao	資源委員會技術人員赴美實習史料

NOTES

Introduction

1. International Energy Agency, "China Overtakes the United States to Become World's Largest Energy Consumer," news release, July 20, 2010.

2. Zhu Yongpeng, "New Energy Sources."

3. Fravel, *Active Defense*, 66–68. Fravel points out that the Communist Party established control over the army after the Gutian Conference in 1929. The understanding of the people's war was modified to include the mass mobilization of China's population and resources in the event of armed conflict.

4. Kasza, *The Conscription Society*, 1.

5. Yuen Yuen Ang, *How China Escaped the Poverty Trap*, 17.

6. Nathan, "China's Changing of the Guard."

7. Huang Xiaoming, *The Rise and Fall of the East Asian Growth System*, 48. See also Suehiro, *Catch-up Industrialization*. Suehiro argues that economically backward countries can achieve compressed development by importing existing technologies of industrialized nations through import substitution, thus short-circuiting the costly R&D process. Suehiro, however, notes that the recipient nation simply follows the path of early adopters of technology and precludes the possibility of late adopters leapfrogging advanced nations.

8. Steffen, Crutzen, and McNeill, "The Anthropocene."

9. Hamilton, "The Anthropocene as Rupture," 100.

10. Hudson, "Placing Asia in the Anthropocene."

11. Wrigley, *Energy and the English Industrial Revolution*.

12. Muscolino, *The Ecology of War*, 11.

13. Smil, *China's Past, China's Future*, 60–62.

14. Latour, *Science in Action*, 215–257.

15. Seow, "Carbon Technocracy," 8. Seow understands technocracy as "involving interrelated ideas about the unquestionable superiority of the scientific method in framing and solving problems of society, and the necessity for central directives and state planning to bring that method to bear on problems." The engineer-bureaucrats featured in this study recognize the limits of scientific methods. They see the rational mindset of profit-optimizing electrical utilities as an impediment to the equitable distribution of electrical power. The "technocracy" that emerges in chapter 7 of this study does not reflect the preeminence of "scientific rationality" but results from scientific experts relinquishing control over power production decisions to party cadres.

16. Yun Zhen, "Yun Zhen zizhuan" [Autobiography of Yun Zhen], February 19, 1944, 129–000000–087A, AH, New Taipei City, Taiwan. The autobiography is filed with the intelligence reports of Chiang Kai Shek's aide-de-camp. Chiang's closest aides screened the engineers of the National Resources Commission (NRC) for their political views and monitored their performance in the party education camps.

17. Dodgen, *Controlling the Dragon*, 7.

18. Elvin, "Three Thousand Years of Unsustainable Growth," 38.

19. A resurgence in scholarship about the NRC appeared around the early 2000s, as China experts became interested in the dominance of state-owned enterprises. An example of this work is Cheng Linsun, "Zhongguo jihua jingji." The most comprehensive English-language monograph on the legacy of NRC's wartime industrialization is Morris Bian's *The Making of the State Enterprise System in Modern China*.

20. Survey histories of China's electrical industries in the Chinese language are collective scholarly endeavors completed with state sponsorship. Examples include Li Daigeng, *Xin Zhongguo dianli gongye fazhan shilue*; Huang Xi, *Zhongguo jinxiandai dianli jishu fazhan shi*. Li Daigeng was deputy minister of water resources and electrical power. He thus emphasized the achievements of post-1949 China. Huang Xi's work largely repeats many of the insights from earlier state-sponsored histories and draws heavily from Li's earlier work. Huang was diagnosed with cancer and suffered a stroke shortly after completing the manuscript in 2002.

21. R. Bin Wong points out that Chinese and Western scholarship offer different interpretations of the impact of foreigners on the Chinese economy. Chinese Marxists see that "imperialism twisted and distorted China's path of development," while Western-trained scholars emphasize that "foreigners created the opportunities and offered the skills and techniques to build a modern economy in China." See Wong, *China Transformed*, 21.

22. Sun Yat-sen, "A Plea to Li Hung-chang (June 1894)," 9.

23. Li Daigeng, *Zhongguo dianli gongye fazhan shiliao*, 5.

24. See Wang, *Fin-de-siécle Splendor*.

25. Cited in Schwarcz, *The Chinese Enlightenment*, 285.

26. "Riben quanguo dianli tongji," 1507; Jianshe weiyuanhui, *Quanguo dianchang tongji*, 7.

27. Mitter, *Forgotten Ally*, 182. The figure is from Kirby, "The Chinese War Economy," 191.

28. Mitter, *Forgotten Ally*, 191.

29. Zhongyang ranliao gongyebu bianyi shi dianye zu, "Xiang Sulian zhuanjia men xuexi."

30. Pearson, "Past, Present and Prospective Energy Transitions."

31. Yu-kwei Cheng, *Foreign Trade and Industrial Development of China*, 227–237. After completing his stint at the Brookings Institute, Zheng left his wife and child in the United States and returned to China to continue his research into Chinese economic history at the Shanghai Academy of Social Sciences (SASS). Zheng Youkui, who had been branded an American spy during the Cultural Revolution, was rehabilitated and returned to the SASS in 1979.

32. Yeh, *Electric Power Development in Mainland China*.

33. Founded in 1951, the Union Research Institute collected newspaper clippings and monitored broadcasts from Communist China for Western intelligence agencies.

34. Cheng Linsun, "Lun kangri zhanzheng qian Ziyuan"; Zheng Youkui, Cheng Linsun and Zhang Chuanhong, *Jiu Zhongguo ziyuan weiyuanhui*, 3–4.

35. Li Daigeng, *Xin Zhongguo dianli gongye*, preface.

36. Joint Economic Committee, *Chinese Economy Post-Mao*, ix.

37. Clarke, "China's Electric Power Industry," 404.

38. Li Cheng and White, "Elite Transformation and Modern Change."

39. Kirby, "Engineering China."

40. Kirby, "Technocratic Organization in China"; Kirby, "Continuation and Change."

41. This phenomenon is not unique to Chinese history. Alain Beltran notes that energy is pervasive, "And yet, is energy the forgotten item in the history of human society?" Beltran, "Introduction."

42. Hughes, *Networks of Power*.

43. Nye, *Electrifying America*, 27.

44. Larkin, "The Politics and Poetics of Infrastructure."

45. Coopersmith, *The Electrification of Russia*, 4.

46. Edgerton, *The Shock of the Old*, xiv.

47. Ekbladh, *The Great American Mission*; Sneddon, *Concrete Revolution*.

48. Mitchell, *Carbon Democracy*.

49. Kale, *Electrifying India*; Meiton, *Electrical Palestine*. Meiton points out that he is not the first historian to note the mutual influence between energy and politics. See page 226 note 22 for an annotated bibliography on the technopolitics of energy.

50. Wright, *Coal Mining in China's Economy and Society*; Wu, *Empires of Coal*, 3.

51. Esherick and Rankin, eds., *Chinese Local Elites and Patterns of Dominance*.

52. Tajima Toshio, ed., *Gendai chūgoku no denryoku sangyō*. Tajima Toshio was also one of the contributors to Zhu Yingui and Yang Daqing, eds., *Shijie nengyuan shi zhong de Zhongguo*.

53. Muscolino, *The Ecology of War*, 6–7.

54. Kinzley, *Natural Resources and the New Frontier*, 10–13.

55. Pomeranz, *The Great Divergence*, 64.

56. Esherick and Rankin, eds., *Chinese Local Elites and Patterns of Dominance*.

57. Kirby, "Engineering China." See especially 141–143.

Chapter 1

1. Broggi, *Trade and Technology Networks*, 36. Broggi points out that the self-strengthening reformers saw the establishment of textile mills as part of "'commercial warfare' (*shangzhan*) that was taking place in the market of consumer goods."

2. Jackson, *Shaping Modern Shanghai*, 8.

3. Beckert, *Empires of Cotton*, xiv.

4. Ghosh, *The Great Derangement*, 110.

5. Shanghai shi dianli gongyeju shizhi bianzuan weiyuanhui, "Shanghai shi dianli gongyezhi-gaishu" [Gazetteer of Shanghai's Electrical Industries—Introduction], last modified February 27, 2003, accessed October 12, 2017, http://www.shtong.gov.cn/newsite/node2/node2245/node4441/node58149/index.html.

6. Chūgoku tsūshinsha chōsa bu, *Shanhai*.

7. Yang Yan, "Gongbuju," 64.

8. Chūgoku tsūshinsha chōsa bu, *Shanhai*.

9. Schivelbusch, *Disenchanted Night*; Freeberg, *The Age of Edison*, 189–190.

10. Shanghai Municipal Council, *Report for the Year 1907*, 200. Accessed through HathiTrust.

11. Yang Yan, "Gongbuju," 74. Based on the *Annual Report of Shanghai Municipal Council (1882–1894)*, U1-1-907, Shanghai Municipal Archives, Shanghai, China.

12. Yang Yan, "Gongbuju," 75.

13. Shanghai Municipal Council, *Report for the Year 1907*, 201.

14. Shanghai Municipal Council, *Report for the Year 1907*, 201, 205.

15. Shanghai Municipal Council, *Report for the Year 1907*, 202.

16. Shanghai Municipal Council, *Report for the Year 1907*, 204.

17. Shanghai Municipal Council, *Report for the Year 1907*, 204.

18. Shanghai Municipal Council, *Report for the Year 1907*, 213.

19. Li Daigeng, *Zhongguo dianli gongye fazhan shiliao*, 9–10.

20. Chen Weiguo et al., eds., *Zhongguo jindai mingren*, 51.

21. Shiroyama, *China during the Great Depression*, 43–45.

22. Liu Taotian, "Fangzhiye gaikuang diaocha."

23. Faure, "The Control of Equity in Chinese Firms."

24. Köll, *From Cotton Mill to Business Empire*.

25. Chūgoku tsūshinsha chōsa bu, *Shanhai*.

26. Huashang shachang lianhe hui, *Zhongguo shachang yilanbiao*.

27. Lieu, *The Silk Reeling Industry in Shanghai*, 69. Lieu described the work of laogui in silk filatures in his research. The practice of hiring a subcontractor to manage the boiler room began in British-owned cotton mills.

28. Faure, "The Control of Equity in Chinese Firms," 60.

29. Zhu Youzhi and Guo Qin, *Hunan jindai shiye renwu zhuanlue*.

30. The scale of the Shanghai Spinning Factory at its inception is obtained from the table of cotton of mills in China published in 1933. Power consumption data is from the statistical investigation of electrical power industries by the NCC. By then, only four out of nine Japanese textile mills reported using self-supplied power.

31. Tang Mengxiong, "Yingguo."

32. Duus, "Zaikabo," 84.

33. Shanghai Municipal Council, *Report for the Year 1907*, 213–214.

34. *Annual Report of the Shanghai Municipal Council 1927*, 370. Generating capacity doubled between 1912 and 1914 following the completion of the Riverside Power Station,

while the number of units sold for power and heating increased by 544 percent over the same period.

35. Jianshe weiyuanhui, *Quanguo fadianchang diaocha biao.*

36. Huashang shachang lianhe hui, *Zhongguo shachang yilanbiao.*

37. Zhang Wangliang, "Shachang dianqi," 1.

38. Shanghai shi dianli gongyeju shi zhi bianzuan weiyuanhui, ed., *Shanghai shi dianli gongye*, 25–26.

39. Huashang shachang lianhe hui, *Zhongguo shachang yilanbiao.*

40. Jianshe weiyuanhui, *Quanguo fadianchang diaocha biao*, 63–65.

41. *The International Library of Technology* textbook on cotton milling states that cotton yarn is "divided into coarse, medium, and fine, according to the thickness of the thread, which in turn is determined by the number of hanks to the pound. A hank of cotton yard contains 840 yards, and the size of the yarn is indicated by the number of hanks required to weigh one pound; thus 10s yarn would contain ten hanks, or 10 × 840 yards, making 8,400 yards, in a pound." There is also no rule that defines which yarn is coarse, medium, or fine. As a rule of thumb, anything below 30s is considered coarse.

42. He Da, "Miansha fangzhi."

43. Li Shizhong, "Fangzhi yuanliao," 6. See also Shiroyama, *China during the Great Depression*, 45.

44. Shanghai Municipal Council, *Report for the Year 1907*, 213–214.

45. "Xihe shachang dianji zhahui" [Generator of Xihe Cotton Mill destroyed in explosion], *Fangzhi shibao*, May 31, 1926.

46. Perry, *Shanghai on Strike*, 82.

47. Zhang Qian, ed., "Minutes for the meeting on June 22, 1925," in *The Minutes of Shanghai Municipal Council*, 105. Hereafter abbreviated as *Minutes of the SMC* with volume number, meeting date, and page number specified.

48. Goto-Shibata, *Japan and Britain in Shanghai*, 26.

49. Zhang Qian, ed., *Minutes of the SMC*, vol. 23, July 6, 1925, 118.

50. "Shanghai sha hua shikuang" [Conditions of Shanghai's yarn and cotton market], *Fangzhi shibao*, July 13, 1925.

51. Goto-Shibata, *Japan and Britain in Shanghai*, 27.

52. Zhang Qian, ed., *Minutes of the SMC*, vol. 23, July 31, 1925, 140.

53. Goto-Shibata, *Japan and Britain in Shanghai*, 27.

54. Zhang Qian, ed., *Minutes of the SMC*, vol. 23, August 13, 1925, 148.

55. Robert L. Jarman, ed., "Despatch no. 143 dated 21st August 1925 from the British consul general in Shanghai to the British Legation in Peking," in *Shanghai: Political & Economic Reports, 1842–1943 British Government Records from the International City*, vol. 14 (London: Archive Editions, 2008), 248.

56. Zhang Qian, ed., *Minutes of the SMC*, vol. 23, August 27, 1925, 153.

57. Zhang Qian, ed., *Minutes of the SMC*, vol. 23, August 27, 1925, 153.

58. Zhang Qian, ed., *Minutes of the SMC*, vol. 23, September 1, 1925, 158; Goto-Shibata, *Japan and Britain in Shanghai*, 30.

59. Shiroyama, *China during the Great Depression*, 45. Shiroyama cites from Shanghai Cotton Textile Planning Committee, whose figures are largely consistent with the overview of cotton industries compiled by the National Confederation of Chinese Cotton Mill Owners.

60. Zhang Qian, ed., *Minutes of the SMC*, vol. 24, March 25, 1929, 354.

61. Shanghai shi dianli gongyeju, *Shanghai shi dianli gongye zhi*, 26.

62. Huashang shachang lianhe hui, *Zhongguo shachang yilanbiao*.

63. "Shanghai Power Company advertisement," *Huashang shachang lianhehui jikan* 10:1 (June 1932): cover.

64. "News Bulletin," *Fangzhi zhoukan* 2:25 (1932): 688.

65. Wugong, "Longmao shachang tingye pingyi," 1.

66. Founding date of Tōka Number 1 is obtained from *Zhongguo shachang yilanbiao* (1933), while power output information is contained in the NCC's statistical investigation of power plants (1929).

67. Wugong, "Longmao shachang tingye pingyi," 1.

68. Broggi, *Trade and Technology*, 57.

69. Lieu, *Preliminary Investigation on Industrialization*, Tables 4 and 5. Lieu stated that the investigators calculated the "amount of power employed in the Shanghai factories and workshops." They added the figures for steam engines, steam turbines, petrol engines, electric generators and rented electricity from the power station. To obtain the net power output, I subtracted the 24,248 kW of self-supplied power in the electrical power industries.

70. Cotton mill data extracted from *Zhongguo shachang yilanbiao*; motive power data from the NCC's statistical investigation of power plants (1929).

71. "Qingdao diandengchang" [Qingdao Electric Works], Qingdao Gazetteer Research Institute, accessed April 4, 2019, http://qdsq.qingdao.gov.cn/n15752132/n20546827/n20552822/n20553782/n20554016/151215024046156842.html.

72. Köll, *From Cotton Mill to Business Empire*; Shao, *Culturing Modernity*.

73. Shao, *Culturing Modernity*, 71.

74. Dasheng Cotton Mill report to National Construction Commission (NCC), September 1933, 23-25-11-029-01, NCC Papers, AS, Taipei, Taiwan. *Jianshe weiyuanhui* has also been translated as the National Reconstruction Commission. I am using the translation that the agency used in its official communication.

75. "Nantong shachangye jiang lianhe sheli da fadian chang," *Xin dian jie* 18 (1932): 4.

76. Huang Hui, "Inspection Report," February 1935, 23-25-11-029-01, NCC Papers, AS, Taipei, Taiwan.

77. Huang Hui, "Inspection Report," February 1935, 23-25-11-029-01.

78. Huang Hui, "Inspection Report," February 1935, 23-25-11-029-01.

79. Wright, *Coal Mining in China's Economy and Society*, 2. Wugong, "Longmao shachang tingye pingyi."

80. Cheng, Foreign Trade and Industrial Development in China, 252.

Chapter 2

1. Lieu, *The Silk Reeling Industry*, 69.
2. Lieu, *The Silk Reeling Industry*, 71.
3. Shiroyama, *China during the Great Depression*, 58.
4. Bell, "From Comprador to County Magistrate," 138.
5. Jianshe weiyuanhui, *Quanguo dianchang tongji*, 27, 37.
6. Yao Yuming, "Lue lun jindai Zhejiang."
7. Shiroyama, *China during the Great Depression*, 53–54.
8. Zhu Xinyu, *Zhejiang sichou shi*, 143.
9. Li, *China's Silk Trade*, 79.
10. Feng Zizai, "Zhejiang Wuxing sichouye gaikuang," 96.
11. Feng Zizai, "Zhejiang Wuxing sichouye gaikuang"; Wang, "Xinhai geming Jiang-zhe sizhiye"; Zhu Xinyu, *Zhejiang sichou shi*, 198.
12. Huzhou difangzhi bianzuan weiyuanhui, *Huzhou shizhi*, 935–936.
13. Huzhou difangzhi, *Huzhou shizhi*, 938.
14. "Huzhou Dachang sizhi chang jinggao gejie shu," *Huzhou Gongbao*, August 6, 1935.
15. Sun Yat-sen, "How China's Industry Should Be Developed, 1919," 240.
16. Jiao tong da xue xiao shi bian xian zu, *Jiaotong Daxue xiaoshi*, 58. See Köll, *Railroads*, 169–170 for more details on engineering education in Jiaotong University.
17. "Alumni Notes," *The Wisconsin Engineer* 22 (1918): 294.
18. Shen Sifang, "Dianli guantian."
19. Yun Zhen, "Yun Zhen zizhuan" [Autobiography of Yun Zhen], February 19, 1944, 129-000000-087A, Attendants Office of the Guomindang Military Commission papers, AH, New Taipei City, Taiwan.
20. Yun Zhen, "Xuesheng yundong." Yun's article was published in a special edition about new realism.
21. "Si tai yin mai zi."
22. Kline, *Steinmetz*, 9.
23. "Si tai yin mai zi."
24. Yun Zhen, "Yun Zhen zizhuan," February 19, 1944, 129-000000-087A, AH, New Taipei City, Taiwan.
25. Shen Sifang, "Dianli guantian."
26. Zhejiang sheng dianli gongyezhi bianzuan weiyuanhui, eds., *Zhejiang sheng dianli gongye zhi*, 281.
27. Yun Zhen's autobiography was published in Zhang Baichun, ed., "Kangzhan qian de diangong jishu," 161–162.
28. Kirby, "Engineering China," 142.
29. Shen Sifang, "Zhengli shoudu dianchang."
30. Musgrove, "Building a Dream."
31. Shen Sifang and Li Yanshi, "Riben chansi ye." Translated from a Japanese article by Genjiro Tada, an electrical engineer in the Bureau of Electricity in Japan's Ministry of Communications.

32. Jianshe weiyuanhui, *Dianchang tongji*, 3.

33. Jianshe weiyuanhui, *Dianchang tongji* (December 1929), 2.

34. Bao Guobao, "Shoudu dianchang zhi zhengli." For more details on the repairs of the Nanjing power plant, see Tan, "Repairing China's Power Grid."

35. Jianshe weiyuanhui, *Dianchang tongji* (December 1929), 11. There was a numerical error in the tabulation of the fuel mix in the 1929 report. The authors erroneously listed the total generating capacity of diesel generators as 7,500 kW. This could not have been the case, as the diesel generator at the French Concession Power Station was 15,000 kW. There were another 107 diesel-powered generators. The sum total should be 43,315 kW.

36. World Power Conference, *Transactions Third World Power Conference*, vol. 4, 415.

37. Jianshe weiyuanhui, *Dianchang tongji* (December 1929), 75–76.

38. "Zhi Henan sheng minzheng ting Zhang Boying," 78.

39. "Quanguo minying dianye qingyuan," 6.

40. "Quanguo minying dianye qingyuan," 6.

41. "Quanguo minying dianye qingyuan," 7.

42. In a full-text search on the term *minying* on the Chinese Periodical Full-Text database spanning between 1890 and 1930, most of the hits come from the *Dianye jikan*, the journal of the Association of Private Chinese Electric Corporations.

43. "Dianqi shiye qudi guize cao'an."

44. World Power Conference, *Transactions Third World Power Conference*, vol. 4, 411.

45. The legal scholar Lou Tongsun also graced the inaugural issue with the words, "The modern world will not be bright without electricity, new railroads will not run without electricity. Cure illnesses and irrigate fields with half the effort, the essential path to national salvation and the core of reconstruction."

46. Guo Zhicheng, "Fakanci."

47. Shen Sifang, "Tongye cihou."

48. Shen Sifang and Li Yanshi, "Benhui chuxi Deguo."

49. Shen Sifang and Li Yanshi, "Li Yanshi Shen Sifang er jun," 6. Li Yanshi uses the term *shangban* (merchant-run) rather than *minying* (operated by the people) to refer to private power plants.

50. Shen Sifang and Li Yanshi, "Li Yanshi Shen Sifang er jun," 10.

51. Dong Shu, trans., "Minying dianye."

52. World Power Conference, *Transactions Third World Power Conference*, vol. 1, 32.

53. Huzhou difangzhi bianzuan weiyuanhui, *Huzhou shizhi*, 935–936.

54. World Power Conference, *Transactions Third World Power Conference*, vol. 1, 108.

55. World Power Conference, *Transactions Third World Power Conference*, vol. 1, 117.

56. World Power Conference, *Transactions Third World Power Conference*, vol. 1, 53. The program directors for the Washington conference had a distinguished record in public service. Ely Sumner, a mechanical engineering professor at Carnegie Institute of Technology (CIT), served as CIT's power plant supervisor and later joined Pittsburgh's Department of Public Health to work on smoke prevention. A. C. Jewett built the first hydropower dam for the emir of Afghanistan.

57. Contract between Dachang and Woo-shing Electric, January 1929, 23-25-13-014-01, NCC Papers, AS, Taipei, Taiwan.

58. Arrears, July 1935, NCC Papers, 23-25-13-014-02, AS, Taipei, Taiwan.

59. "Wuxing dianqi youxian gongsi duiyu Wuxing ji sizhi ye tongye gonghui wei jianrang dianfei jingguo shishi xuanyan zhi dafu," [Reply from Woo-shing Electric in response to the silk weavers' association's request for the reduction of electric tariffs], August 7, 1935, *Huzhou Gongbao*.

60. "Wuxing dianqi youxian gongsi duiyu Wuxing," August 7, 1935, *Huzhou Gongbao*.

61. "Wuxing dianqi gufen gongsi bofu Wuxing," August 16, 1935, *Huzhou Gongbao*.

62. Minutes of meeting between electric company and silk industries, August 21, 1935, 23-25-13-014-01-246, NCC Papers, AS, Taipei, Taiwan.

63. Zhang Renjie, "Benhui wei jiejue Beiping."

64. Woo-shing Electric Accounts Receivable, September 1935, 23-25-13-014-01-246, NCC Papers, AS, Taipei, Taiwan.

65. World Power Conference, *Transactions Third World Power Conference*, vol. 4, 664. Zimmerman explains that the rate-base rule had been established in 1898 after disputes over railway freight pricing. The public believed that railway companies had set exorbitant rates based on fictitious capitalization. The Supreme Court ruled that return to value had to be calculated based on the present value of the property rather than the investment on the business. This was later extended in 1907 over gas and electric utilities.

66. World Power Conference, *Transactions Third World Power Conference*, vol. 4, 668.

67. World Power Conference, *Transactions Third World Power Conference*, vol. 4, 35.

68. Ang, *How China Escaped the Poverty Trap*, 69.

69. World Power Conference, Transactions Third World Power Conference, vol. 1, 117.

Chapter 3

1. Chen Zhongxi, "Dianqi shiye," 4–5.

2. Kanemaru Yūichi, "Shina jihen chokugo"; Tajima Toshio, ed., *Gendai chūgoku no denryoku sangyō*. Kanemaru has written extensively about the history of Shanghai's electrical industries for three decades. His 2010 article is a compilation of newly discovered sources tucked away in manuscript collections across Japan. Tajima is a professor with the Social Science Research Institute at the University of Tokyo. Drawing mostly from published materials such as industrial gazetteers and statistical digests, Tajima's research team began their work of studying the electrical industries in contemporary China in 2004.

3. Chen Zhongxi, "Dianqi shiye," 15–16. Chen Zhongxi compared the cost of a 5,000 kW hydroelectric plant with a coal-fired plant in September 1944. The construction cost of hydropower, calculated in terms of average cost per kWh, is 100 CNC dollars higher than coal-fired plants. Interest and depreciation of the hydropower dams came up to 12 CNC dollars per kWh. The average cost per kWh for coal-fired power is two CNC dollars.

4. Hughes, *Networks of Power*, 286.

5. For a chronology on the founding of the National Resources Commission, see Bian, *The Making of the State Enterprise System*, 45–53. Bian argues that the NRC's establishment was not just a name change, but marked the transformation of the agency into Chiang Kai Shek's brain trust for industrial development.

6. Suzuki Jun, *Nihon no kindai*, 265.

7. Bao Guobao and Yun Zhen, "Dianqi shiye gailun," 14–15.

8. Chen Zhongxi, "Dianqi shiye yu dianqi shiye jianshe," 1.

9. Zheng Youkui, Cheng Linsun, and Zhang Chuanhong, *Jiu Zhongguo ziyuan wei-yuanhui*, 24.

10. Mantetsu keizai chōsa kai, *Shina denki jigyō chōsa shiryō* [Materials on the investigation of electrical industries in China], October 1935, DL 177-22, NDL, Tokyo, Japan.

11. Mantetsu keizai chōsa kai, *Shina denki*, 1–18.

12. Ide Taijiro, "Shina senryō chi."

13. "Hokushi Keizai kōsaku taikō," circa July 1937 in "Hokushi keizai kōsaku tō shiryō" [Materials related to the North China economic project], 90:2:H, University of Tokyo Economics Library, Tokyo, Japan. The folder contains a collection of newspaper clippings, electrical engineering journal articles, and cabinet reports of the Manchukuo Ministry of Industries between 1937 and 1938.

14. "Doitsu denki jigyō."

15. Hein, *Fueling Growth*, 45–46.

16. Liang Mu, "Riben zhi guoying dianli."

17. "Hokushi Keizai."

18. "Hokushi Keizai."

19. Tajima Toshio, "Kahoku ni okeru kōiki denryoku," 116.

20. "Huabei dianye gongsi Beiping fengongsi shenqing goumai wuzi he fafang xuke-zheng ji wuzi xuyao de diaocha deng" [Materials on the applications for the purchase of materials and release of resources], J006-003-00001, BMA, China. The power plant's general manager forwarded its requests for raw materials such as power-generating equipment, rubber, and wires, to the North China Development Company for approval.

21. Denki shinpō sha, *Hoku chūshi denki*, 34.

22. Tōkyō dentō kabushiki kaishashi iinkai, ed., *Tōkyō dentō kabushiki kaishashi* [The history of the Tokyo electric light company] (Tokyo: Tokyo Electric, 1956), 140. Quoted by Kanemaru Yūichi, "Kachū denryoku" (2010): 154.

23. Denki shinpō sha, *Hoku chūshi denki*, 35.

24. Sogō Shinji, ed., *Shigen kaihatsu hokushi*, 110, 115.

25. Hein, *Fueling Growth*, 46.

26. Denki shinpō sha, *Hoku chūshi denki*, 1–5.

27. Sogō Shinji, ed., *Shigen kaihatsu hokushi*, 117.

28. Denki shinpō sha, *Hoku chūshi denki*, 24–25.

29. Materials on the third anniversary of North China Electric, February 1942, J006-003-0024, BMA, Beijing, China.

30. Tajima Toshio, "Kahoku denryoku," 116.

31. Power-generating figures for Shijingshan Power Station and Beijing Power Station, May 1940, J006-001-0003, BMA, Beijing, China.

32. Figures based on 1936 statistical survey of China's Electrical Industries by the Guomindang's National Construction Commission. This excludes the three northeastern provinces held by the Japanese. Cited in Chen Zhongxi, "Dianqi shiye yu dianqi shiye jianshe," 10–11.

33. Kanemaru Yūichi, "Kachū denryoku," 158.

34. Ishikawa Yoshijiro, *Naka shina.*

35. Ishikawa Yoshijiro, *Naka shina,* 4.

36. Ishikawa Yoshijiro, *Naka shina,* 20.

37. Ishikawa Yoshijiro, *Naka shina,* 21.

38. Kanemaru Yūichi, "Kachū denryoku," 159.

39. "Huazhong shuiden zhushi huishe sheli yaogang" [Outline on the establishment of Central China Water and Electric Company], June 30, 1938, R1-10-821, SMA, Shanghai, China.

40. Kanemaru Yūichi, "Cong pohuai dao fuxing," 859. Original article published in Chinese.

41. Kanemaru Yūichi, "Cong pohuai dao fuxing," 860.

42. "Shiyebu deng guanyu zhun Rijun yaoqiu jinzhi wang Shanghai ji zujie neiwai shusong wupin ziling" [Order approving the Japanese Army's ban on the shipment of material to Shanghai and foreign concessions], December 3, 1938, in *Riwei Shanghai,* ed. Shanghai shi dang'an guan, 472–474.

43. Ishikawa Yoshijiro, *Naka shina,* 25.

44. "The Chinese Electric Power Co. Ltd: A Brief Report of Its Post-War Preliminary Reconstruction Efforts," April 1949, Q578-1-94, SMA, Shanghai, China.

45. The claims of the Chinese Electric Company are further supported by the records of the American-owned Shanghai Power Company, which states that it had to supply electricity to the service areas outside the foreign concessions.

46. Letter from Reformed Government of Shanghai to Central China Water and Electric, November 1, 1939, R1-10-823, SMA, Shanghai, China.

47. "Sekitan haikyū tōsei shikō rei."

48. Minutes of Council Meeting for May 15, 1940, *The Minutes of the Shanghai Municipal Council,* vol. 28 (Shanghai: Shanghai Classics Publishing House), 54.

49. Coal Conservation report, July 1944, R13-1-160-77, SMA, Shanghai, China. The report lists the conservation targets for each category in tons of coal equivalent. Counting backward, I obtained the estimated total coal consumption for each sector. For the month of July 1944, 2,400 tons of coal had to be burned for electric lighting out of an estimated total consumption of 21,719 tons.

50. Minutes of Council Meeting for March 5, 1941, *The Minutes of the Shanghai Municipal Council,* vol. 28, 187.

51. Keswick survived an assassination attempt in January 1941, when he was shot in the shoulder by the chairman of the Japanese Street Union at the Shanghai Racecourse.

52. Minutes of Council Meeting for March 5, 1941, *The Minutes of the Shanghai Municipal Council*, vol. 28, 191.

53. "Gongye ju qiewang dangju gongyun fenpei dianliu" [Industries call on Municipal Council to publicly discuss power allocation plans], March 14, 1941, *Shenbao*.

54. Minutes of Council Meeting for March 27, 1941, *The Minutes of the Shanghai Municipal Council*, vol. 28, 206.

55. Minutes of Council Meeting for March 27, 1941, *The Minutes of the Shanghai Municipal Council*, vol. 28, 259.

56. Minutes of Council Meeting for May 16, 1942, *The Minutes of the Shanghai Municipal Council*, vol. 28, 466. Okazaki Katsuo eventually became the Japanese foreign minister and was present on board the USS *Missouri* for the signing of the surrender agreement.

57. "Xingya yuan huazhong lianluo bu guanyu tongzhi Shanghai diqu zhongyao wuzi yidong han ji shifu hanling" [Orders on the restriction of important resources in the Shanghai area issued by the East Asia Development Board], April 1942, in *Riwei Shanghai*, ed. Shanghai shi dang'an guan, 570–574.

58. "Meijin jiang zi Huabei yun hu dianliu gongji wuyu" [Coal to be shipped from North China to Shanghai, no concerns for electric power supply], *Shenbao*, March 4, 1942.

59. Minutes of Council Meeting for May 16, 1942, *The Minutes of the Shanghai Municipal Council*, vol. 28, 468.

60. Minutes of the first Industry Fuel Economy Committee, November 29, 1943, R22-2-135, SMA, Shanghai, China.

61. Curtailment of Unnecessary Consumption of Electricity, December 31, 1943, R22-2-198, SMA, Shanghai, China.

62. Suggestion for Economy in Fuel and Power, February 1944, R22-2-135, SMA, Shanghai, China.

63. Letter from Shanghai Miscellaneous Grains Processing Association to Shanghai Public Works, November 28, 1944, R47-2-71, SMA, Shanghai, China.

64. Coal Conservation report, July 1944, R13-1-160-77, SMA, Shanghai, China. The report lists the conservation targets for each category in tons of coal equivalent.

65. Bill to suppress power wastage proposed by Shanghai Provisional Senate, August 1945, Q109-1-1999, SMA, Shanghai, China. Cited in Gao Ming, "Yijiusiwu zhi yijiuliuwu Shanghai."

66. Kajima Jun, "Shanhai denryoku sangyō," 95. Kajima cites figures compiled by the Shanghai Bureau of Statistics in 1947. Gao Ming derives the same figures from the annual reports at the Shanghai Municipal Archives.

67. Zhang Baichun, ed., "Kangzhan qian de diangong jishu," 163.

68. Letter to National Construction Commission to fifteen power stations, August 26, 1937, 23-25-00-020-02, NCC Papers, AS, Taipei, Taiwan.

69. Huang Hui, "Zhongguo zhi shuili ziyuan ji shuili fadian zhi zhanwang."

70. Liu Jinyu, "Kunhu dianchang choubei jingguo."

71. Zhang Baichun, ed., "Kangzhan qian de diangong jishu," 185.

72. Liu Jinyu, "Kunhu dianchang choubei jingguo," 318.

73. Cheng Yufeng and Cheng Yuhuang, eds., *Ziyuan weiyuanhui dangan shiliao chubian*, 1:1, 30. On paper, the Industrial Office oversaw all manufacturing activities, which includes the Electrical Works. The Electrical Power Office managed the power stations.

74. Zhang Baichun, ed., "Kangzhan qian de diangong jishu," 185. Other components for the Kunming Lakeside Electric Works also traveled circuitous paths. A diesel engine transported from the Hanyeping company in Hubei sat in the Kunming Railway Station for two months before it finally made its way to Majiezi. A fifteen-ton flywheel was also too heavy for a truck to transport and had to be manually hauled to higher ground with steel cables. See *Kangzhan shiqi neiqian xinan*, 196.

75. Liu Jinyu, "Kunhu dianchang choubei jingguo," 314.

76. "Tongji."

77. *Kangzhan shiqi neiqian xinan*, 193–194.

78. *Kangzhan shiqi neiqian xinan*, 193.

79. Lu Yuezhang, "Zhongguo zhi ranliao ziyuan."

80. Annual report for Kunming Lakeside Electrical Works, January–December 1940, 003-010301-0581, NRC Papers, AH, Taipei, Taiwan.

81. Annual report for Kunming Lakeside Electrical Works, January–December 1940, 003-010301-0581, NRC Papers, AH, Taipei, Taiwan.

82. Tao and Liu, "Kunhu Penshuidong," 133.

83. Yunnan sheng difangzhi bianzuan weiyuanhui, eds., *Yunnan shengzhi*, vol. 37, 201–208.

84. Tao and Liu, "Kunhu Penshuidong," 132.

85. Liu Jinyu, "Kunhu dianchang choubei jingguo," 132. The engineers who completed the tunneling had worked on the Guangzhou-Wuhan railway between 1930 and 1936. They had burrowed through 262 miles of mountainous territory between Zhuzhou in Hunan and Shaoguan in Guangdong Province to connect two ends of the Guangzhou-Wuhan railway. The cave complied with construction standards established for the Guangzhou-Wuhan railway project.

86. Liu Jinyu, "Kunhu dianchang," 131.

87. Gui Naihuang, Liu Jinyu, and Yang Guohua, "Tanglangchuan jihua ji Dianbei shuili fadian zhi zhanwang."

88. Carin, *Power Industry in Communist China*, 15.

89. Carin, 14.

Chapter 4

1. Services of Supply, China Theater to NRC, December 25, 1944, RG 0493, US Forces in the China-Burma-India Theater, UD-UP 402, NARA, College Park, Maryland, United States.

2. Yun Zhen, "Huadong gongyebu."

3. Shen, *Unearthing the Nation*, 147.

4. Zhang Baichun, ed., "Kangzhan qian de diangong jishu," 189.

5. Xu Ying, "Yigeren tan diangong shiye: jintian shi tamen fenbie duli chengzhang de rizi" [Interview with Yun Zhen: Today marks the day when they will thrive and grow individually], *Dagongbao* [L'Impartial], July 1, 1948. This article was a commemorative piece for the tenth anniversary of the Central Electrical Manufacturing Works.

6. Yun Zhen, "Diangong qicaichang."

7. Yun Zhen, "Sanshinian lai Zhongguo," 255.

8. Nakashi kensetsu seibi iinkai, 42.

9. Chiah-sing Ho, "Dongli jiqi jinkou."

10. Zhang Chenghu, "Zhongguo dianxian gongye," 196. The price data cited in this article came from Zhang's investigation before he set off for England in 1936.

11. Hughes, *Networks of Power*, 324–325. Hughes also compares Britain's standardization problem to "a Chinese box—problems within problems" (358).

12. See Argersinger, "Voltage Standardization."

13. Jianshe weiyuanhui, "Jianshe: Ling fadian zhoulü," 15.

14. The currency unit of China after 1935 is denoted in CNC dollars, which stands for Chinese National Currency. It is also known as the *fabi*. For a discussion about the five-year plan, see Zheng Youkui et al., *Jiu Zhongguo ziyuan weiyuanhui*, 24. Zheng cites the National Resources Commission papers (Record number: 28-5965) in the Number Two Archives in Nanjing. The petroleum and steel industries received the bulk of the capital allocation with 86.3 million CNC dollars and 80 million CNC dollars respectively.

15. Zheng Youkui et al., *Jiu Zhongguo ziyuan weiyuanhui*, 32.

16. Zhang Baichun, ed., "Kangzhan qian de diangong jishu," 179.

17. Zhang Baichun, ed., 179.

18. Zhang Chenghu, "Zhongguo dianxian gongye," 196.

19. Zhongguo dianqi gongye fazhanshi bianzuan weiyuanhui, *Zhongguo dianqi gongye fazhanshi*, 32–33.

20. Zhang Baichun, ed., "Kangzhan qian de diangong jishu," 184.

21. Zheng Youkui et al., *Jiu Zhongguo de ziyuan weiyuanhui*, 37.

22. Guo Dewen and Sun Keming, "Kangzhan banian lai zhi dianqi gongye."

23. Zhang Baichun, ed., "Kangzhan qian de diangong jishu," 179–180. In his autobiography republished in Zhang Baichun's volume, Yun Zhen described how the technology transfer for field telephones was complicated by domestic politics and increasing tensions in Sino-German relations. He recalled that T. V. Soong, then the managing director of the Central Bank of China, had personal connections to the Chicago Telephone Company and pressured the deputy minister for the NRC Qian Changzhao to purchase the patents from the Americans. The first shipment of German field telephones, which served as the models for the telephones to be assembled by the NRC, nearly did not make it to China. Unable to receive the shipments in the port of Haiphong, as the colonial government in French Indo-China imposed an embargo on German goods by 1938, the Electrical Works diverted the shipment to their temporary assembly facility in the British colony of Hong Kong. The manufacturing operation was transferred to Kunming in early 1939.

24. Xu Yingqi, "Zhongyang diangong," 22.

25. Xiangtan dianji changzhi bianzuan weiyuanhui, *Xiangtan dianji changzhi*. After 1949, the Central Electrical Manufacturing Works underwent reorganization and was renamed as the Xiangtan Electrical Manufacturing Works. It is still one of the top five hundred state-owned enterprises with assets totaling 1.63 billion RMB.

26. *Xiangtan dianji changzhi*, 5.

27. Ministry of Economic Affairs to National Resources Commission, "Jingjibu fengfa feichang shiqi guanli nong kuang gong shang guanli tiaoli zhi ziyuan weiyuan-hui xunling" [Management guidelines for agriculture, mining, industry and commerce during extraordinary times], October 18, 1938, *Zhonghua minguo dangan ziliao huibian*, part v, vol. 2, 11.

28. *Xiangtan dianjichang zhi*, 734.

29. Zhongyang diangong qicaichang [Central Electric Corporation], "*Shizhounian jiniance*" [Tenth anniversary commemorative booklet], 1948, 24-16-03-11-6, NRC Papers, AS, Taipei, Taiwan. The tenth anniversary booklet is a valuable source that summarizes the key milestones of the Electrical Works, and several articles about archival collections on the Electrical Works repeat the material of this commemorative publication.

30. Yun Zhen, "Zhongyang diangong."

31. Yun Zhen, "Huadong gongyebu."

32. Yun Zhen, Zhongyang diangong," 12.

33. Tawney, *Land and Labor in China*, 77.

34. Yun Zhen, Zhongyang diangong," 17.

35. "Caizheng bu ziqing jingji bu hefu jingji sannian jihua shishi banfa an" [The Ministry of Finance requests the Ministry of Economic Affairs to ratify the implementation of the Three-Year economic plan], April 26, 1940, in *Ziyuan weiyuanhui dang'an shiliao*, 74.

36. Yun Zhen to Weng Wenhao, "Zhongyang diangong qicaichang zaoshou zhanshi caiwu sunshi qingxing" [War damage sustained by the Central Electrical Manufacturing Works], October 21, 1941, 003-010309-0279-0006a, NRC Papers, AH, Taipei, Taiwan

37. Yun Zhen, Zhongyang diangong," 12.

38. Yun Zhen, "War damage," 003-010309-0279-0006a, AH, Taipei, Taiwan.

39. Yun Zhen, "War damage," 003-010309-0279-0007a, AH, Taipei, Taiwan.

40. The unscheduled relocation was costly. The Electrical Works still owed its contractors about 480,000 CNC dollars for earlier construction projects, incurred 3.45 million CNC dollars for the works in progress from 1941, and had to fork out 2.82 million CNC dollars to cover the cost of the relocation in 1942. "Zhongyang diangong qicaichang chuangye yusuan gaisuan ji jisuan shubiao" [Start-up budget, estimates, and accounts for the Central Electrical Manufacturing Works], January 1942, 24-16-03-9-1, NRC Papers, AS, Taipei, Taiwan.

41. Ministry of Economic Affairs, Minutes of meeting with Central Electrical Manufacturing Works, March 12, 1941, NRC Papers, 003-010303-0449-0024, AH, Taipei, Taiwan.

42. Central Electrical Manufacturing Works to Ordnance Department, date not specified (second half of 1941), 003-010303-0451-0020a, NRC Papers, AH, Taipei, Taiwan.

43. Maochun Yu, *The Dragon's War*, 93. Yu points out that Treasury Department officers placed restrictions on the lend-lease loans to the Chinese. At the end of the war, the Chinese were only able to use half of the loans.

44. "Summary Sheets for N.R.C. Order No. Machineries and Materials to be Purchased from U.S.A. for Central Electrical Manufacturing Works National Resources Commission," April 1942, NRC Papers, 003-020500-0222-0026a, AH, Taipei, Taiwan.

45. Guo Dewen and Sun Keming, "Kangzhan banian lai," 126.

46. "Summary Sheets," NRC Papers, 003-020500-0222-0012a and 003-020500-0222-0027a, AH, Taipei, Taiwan.

47. Zhang Baichun, "Yun Zhen Xiansheng," 79.

48. National Resources Commission, "Letters and Purchase Orders of Central Electrical Manufacturing Works," 1942–1947, NRC Papers, 003-020500-0222, AH, Taipei, Taiwan; National Resources Commission, "Zhongyang diangong qicaichang: Ying daikuan dinggou qicai an" [Central Electrical Manufacturing Works: Equipment purchased with British loans], September 1942 to November 1943, 24-16-03-6-2, NRC Papers, AS, Taipei, Taiwan.

49. "Zai Yin qicai," *Ziyuan weiyuanhui gongbao* 4:2 (February 1943): 22–23.

50. Letter from L. F. Chen to National Resources Commission (Chongqing), August 31, 1942, 003-020500-0222-0269a, NRC Papers, AH, Taipei, Taiwan.

51. Letter from L. F. Chen to Universal Trading Company, November 12, 1942, 003-020500-0222-0269a, NRC Papers, AH, Taipei, Taiwan.

52. "Machineries and Materials to be Purchased from England N.R.C. Electrical Manufacturing Works N.R.C. List 304," 1942, 24-16-03-6-2, NRC Papers, AS, Taipei, Taiwan.

53. Letter from L. Hammond & Co. Ltd. to China Purchasing Agency (London), February 23, 1943, 24-16-03-6-2, NRC Papers, AS, Taipei, Taiwan.

54. Letter from Pekin Syndicate to Resources Commission, September 18, 1942, 24-16-03-6-2, NRC Papers, AS, Taipei, Taiwan.

55. Letter from Pekin Syndicate to H. C. Hsia of the Ministry of Economic Affairs Office in Calcutta, October 14, 1942, 24-16-03-6-2, NRC Papers, AS, Taipei, Taiwan.

56. Central Electric Corporation, *Shizhounian jiniance* [Tenth anniversary commemorative booklet] 1948, 3.

57. Lassman, "Industrial Research Transformed." Lassman's article looks at how Edward Condon, a theoretical physicist, introduced research in fundamental research to Westinghouse in 1935.

58. "Ziyuan weiyuanhui zhongyang diangong qicaichang zongbanshichu banshi xize" [Administrative details for the head office of the National Resources Commission Central Electrical Manufacturing Works], 1942, 003-010101-0151-0063x, NRC Papers, AH, Taipei, Taiwan.

59. Kunming Copper Smelting Plant to National Resources Secretariat, Notification of successful replication of tungsten powder, May 22, 1943, 003-010201-0157-0052a~003-010201-0157-0056a, NRC Papers, AH, Taipei, Taiwan.

60. Central Radio Manufacturing Works to National Resources Commission, Patent Application, February 1944, 003-010305-0087-0024a, NRC Papers, AH, Taipei, Taiwan.

61. "Gongye jishu gaijin," 48–49.

62. Ge Zuhui, "Ziyuan weiyuanhui," 51–52.

63. Zhang Chenghu, "Guoying dianli," 216.

64. Zhang Chenghu, "Guoying dianli," 214.

65. Zhang Chenghu, "Guoying dianli," 214.

66. Ministry of Economic Affairs to National Resources Commission, "Jingjibu guanyu 1945 shangban nian gongzuo baogao" [The Ministry of Economic Affairs work report on the first half of 1945], July 1945, *Zhonghua minguo dangan ziliao huibian*, part v, vol. 2, 401.

67. Zhang Chenghu, "Guoying dianli," 213.

68. Zhang Chenghu, "Guoying dianli," 216.

69. Yun Zhen, "Dianqi gongye jianshe," 41–42.

70. "Comparison of Voltage Standards," October 1941, 003-020700-0118-0160a, 003-020700-0118-0186a, 003-020700-0118-0209a, NRC Papers, AH, Taipei, Taiwan.

71. Wang Shoutai, "Guanyu zhiding."

72. "Dianqi gongye biaozhun," 2.

73. Zhang Baichun, ed., "Kangzhan qian de diangong jishu," 174–175.

74. Letter from Robert Russell of Westinghouse to Universal Trading Company, October 23, 1941, 003-010310-0032-0009a, NRC Papers, AH, Taipei, Taiwan.

75. Westinghouse Electric International Company, "A Plan for the Development of the Electrical Manufacturing Industry of China Presented to the Republic of China," March 1944, box 38, folder 8, Westinghouse Papers, Heinz Historical Center, Pittsburgh, PA, United States.

76. S. V. Falinsky to Yun Zhen, Letter, May 22, 1942, 003-020700-0118-0145a, NRC Papers, AH, Taipei, Taiwan.

77. Yun Zhen to S. V. Falinsky, September 24, 1942, 003-020600-1968-0008a, NRC Papers, AH, Taipei, Taiwan.

78. C. A Powel, "American Standards Association Letter to J. W. McNair on Voltage Standards in China," January 14, 1943, 003-020700-0118-0202a, 003-020700-0118-0196a, NRC Papers, AH, Taipei, Taiwan.

79. A. C. Monteith to Chen Liangfu, Letter, December 8, 1943, 003-020700-0118-0041a, NRC Papers, AH, Taipei, Taiwan.

80. Chen Liangfu to P. G. Agnew, Letter, August 2, 1943, 003-020700-0118-0125a, NRC Papers, AH, Taipei, Taiwan. Chen is citing the fable of *xuezu shilü* (Cut the feet to fit one's shoes) from *Huainanzi*.

81. Chen Liangfu to Westinghouse, Letter, September 7, 1944, 003-020600-2377-0352a, NRC Papers, AH, Taipei, Taiwan.

82. Weng Wenhao to Chen Liangfu, Letter, September 1, 1942, 003-02000-0118-0207a, NRC Papers, AH, Taipei, Taiwan.

83. Westinghouse International, "Plan," 3.

84. Westinghouse International, "Plan," I-A-3.

85. Westinghouse International, "Plan," I-A-3.

86. Yun Zhen, "Yun Zhen zizhuan," February 19, 1944, 129-000000-087A, AH, New Taipei City, Taiwan.

87. Zhang Baichun, ed., "Kangzhan qian de diangong jishu," 195.

88. Yun Zhen to R. D. McManigal, Letter, November 1, 1944, 003-020600-2377-0328a, NRC Papers, AH, Taipei, Taiwan.

89. Contract between Central Electrical Manufacturing Works of China and Technical Advisor, November 27, 1944, 003-020600-2377-0319a, NRC Papers, AH, Taipei, Taiwan.

90. Zhang Baichun, ed., "Kangzhan qian de diangong jishu," 201–202; Letter from Yun Zhen to Universal Trading Company, December 20, 1944, 003-020600-2377-0309a, NRC Papers, AH, Taipei, Taiwan.

91. Chen Liangfu to F. Chang, Letter, December 5, 1944, 003-020600-2377-0310a, NRC Papers, AH, Taipei, Taiwan.

92. Zhang Baichun, ed., "Kangzhan qian de diangong jishu," 199.

93. Zhang Baichun, ed., "Kangzhan qian de diangong jishu," 199.

94. Xu Ying, "Interview with Yun Zhen," *Ta Kung Pao*, July 1, 1948.

95. Zhang Chenghu, "Guoying dianli," 212.

96. Xu Ying, "Interview with Yun Zhen," *Ta Kung Pao*, July 1, 1948.

97. Yun Zhen, "Dianli jianshe zhi fangzhen."

98. Steffen, Crutzen, and McNeill, "The Anthropocene," 618.

Chapter 5

1. Fleming, "Damming the Yangtze Gorge." Accessed from box 4, John Lucian Savage Papers, AHC, Laramie, WY.

2. J. L. Savage, "Excerpts from Preliminary Report on Yangtze Gorges Project, Chungking, China," November 9, 1944, RG 115, Yangtze Gorge Project (1943–1949), box 10, NARA (Denver), 2.

3. "Yangtze Dam Would Make Boulder Dam Look Like Mud Pie," March 13, 1945, *Christian Science Monitor*. Accessed through RG 115, Yangtze Gorges Project (1943–1949), box 9, NARA (Denver).

4. Savage, "Excerpts from Preliminary Report," 12. See also Byrnes, *Fixing Landscape*, 93.

5. Cheng Yu-feng and Cheng Yu-huang eds., *Ziyuan weiyuanhui jishu renyuan fu mei shixi shiliao*, 163. This is a compilation that faithfully reproduces the reports and documents of the engineers sent for advanced training in the United States. Hereafter referred to as FMSXSL.

6. Ekbladh, *The Great American Mission*, 51.

7. Sneddon, *Concrete Revolution*, 28.

8. Ekbladh, *The Great American Mission*, 48. Ekbladh points out that the TVA presented itself as a model of liberal development and promoted itself as such to the visiting Chinese engineers. Ekbladh largely relies on the papers of David Lilienthal and other TVA officials.

9. "Hu Shih Discounts U.S. Dictator Fear," September 20, 1940, *New York Times*, 15. Clipping in box 94, David Lilienthal Papers, SGML, Princeton, NJ.

10. Lilienthal to Hu Shi, October 1, 1940, box 94, David E. Lilienthal Papers, SGML, Princeton, NJ.

11. Huang Hui to Lilienthal, circa July 1943, box 103, David E. Lilienthal Papers, SGML, Princeton, NJ.

12. FMSXSL, 61.

13. FMSXSL, 59.

14. FMSXSL, 56.

15. Yun Zhen, "Sun Yunxuan xiansheng zai dalu qijian."

16. Scholars in China have recognized the significance of the engineering mission in the TVA and produced voluminous accounts of the event. The most accurate overview of the event appears in Lin Lanfang, "Zhanhou chuqi ziyuan weiyuanhui." Lin uses the events of the engineering mission as background to understand why the Nationalists in Taiwan continued to employ Japanese engineers during the initial takeover.

17. FMSXSL, 162–165.

18. FMSXSL, 163.

19. Sun Yun-suan, "Travel Report: Traveling from Chung-king to New-York City U.S.A.," March 6, 1943 to May 31, 1943, 003-020300-0791-0158, 003-020300-0791-0159, NRC Papers, AH, Taipei, Taiwan; Cheng Yufeng, "Sun Yunxuan shixi riji" [Sun Yunxuan's diary on his advanced training], in *Wo suo renshi de Sun Yunxuan*, 68–76. Cheng Yufeng interviewed Sun Yun-suan during the 1980s as part of a larger project to compile sources for Academia Historica. In one of the interviews, Sun's wife showed Sun Yun-suan's diaries to Cheng. Cheng published a selection of the diaries in a commemorative volume for Sun's eightieth birthday.

20. Cheng Yufeng, "Sun Yunxuan shixi riji," 73.

21. Mitchell, *Rule of Experts*, 19. Timothy Mitchell notes that the anopheles gambiae had already invaded Egypt around the summer of 1942.

22. Sun Yun-suan diary, digital reproduction at Sun Yun-suan Memorial, Taipei, accessed on July 7, 2017.

23. Sun Yun-suan, "Travel Report."

24. Zhang Guangdou to Chen Liangfu, Letter, May 20, 1943, 003-020600-1076-0068, NRC Papers, AH, Taipei, Taiwan.

25. *FMSXSL*, 28.

26. *FMSXSL*, 410.

27. *FMSXSL*, 725.

28. *FMSXSL*, 743.

29. *FMSXSL*, 103.

30. *FMSXSL*, 547.

31. Zhang Guangdou, "Memorandum to Chen Liangfu," July 18, 1943, 003-020700-00885-0072a, NRC Papers, AH, Taipei, Taiwan.

32. Yu En-ying, "Wo de jidian jiyi" [My few recollections], in *Wo suo renshi de Sun Yunxuan*, 64. Yu En-ying was Sun Yun-suan's brother in law. He introduced Sun Yun-suan to his sister after they returned to China after the war.

33. Sun Yunxuan, "Letter to Chen Haomin (Chen Zhongxi)," August 23, 1944, 003-010101-0907-0067a, NRC Papers, AH, Taipei, Taiwan.

34. FMSXSL, 513.

35. David Lilienthal to Chen Liangfu, Letter, October 13, 1943, 003-020700-0885-0085a, NRC Papers, AH, Taipei, Taiwan.

36. FMSXSL, 732.

37. FMSXSL, 797.

38. FMSXSL, 746.

39. FMSXSL, 747.

40. FMSXSL, 792–798.

41. FMSXSL, 531; see also Lin Lanfang, "Zhanhou chuqi Ziyuan weiyuanhui," 98.

42. Liu Jinyu letter to NRC Technical Office, June 1944, 003-020100-0150-0038a, NRC Papers, AH, Taipei, Taiwan.

43. Liu Jinyu letter to NRC Technical Office, August 1944, 003-020100-0150-0024a, NRC Papers, AH, Taipei, Taiwan.

44. Arthur Davies to L. F. Chen, Letter, November 16, 1944, 003-020400-0176-0076a, NRC Papers, AH, Taipei, Taiwan.

45. "Map and Profile of the Tennessee River System," December 1944, 003-020400-0176-0022a, NRC Papers, AH, Taipei, Taiwan.

46. Yun Zhen to David Lilienthal, Letter, December 21, 1944, 003-020400-0176-0011a, NRC Papers, AH, Taipei, Taiwan.

47. G. H. Shaw, letter to John L. Savage, December 13, 1943, RG 115, Yangtze Gorge Project (1943–1949), box 3, NARA (Denver). Also cited in Sneddon, *Concrete Revolution*, 37. C.V. Davis, project design engineer of TVA, also expressed interest in working in China. See Xiangli Ding, "Transforming Waters," 68.

48. Sneddon, *Concrete Revolution*, 36.

49. "Post-War Plans," September 27, 1943, 003-020700-0885-0085a, NRC Papers, AH, Taipei, Taiwan.

50. "Post-War Plans," September 27, 1943.

51. Letter by J. L. Savage to Zhang Guangdou, October 5, 1943, 003-020700-0885-0095a, NRC Papers, AH, Taipei, Taiwan.

52. Chen Liangfu to J. L. Savage, Letter, October 13, 1943, 003-020700-0885-0092, NRC Papers, AH, Taipei, Taiwan.

53. FMSXSL, 230. See also Lin Lanfang, "Zhanhou chuqi Ziyuan weiyuanhui," 97.

54. Zhang Guangdou memorandum to Chen Liangfu, March 22, 1944, 003-020600-1076-0157a, NRC Papers, AH, Taipei, Taiwan.

55. C. E. Blee to Zhang Guangdou, October 6, 1944, 003-020600-1076-0098a, NRC Papers, AH, Taipei, Taiwan.

56. Zeng Zhaojian, "Ji Sanfanqi xiansheng."

57. Fleming, "Damming the Yangtze Gorge," 101; "Who's Who in the News," *Current Events*, March 25–29, 1946. Accessed from box 4, John Lucian Savage Papers, AHC, Laramie, WY.

58. Min Jiangyue, "Safanqi kaocha changjiang sanxia qianhou." In 1997, the Chinese People's Political Consultative Conference published a collection of archival materials about the history of the Three Gorges. The compilers constructed a chronology of events from documents in the Number Two Historical Archives.

59. Yun Zhen, "Ziyuan weiyuanhui yu meiguo."

60. J. L. Savage, "Excerpts from Preliminary Report on Yangtze Gorges Project, Chungking, China," November 9, 1944, RG 115, Yangtze Gorge Project (1943–1949), box 10, NARA (Denver); Savage, "Yangzijiang sanxia jihua chubu baogao (shang)," 27. The editors translated and republished the Chinese version of Savage's report filed in the Number Two Historical Archives.

61. J. L. Savage Preliminary Report, November 9, 1944, RG 115, box 10, NARA (Denver). Also cited in Sneddon, *Concrete Revolution*, 40.

62. John Reed Paschal, "The Building of Hydroelectric Power Plants in China under US Government Loan and a Suggested Mode of Repayment Therefor," August 28, 1944, 003-020100-0217-0004a, NRC Papers, AH, Taipei, Taiwan.

63. Savage, "Yangzijiang sanxia jihua chubu baogao (xia)," 41.

64. Savage, "Yangzijiang sanxia jihua chubu baogao (xia)," 43.

65. J. L. Savage Preliminary Report, November 9, 1944, RG 115, box 10, NARA (Denver).

66. Drawing for Preliminary Report on Ta-Tu-Ho and Ma-Pien-Ho Projects, Upper Ming-Kiang and Kwan-Hsien Projects, Lung-chi-ho Projects, Tang-Lang-Chuan Projects, September 10, 1945, John Lucian Savage Papers, 02852, box 6, American Heritage Center, Laramie, WY. The drawings were dated between September and November 1944 with a letter of transmittal dated September 10, 1945.

67. John Lucian Savage, Preliminary Report on Ta-Tu-Ho and Ma-Pien-Ho Projects, November 9, 1944, 003-020200-0501, NRC Papers, AH, Taipei, Taiwan.

68. Memorandum for Assistant Secretary Strauss, "Chronology of Events Relating to the Proposed Agreement between the Bureau of Reclamation and the National Resources Commission of China for Preparation of the Yangtze Gorges and Tributary Reports," June 6, 1945, 1, RG 115, Yangtze Gorges Project (1943–1949), box 9, NARA (Denver).

69. Sneddon, *Concrete Revolution*, 41.

70. Minutes of Conference at Bureau of Reclamation, March 10, 1945, RG 115, Yangtze Gorges Project (1943–1949), box 9, NARA (Denver).

71. Memo from Strauss to Secretary Chapman, May 21, 1945, RG 115, Yangtze Gorges Project (1943–1949), box 9, NARA (Denver).

72. Department of the Interior Advance Release, May 9, 1945, RG 115, Yangtze Gorges Project (1943–1949), box 9, NARA (Denver).

73. Wang Shoujing to John S. Cotton, Employment Contract, December 9, 1944, 003-020100-0211-0140a, NRC Papers, AH, Taipei, Taiwan.

74. Wang Shoujing letter to John S. Cotton, March 12, 1945, 003-020400-0053-0116a, NRC Papers, AH, Taipei, Taiwan.

75. Letter from Victor Kwong to W. E Dwyer, March 23, 1945, 003-020100-0211-0082a, NRC Papers, AH, Taipei, Taiwan.

76. Letter from Wang Shoujing to Mrs. Cotton, April 10, 1945, 003-020100-0211-0006a, NRC Papers, AH, Taipei, Taiwan.

77. Tentative Outline, May 18, 1945, 003-020100-0211-0056a, NRC Papers, AH, Taipei, Taiwan.

78. Memorandum Report on Power Development in the Vicinity of I-chang, August 20, 1945, ACC 6499, box 93, John S. Cotton Collection, AHC, Laramie, WY.

79. Huang Yuxian letter to Weng Wenhao and Qian Changzhao, June 27, 1945, 003-010602-0164, NRC Papers, AH, Taipei, Taiwan.

80. Aviation Commission letter to Weng Wenhao, August 22, 1945, 003-020200-0682-0008a, NRC Papers, AH, Taipei, Taiwan.

81. TVA letter to John Cotton, October 22, 1945, 003-020200-0682-0026a, NRC Papers, AH, Taipei, Taiwan.

82. Status Report Yangtze Gorges and Tributary Projects, August 15, 1947, 1–2, ACC 6499, box 28, John S. Cotton Collection, AHC, Laramie, WY.

83. Savage letter to Qian Changzhao, May 17, 1946, RG 115, Project Plans and Feasibility Reports Relating to Yangtze Gorge Project, box 2, NARA (Denver).

84. Min Jiangyue, "Safanqi kaocha,"11.

85. National Hydroelectric Engineering Bureau, "Planning Report on the Lung Chi Ho Development and in particular the Initial Development Lion Rapid Project Upper Tsing Yuan Tung Project," September 28, 1945, 003-020200-0152, NRC Papers, AH, Taipei, Taiwan. See Ding, "Transforming Waters," 57-82 for development of hydropower in wartime Chongqing.

86. Zhou Zhirou petition, October 3, 1945, NRC Papers, 003-010309-0370, NRC Papers, AH, Taipei, Taiwan.

87. Huang Yuxian report, December 22, 1945, 003-010309-0370 NRC Papers, AH, Taipei, Taiwan.

88. Resources Commission memorandum on flooding incident, May 10, 1946, 003-010309-0370, NRC Papers, AH, Taipei, Taiwan.

89. National Hydroelectric Engineering Bureau, "Invitations for Bids, Schedules, Specifications and Drawings: Hydraulic Turbines, Governors, and Butterfly Valves for Lion Rapids Plant and Hydraulic Turbines and Governor for Upper Tsing Yuan Tung Plant," fourth quarter of 1945, 003-020200-0125, NRC Papers, AH, Taipei, Taiwan.

90. National Resources Commission memorandum, January 14, 1946, 003-010602-0078, NRC Papers, AH, Taipei, Taiwan.

91. Savage letter to Qian Changzhao, May 17, 1946, RG 115, Project Plans and Feasibility Reports Relating to Yangtze Gorge Project, box 2, NARA (Denver).

92. Corfitzen, "It Works Both Ways."

93. Savage letter to Qian Changzhao, May 17, 1946, RG 115, Feasibility Reports, box 2, NARA (Denver).

94. Sun Danchen, "Safanqi yu Li Bing fuzi."

95. "YVA jihua de shijian."

96. Ekbladh, *The Great American Mission*, 49.

97. Walker Young letter to W. H. Halder et al., April 2, 1946, RG 115, Project Plans and Feasibility Reports Relating to Yangtze Gorge Project, box 2, NARA (Denver).

98. C. Y. Pan to Savage, June 13, 1946, RG 115, Project Plans and Feasibility Reports Relating to Yangtze Gorge Project, box 2, NARA (Denver).

99. Status Report Yangtze Gorges and Tributary Projects, August 15, 1947, ii, ACC 6499, box 28, John S. Cotton Collection, AHC, Laramie, WY.

100. W. C. Beatty to R. E. Krueger, "Yangtze Power Plant-Estimate of Number of Units," August 26, 1945, RG 115, Project Plans and Feasibility Reports Relating to Yangtze Gorge Project, box 1, NARA (Denver).

101. "Yangtze Power Plant Turbine Study," January 3, 1947, RG 115, Project Plans and Feasibility Reports Relating to Yangtze Gorge Project, box 23, NARA (Denver).

102. General Cable Corporation to L. M. McClellan, December 3, 1946, RG 115, Project Plans and Feasibility Reports Relating to Yangtze Gorge Project, box 2, NARA (Denver).

103. Hou Defeng et al., "Yangzijiang fadian gongcheng dizhi wenti zhi jiantao."

104. E. W. Lane, "Preliminary estimate of sediment load in Yangtze River," July 3, 1947, RG 115, Project Plans and Feasibility Reports Relating to Yangtze Gorge Project, box 5, NARA (Denver).

105. Cost Estimate from W. G. Beatty to J. J. Hammond, May 6, 1947, RG 115, Project Plans and Feasibility Reports Relating to Yangtze Gorge Project, box 4, NARA (Denver).

106. Financial Study of the Power Production Plant Yangtze Gorges Project by Wan-Shih Feng, May 12 1947, RG 115, Project Plans and Feasibility Reports Relating to Yangtze Gorge Project, box 2, NARA (Denver).

107. The Dajia River project proposed in 1941 cost 875 yen per kW (206.85 US dollars) in 1940 prices.

108. John S. Cotton to Qian Changzhao, Telegraphic Wire, March 13, 1947, 003-020200-0683-0052a, NRC Papers, AH, Taipei, Taiwan.

109. Status Report Yangtze Gorges and Tributary Projects, August 15, 1947, 1–2, ACC 6499, box 28, John S. Cotton Collection, AHC, Laramie, WY.

110. "Billion Dollar Engineer," March 28, 1948, *Rocky Mountain Empire Magazine*, 8.

111. Cotton, "Yangzi jiang sanxia shuilifadian jihua (futu)."

112. Hsu Huai-yun to Zhang Guangdou, Letter, April 26, 1948, 003-020200-0689-0015a, NRC Papers, AH, Taipei, Taiwan.

113. Hydroelectric bureau headcount, August 25, 1947, 003-020200-0682-0026a, NRC Papers, AH, Taipei, Taiwan.

114. John S. Cotton letter to Huang Yuxian, September 27, 1946, 003-010201-0087, NRC Papers, AH, Taipei, Taiwan.

115. Memoranda on shipment of four turbines, November 11, 1946, 003-010602-0078, NRC Papers, AH, Taipei, Taiwan.

116. Concrete Testing Report, September 1947, 003-010201-0087, NRC Papers, AH, Taipei, Taiwan.

117. Work Report of the National Hydroelectric Engineering Bureau, July 1948, 003-010301-0538-0081a, NRC Papers, AH, Taipei, Taiwan.

118. Planning Report on Prospective Wong Kiang Project, April 1948, ACC 6499, box 95, John S. Cotton Collection, AHC, Laramie, WY.

119. John Lucian Savage was consulting engineer for hydropower projects such as Shimen in Taoyuan, Chukeng near Taipei, Tongmen in Hualian, and Dajian along the Dajia River in Taichung. See box 6, John Lucien Savage Papers, AHC, Laramie, WY.

Chapter 6

1. Pauley, *Report on Japanese Assets*, 54–71.

2. Pauley, *Report on Japanese Assets*, 54–55.

3. Cited in "Ranliao gongyebu."

4. Yu Depei, ed., *Taiwan dianyuan kaifashi*, 3–70. In an oral interview completed in Bethesda, Maryland, on August 12, 1996, Huang Hui vouched for Liu Jinyu's innocence. He confronted Premier Chen Cheng and denied that Liu had any connections with the Communist Party. He later claimed that he had health issues that prevented him from taking over as Taipower's general manager but still reluctantly accepted the appointment. The inclusion of Huang's interview on the Liu incident suggests that Taipower executives knew that Liu's execution was a miscarriage of justice.

5. Xu Ying, *Dangdai Zhongguo shiye*, 191.

6. "Manchuria (including Jehol) Principal Electric Power Facilities as of July 1945," March 1946, box 4, John Lucian Savage Papers, AHC, Laramie, WY. See also Pepper, *Civil War in China*, 213–216.

7. Leshan shi dang'an guan [Leshan Municipal Archives], "Sun Yunxuan zuan."

8. Lin Lanfang, "Zhanhou chuqi ziyuan weiyuanhui."

9. Sun Yun-suan CV, circa 1964, 128-0000595-0006, Personal Papers, AH, New Taipei City, Taiwan.

10. Sun Yun-suan eulogy to his mother, 1964, 128-000595-0002, Personal Papers, AH, New Taipei City, Taiwan.

11. Liu Jinyu, Letter to Chen Zhongxi, January 6, 1946, in *Guancang minguo Taiwan*, ed. Chen Yunlin et al., 170–171.

12. Huang Hui, Qiu Sujun, and Sun Yun-suan, "Taiwan zhi dongli" [Taiwan's power industry], 1947, 003-020200-0374-0038a, NRC Papers, AH, Taipei, Taiwan.

13. Electrical Power Industry in Taiwan, October 1946, 003-020200-0374-0004a, NRC Papers, AH, Taipei, Taiwan. The engineers still used Japanese place names in the documents. Songshan, which is near Songshan airport in Taipei was referred to as Matsuyama, Kao-hsiung as Takao, and Sun-Moon Lake as Jitsugetsudan.

14. Activities of the Taiwan Power Company in the Past, Present and Future, May 1946, 5.

15. J. G. White Engineering Corporation, 1946, "Survey of the Facilities of the Taiwan Electric Company, Ltd. Formosa: Report to National Resources Commission of

China," 207, Lin Pin-yen Collection, Academia Sinica, Taipei, Taiwan. Substations were installed outdoors without any protective housing, making them easy targets for aerial bombardment.

16. "Taiwan zhi dongli" [Taiwan's power industry], October 1946, 003-020200-0374-0010a, NRC Papers, AH, Taipei, Taiwan; Activities of the Taiwan Power Company in the Past, Present and Future, May 1946, 25.

17. Liu Jinyu, Report on Chiang Kai-shek's inspection of Sun-Moon Lake, October 25, 1946, 003-010700-0076, NRC Papers, AH, Taipei, Taiwan.

18. "Taiwan zhi dongli," 003-020200-0374-0026a, NRC Papers, AH, Taipei, Taiwan.

19. Lin pin-yen, *Taiwan jingyan de kaiduan*, 177–187.

20. "Manchuria (including Jehol) Principal Electric Power Facilities as of July 1945," March 1946, box 4, John Lucian Savage Papers, AHC, Laramie, WY. According to the March 1946 addendum, the Pauley mission did not directly observe all the removals by the Soviets. The mission's coverage was also limited, as it omitted a number of power stations amounting to a generating capacity of 404,000 kW in their initial calculations. The extent of Soviet removals is unclear. Robert Carin's 1964 report underestimated the prewar generating levels and concluded that Soviet removals led to a 26.7 percent drop in generating capacity. See Carin, *Power Industry in Communist China*, 15.

21. "National Resources Commission Report on the Sungari (Ta-Fung-Man) Multiple Purpose Project," November 1946, ACC 6499, box 95, John S. Cotton Collection, AHC, Laramie, WY.

22. Pepper, *Civil War in China*, 215.

23. Li Nanyang, ed., *Fumu zuori shu*, 433. Li Rui was persecuted after the Lushan plenum and sent for twenty years of hard labor. His letters and diaries were thought to be destroyed during the Cultural Revolution. Some time in the late 1990s, Li Rui's daughter Li Nanyang found her parents' diaries and letters in a safe in the basement of the Ministry of Hydraulic Power in Beijing. She went on to write a biography of her parents and published their letters and diaries.

24. Xu Ying, *Dangdai Zhongguo shiye*, 191–202. The biographies of these industrialists were published in the *Dagongbao* newspaper and *Xin Zhonghua* semi-monthly magazine of the Zhonghua bookstore and circulated widely. Li Rui was also in close contact with Xu Ying. By 1948, the *Dagongbao* abandoned its position of neutrality and openly expressed its support for the Communists.

25. Xu Ying, *Dangdai Zhongguo shiye*, 198.

26. Weng Wenhao letter to W. S. Finley, November 26, 1947, 003-020400-0321-0014a, NRC Papers, AH, Taipei, Taiwan.

27. Xu Ying, *Dangdai Zhongguo shiye*, 191.

28. Xu Ying, *Dangdai Zhongguo shiye*, 193–194.

29. Company records of North Hebei Electric Company, May 1946 to July 1946, 18-31-02-066-02, Ministry of Economic Affairs Papers, AS, Taipei, Taiwan.

30. Company Records, December 1946, J006-001-00326, North Hebei Electric Company Papers, BMA, Beijing, China.

31. Balance sheet for period between March 1, 1946, and October 31, 1946, February 19, 1947, 003-010101-0748-0016x, NRC Papers, AH, Taipei, Taiwan. The balance sheet was attached to the documents for the handover of the company from Zhang Jiazhi to Bao Guobao. The revenue stands at 10,891,810.71 CNC dollars, expenses 3,919,293,108.22 CNC dollars, and loss after cost allocation 3,194,988,760.05 CNC dollars. The exchange rate of 700 CNC dollars to one US dollar published in a Chongqing newspaper is cited in Pepper, *Civil War in China*, 34.

32. Xiong Bin to National Resources Commission, August 1, 1946, 003-010303-0508-0043x, NRC Papers, AH, Taipei, Taiwan.

33. "Ziyuan weiyuanhui suo shu Tianjin ge gongchang" [Detailed table on the arrears owed by the National Resources Commission's factories at Tianjin], December 3, 1946, 003-010303-0508-0022x, NRC Papers, AH, Taipei, Taiwan.

34. Rolling outage roster, November 15, 1946, 003-010303-0508-0240x, NRC Papers, AH, Taipei, Taiwan.

35. Central Machinery Works in Tianjin to National Resources Commission, December 1946, 003-010303-0508-0215x, NRC Papers, AH, Taipei, Taiwan.

36. "Jibei dianli gongsi Pingjintangcha."

37. "Ziyuan weiyuanhui zhongyang jiqi youxian gongsi jieqian Jibei dianli gongsi dianli fei mingxi biao [Detailed table on the arrears of electric power bills owed by the Central Machinery Corporation to the North Hebei Power Company], February 1947, 003-010303-0508-0032x, NRC Papers, AH, Taipei, Taiwan.

38. Shanghai Power Company Report on Utility Rates, October 14, 1947, Q5-3-4662, Shanghai Public Utilities Bureau Papers, SMA, Shanghai, China.

39. Salary figures from Pepper, *Civil War in China*, 128.

40. Letter from Wu Guozhen to Shanghai Power Company, September 23, 1947, Q109-1-499-35, Shanghai Municipal Assembly Papers, SMA, Shanghai, China.

41. Suzanne Pepper describes Kalgan as the CCP's experiment with urban administration.

42. Zhangjiakou diangongju, ed., *Zhangjiakou dianye zhi*. The gazetteer covers the period between 1917 and 1988. The industrial gazetteer is silent about the performance of the power industries of Zhangjiakou (Kalgan) during the two-year period referred to as "The Kalgan Experiment."

43. Damage Report on Tongzhou County, February 1947, J006-003-00285, North Hebei Electric Company Papers, BMA, Beijing, China.

44. Li Daigeng, *Zhongguo dianli gongye fazhan shiliao*, 163.

45. Name list, August 1949, Shanghai CCP Public Utilities Committee Papers, A59-1-227, SMA, Shanghai, China. The nominal roll lists 102 party members, but only 48 of them were party members before August 1948. Party membership doubled after the Nationalist defeat at the Huaihai Campaign in January 1949.

46. Pepper, *Civil War in China*, 332.

47. *China: The Roots of Madness, 1967*, directed by Mel Stuart, screenplay by Theodore H. White (National Archives and Records Service, 2012).

48. Guanying, "Haerbin fadianchang" [Harbin power plant], May 1, 1948, *Renmin ribao*. Accessed from the *Renmin ribao* database. The report also states that the power plant manager was an ordinary worker and holds up the Harbin Power Plant as a model for workers' commune. For a more detailed discussion about heroic repairs, see Tan, "Repairing China's Power Grid."

49. Tan, "Repairing China's Power Grid."

50. Tan, "Repairing China's Power Grid."

51. Volti, "Worker Innovation."

52. "Heping zhixu xunsu huifu Tai-an Shuo xian huanqing jiefang" [Peace and order quickly restored Shuo county Tai-an celebrates liberation], July 14, 1946, *Renmin ribao*. Accessed from the *Renmin ribao* database.

53. "Dongbei shengchan jianshe jianxun" [Bulletin on production and reconstruction of the northeast], November 11, 1947, *Renmin ribao*.

54. "Zhengduo diandeng chang gongke Linfen zhandou tongxun" [Seize the electric plant: chronicles of the battle of Linfen], May 27, 1948, *Renmin ribao*.

55. Cited in "Ranliao gongyebu: diyi ci quanguo dianye huiyi," 209.

56. Chen Yun, "Jieshou Shenyang zhi jingyan," 375.

57. "Changchun ting zai 'liudianban zhong'" [Changchun stops at half past six], October 25, 1948, *Renmin ribao* database. I would like to thank James Gao for pointing this out in his article presented at the Civil War workshop at Yale held in May 2016.

58. Peng Zhen, "Zuohao chengshi gongzuo."

59. Wang Hongchao and Li Zhongzhi, eds., *Beijing shi dianli gongye zhi*, 457.

60. Minutes of the Safety Committee of the Central Electrical Manufacturing Works in Shanghai, February 1949–May 1949, Q452-1-19, Central Electrical Works (Kunming) Papers, SMA, Shanghai, China.

61. Senior engineers like Yun Zhen and Bao Guobao participated in the Guomindang's political indoctrination camps during the War of Anti-Japanese Resistance before they were posted to the United States. The writings and political assessments of these senior engineers were filed with the papers of the Attendants Office of the Guomindang Military Commission. They were vetted by Chiang's personal secretaries, thus indicating that the Guomindang made a concerted effort to influence their political leanings. The folders of the engineers who remained in the mainland were stamped with the label *"toufei"* (defected to the Communist bandits). The National Security Bureau on Taiwan also tried to gather information about these defected engineers but only managed to gather newspaper clippings and biographical dictionary entries.

62. Liu Yenbin, Zhou Wen, and Cha Renbo, "Woguo jiechu de dianli xitong zhuanjia."

63. Beijing gongdianju dangshiju zhi bangongshi, "Jiefang qianxi," 41.

64. Beijing gongdianju dangshiju zhi bangongshi, 39.

65. "Zhonggong zhongyang Huabeiju dui Pingjin."

66. Yingren, "Jieshou Beiping Shijingshan fadianchang" [Taking over the Shijingshan power station of Beiping], January 12, 1949, *Renmin ribao*.

67. Jiang Yan, "Beiping gongren yingjie jiefang"; Li Daigeng, *Zhongguo dianli fazhan shiliao*, 165.

68. Xu Ying, *Dangdai Zhongguo shiye*, 191–202; Xu Ying, *Beiping eryue weicheng*.

69. Xu Ying, *Beiping eryue weicheng*, 23. The diary entry from December 19, 1948, notes that the power outage began at 10 am on December 15, 1948.

70. "Jiajia gua hongdeng, yingjie Mao Zedong Beiping renmin repan Jiefangjun" [Every family hangs red lanterns to welcome Mao Zedong, the people of Beiping earnestly wait for the People's Liberation Army], January 19, 1949, *Renmin ribao*.

71. Xu Ying, *Beiping eryue weicheng*, 38.

72. Beijing gongdianju dangshiju zhi bangongshi, "Jiefang qianxi Beiping dianli gongsi," 40–41.

73. Lapwood and Lapwood, *Through the Chinese Revolution*, 46–47.

74. Peng Zhen Zhuan bianxie zu, ed., *Peng Zhen nianpu*.

75. Xu Ying, *Beiping liangyue weicheng*, 120.

76. Internal document Shanghai Public Works Department, February 3, 1949, Shanghai Garrison Command Papers, Q127-7-15, SMA, Shanghai, China. The Communist Party Committee of Shanghai found this document on August 1955, when they were sorting out the documents of the Shanghai Power Company.

77. Nanjing difangzhi bianzuan weiyuanhui, *Nanjing dianli gongye zhi*.

78. Zhang Baichun, "Kangzhan qian de diangong jishu," 224.

79. Memorandum, May 28, 1949, Q452-1-19, Central Electrical Works (Kunming) Papers, SMA, Shanghai, China.

80. Huang Yuxian Biography, January 6, 1944, 129-00000-0089A, Attendants Office of the Guomindang Military Commission Papers, AH, New Taipei City, Taiwan.

81. National Security Bureau to personnel office of Chiang Kai-shek, May 28, 1963, 129-00000-0089A, Attendants Office of the Guomindang Military Commission Papers, AH, New Taipei City, Taiwan.

82. Zhang Baichun, ed., "Kangzhan qian de diangong jishu," 233.

83. Liu's execution brings to mind the execution of Peter Palchinsky in Graham, *The Ghost of the Executed Engineer*. His execution, like that of Palchinsky, is a "cautionary tale about the fate of engineering that disregards social and human issues." Graham also notes that the execution of Palchinsky caused Soviet engineers to steer away from economic and social questions and continue to focus on narrow technical tasks. In February 2018, the Control Yuan opened an investigation into Liu Jinyu's case as an act of restorative justice, recognizing Liu as a victim of white terror.

84. Court records, July 7, 1950, B3750347701/0039/3132018/18/1/001, Ministry of Defense Judge Advocates Bureau Papers, National Archives Administration, New Taipei City, Taiwan.

85. Xu Junrong, Huang Zhiming, eds., *Baise kongbu mimi dang'an*, 101–113.

86. Xu Junrong, Huang Zhiming, 102.

87. Xu Junrong, Huang Zhiming, 103–104.

88. Taiwan Power Company, "Kan Fengshan weiwen ben gongsi ruying yuangong."

89. National Resources Commission memo to Taipower, May 3, 1950, 24-11-10-003-01-03, Taiwan Power Company Papers, AS, Taipei, Taiwan.

90. Hu Ping, *Haijiao qiying*, 131. The author cites Gong Debo's memoirs.

91. Huang Hui letter to the Resources Commission about the appointment of Sun Yunxuan, May 15, 1950, 24-11-10-003-01-17, Taiwan Power Company Papers, AS, Taipei, Taiwan.

92. Xu Junrong, Huang Zhiming, eds., *Baise kongbu mimi dang'an*, 110.

93. "Gan zuo feidie weihuzuochang Qian taidian gongsi zongjingli jinchen qiangjue" [Holding a candle to the Communist devils, former Taipower general manager executed by firing squad this morning], *Xinshengbao*, July 17, 1950. The newspaper clipping is filed with the Attendants Office of the Guomindang Military Commission papers at the Academia Historica in Taipei, along with biographical materials of all other defected engineers.

94. Zhuge Ming, "Liu Jinyu fufa jixiang" [A detailed record on the execution of Liu Jinyu], *Niusi* 78, July 23, 1950. The pseudonym is a reference to a famous third-century strategist Zhuge Kongming.

95. *Xinshengbao*, July 17, 1950.

96. Zhuge Ming, "Liu Jinyu fufa jixiang."

97. Xinshengbao, July 17, 1950, 1.

Chapter 7

1. Xu Yongfeng, "Yijiu wuling Shanghai erliu hongzha."

2. Centeno, "The New Leviathan."

3. For a more detailed discussion on the reds vs. experts issue, see Andreas, *Rise of the Red Engineers*.

4. Kasza, *The Conscription Society*, 72. Kasza sees "mass war as a central characteristic of twentieth-century politics."

5. Mao Zedong, "Ba jundui bian wei gongzuo duiwu."

6. Shanghai Power Company, "Shanghai dianli gongsi er-liu fan hongzha zongjie" [Concluding Report on anti-bombardment measures after February 6], March 15, 1950, A59-1-23, Shanghai CCP Public Utilities Committee Papers, SMA, Shanghai, China.

7. Report on the Work of the Trade Union after the Air Raid by the Municipal Committee Taskforce, March 15, 1950, A59-1-23, Shanghai CCP Public Utilities Committee Papers, SMA, Shanghai, China.

8. Report on the Work of the Trade Union, March 15, 1950.

9. Lou Xichen report, February 17, 1950, A59-1-22, Shanghai CCP Public Utilities Committee Papers, SMA, Shanghai, China.

10. Shanghai Power Company Concluding Report, March 15, 1950.

11. Shanghai Power Company Concluding Report, March 15, 1950.

12. Overall Report, March 15, 1950, A59-1-23, Shanghai CCP Public Utilities Committee Papers, SMA, Shanghai, China.

13. Report on the Work of the Trade, March 15, 1950.

14. Public Utilities Party Committee Report on the immediate dispatch of sentries for primary Substations, October 28, 1950, A59-1-40-17, Shanghai CCP Public Utilities Committee Papers, SMA, Shanghai, China.

15. Report by the Public Utilities Municipal Party Committee, November 22, 1950, A59-1-32, Shanghai CCP Public Utilities Committee Papers, SMA, Shanghai, China.

16. "Zhengwu yuan guanyu guanzhi Meiguo zai hua caichan dongjie zai hua cun kuan de mingling" [Government Administration Council's order on the regulation of American assets in China and freezing of American savings in China], December 18, 1950, in *Zhonggong dangshi*, vol. 7, ed. Zhonggong zhongyang dangxiao dangshi jiaoyuanshi, 178.

17. Kajima Jun, "Shanhai denryoku," 102–103.

18. In March 1955, the PRC readjusted the value of the Renminbi (RMB) at a conversion rate of 10,000 to 1. The "250 million" donation by the workers was only worth RMB 25,000 in 1955 dollar terms. This was a small amount of money. Back in 1951, the cost of 1 kWh of electricity was 55 RMB (old) and the power station generated 2.4 million kWh of electricity daily. (Figures from *Shanghai dianli gongyezhi*, 44.) The donation was equivalent to two days of Shanghai Power Station's revenue.

19. Concluding Report on the Campaign to Resist American Aggression and Aid Korea, 1951, A59-1-13, Shanghai CCP Public Utilities Committee Papers, SMA, Shanghai, China.

20. Concluding Report on the Campaign to Resist American Aggression.

21. "Souji yongtong, jieyue yongtong" [Collect scrap copper, conserve copper], February 12, 1957, *Renmin ribao*, 2.

22. Yun Zhen speech at the copper conservation campaign, September 27, 1951, C48-2-2412, Shanghai Federation of Commerce and Industry Papers, SMA, Shanghai, China.

23. "Shanghai gongchang zhankai jieyue yongtong yundong" [Shanghai's factories launch copper conservation campaign], October 27, 1951, *Renmin ribao*.

24. Li Fengwu and Li Bin, "Liyong luo gangxian."

25. Shanghai Chinese Communist Party Finance and Economic Committee, Minutes on the Meeting on Power Supply Problems, September 7, 1951, Shanghai Finance and Economic Affairs Committee Papers, B28-2-45, SMA, Shanghai, China.

26. Minutes on the Meeting on Power Supply Problems, September 7, 1951.

27. Minutes on the Meeting on Power Supply Problems, September 7, 1951.

28. Shanghai Power Party Committee Basic Work Report for 1952, 1953, A38-2-417, Shanghai CCP Industrial Production Committee Papers, SMA, Shanghai, China.

29. Shanghai Power Party Committee Basic Work Report for 1952.

30. Gu Zhun, the deputy chairman of Shanghai Chinese Communist Party Finance and Economic Committee and the chair of the Electrical Power Committee, noted that the deputy mayor of Shanghai Pan Hannian was under pressure from Mao to pick out corrupt elements under the Three-anti Campaign and hastily impeached Cheng Wanli.

Many of those present at the September 1951 meeting, such as Xu Dixin and Gu Zhun, would later be persecuted as right-wing opportunists.

31. Shanghai shi dianli gongyeju shi zhi bianzuan weiyuanhui, ed., *Shanghai shi dianli gongye zhi*, 372. The military also contributed to the accelerated growth of railroads. See Köll, *Railroads*, 240–246 for the role of the PLA's railroad army corps in the construction of railroad lines.

32. Shanghai Power Party Committee Basic Work Report for 1952.

33. Letter from Li Daigeng to Public Utilities Party Committee, June 24, 1952, A59-1-147, Shanghai CCP Public Utilities Committee Papers, SMA, Shanghai, China.

34. Gao Ming, "Yijiusiwu zhi yijiuliuwu Shanghai dianli gongye yanjiu," 106. Gao Ming points out that Shanghai only started increasing the generating capacity after 1957, after the city expanded chemical and metal-processing industries.

35. Minutes for the fourth board meeting of the (reincorporated) Nanshi power station, July 2, 1955, Chinese Merchants' Electric Company Papers, Q578-2-1594, SMA, Shanghai, China.

36. Zhongyang renmin zhengfu ranliao gongyebu jihua si, ed., *Dianye jihua jiangxiban*, 70–80.

37. Proposal on the power adjustment plan for Shanghai's industries, June 27, 1951, B1-2-1465, Shanghai Municipal People's Government Papers, SMA, Shanghai, China.

38. Minutes of the Shanghai Chinese Communist Party Finance and Economic Committee meeting, September 7, 1951.

39. Shanghai Power Administration plan on rational use of power forward by Shanghai Communist Party Committee, February 22, 1953, B8-2-37-12, Shanghai Municipal Commission of Urban Administration Papers, SMA, Shanghai, China.

40. Shanghai Power Administration plan on rational use of power, August 1953, B8-2-37-12, Shanghai Municipal Commission of Urban Administration Papers, Shanghai China.

41. Shanghai dianye guanli ju, "Shanghai diqu dianli xitong tiaozheng."

42. Quarterly work orders for East China Power Bureau for 2nd quarter 1954, April 30, 1954, B41-2-25, Shanghai Municipal People's Government Industrial Production Committee Papers, SMA, Shanghai, China.

43. Shanghai Power Bureau Report on Rational Use of Power, September 16, 1953, B8-2-37-12, Shanghai Municipal Commission of Urban Administration Papers, Shanghai, China.

44. Shanghai Power Administration Report on Power Conservation for 1954, date not specified, B109-5-120-209, Shanghai Materials Bureau, SMA, Shanghai, China.

45. Shanghai Power Administration, Proposal on peak load management for 1954, December 1954, Shanghai Municipal People's Government Papers, B41-2-26-16, Shanghai, SMA, China. Figures also cited in Kajima, "Shanhai denryoku," 107.

46. Shanghai Power Bureau Report on Rational Use of Power, September 16, 1953, B8-2-37-12, Shanghai Municipal Commission of Urban Administration Papers, SMA, Shanghai, China.

47. Shanghai Power Administration, Report on Power Conservation for 1954, circa 1955, B109-5-120-209 Shanghai Materials Bureau, SMA, Shanghai, China.

48. Shanghai Power Administration, Report on Power Conservation for 1954.

49. Shanghai Power Administration, Report on Power Conservation for 1954.

50. Zhabei Electric Public-Private Partnership Contract, December 25, 1952, B1-2-1463, Shanghai People's Municipal Government Papers, SMA, Shanghai, China.

51. Pudong Electric Public-Private Partnership Proposal, December 16, 1953, A38-2-277, Shanghai CCP Industrial Production Committee Papers, SMA, Shanghai, China. See also Gao Ming, "Yijiusiwu zhi yijiuliuwu Shanghai dianli gongye yanjiu," 32–49.

52. Gao, "Yijiusiwu zhi yijiuliuwu Shanghai dianli gongye yanjiu," 38–39.

53. Minutes of joint meeting between Nanshi Electric Company (reincorporated) and provisional management committee, August 23, 1954, Q578-2-1595, Chinese Merchants' Electric Company Papers, SMA, Shanghai, China.

54. Minutes of the fourth board meeting, April 5, 1955, Q578-2-1594, Chinese Merchants' Electric Company Papers, SMA, Shanghai, China.

55. Formal application for Public-Private Partnership, circa November 1954, Q578-2-1595, Chinese Merchants' Electric Company Papers, SMA, Shanghai, China.

56. Huadong dianli gongye zhi bianzuan weiyuanhui, *Huadong dianli gongye zhi*, 287.

57. *Huadong dianli gongye zhi*, 132.

58. Yun Zhen, "Dianli diangong zhuanjia Yun Zhen zishu san." The third segment of the biographical account covers Yun Zhen's career from 1953 to 1980.

59. Yun Zhen, "Dianli diangong zhuanjia Yun Zhen zishu san," 172.

60. *Huadong dianli gongye zhi*, 21.

61. Liu Lanbo, "Yikao qunzhong, tigao jishu, gaijin lingdao."

62. Liu Lanbo, "Yikao qunzhong, tigao jishu, gaijin lingdao," 4.

63. "Kewaluofu: Gaishan he tigao dongbei and huabei ge fadianchang zhi jingying he gongzuo xiaolü jianyao cuoshi 1949 nian 11 yue [Ivan Kovalyov: Basic measures to improve and enhance the operations and efficiency of power plants in the Northeast and North China, November 1949]," in *Zhonghua renmin gonghe guo jingji dang'an*, ed. Zhongguo shehui kexueyuan, 762–764.

64. *Sulian dianye zhuanjia* vol.2, 8–9.

65. "San nian li Sulian zhuanjia."

66. Zhang Tianzhu and Ren Jingye, "Tianjin dianye ju diyi fadian."

67. National Electrical Industries Work Report of 1950, March 1951, Finance and Economic Committee Papers, 1-6-446, BMA, Beijing, China.

68. National Electrical Industries Work Report of 1950, March 1951, 1-6-446, Beijing Municipal Finance and Economic Committee Papers, BMA, Beijing, China.

69. Chen Deyu and Chen Dingkun, "Jingjintang dianli wang."

70. Report on Beijing Power Supply 1949–1960, 1963, 133-4-474, Beijing Municipal Bureau of Statistics Papers, BMA, Beijing, China.

71. Beijing Power Supply, April 18, 1956, 005-002-00064, Beijing Urban Planning Committee Papers, BMA, Beijing, China.

72. National Electrical Industries Work Report of 1950, March 1951, 1-6-446, Finance and Economic Committee Papers, BMA, Beijing, China.

73. "Duiyu Shijingshan fadian."

74. Report on Beijing Power Supply, April 18, 1956, 005-002-00064, Beijing Urban Planning Committee Papers, BMA, Beijing, China.

75. Vladimir Lenin, "Report on the Work of the Council of People's Commissars."

76. Zhang Baichun, ed., "Kangzhan qian de diangong jishu," 223.

77. Party Committee report on Yun Zhen, October 1955, Q452-1-26, Central Electrical Works (Kunming) Papers, SMA, Shanghai, China.

78. Speech by Bao Guobao, May 4, 1956, Xinhua News Agency.

79. Xieersufu, *Liening Sidalin de Sulian dianqi hua*, 1. This is the Chinese translation of S. F. Shershov, *Leninsko Stalinskaia elektrifikatsiia SSSR* (Leningrad: Gosenergoizdat, 1951).

80. Liebknecht, *Karl Marx*, 57.

81. Chen Yun, "Gongsi heying zhong ying zhuyi de wenti," 294. See Cochran, ed., *The Capitalist Dilemma* for further studies on the relationship between the Communists and capitalists.

82. Zhongguo shehui kexueyuan, ed., *Zhonghua renmin gonghe guo jingji dang'an*, 954. By 1954, power stations under the command of the Ministry of Fuel Resources accounted for 81.3 percent of total electrical energy output.

83. Steffen, Crutzen, and McNeill, "The Anthropocene," 618.

Conclusion

1. Taoyuan District Prosecutor's Office, "Investigation into the Power Outage Incident at Datan Power Plant," news release, September 21, 2017, http://www.tyc.moj.gov.tw/ct.asp?xItem=488052&ctNode=14411&mp=012.

2. "Quantai da duandian pu guo'an weiji" [All-Taiwan power outage exposes national security crisis], Dai Ligang, *News Tornado*, aired August 16, 2017, CTI News.

3. Huang Shih-hsiu, "Bayiwu da tingdian de wu ge yidian" [Five questionable points of the August 15, 2018 blackout], August 23, 2018, https://cnews.com.tw/126180823a01.

4. Pearson, "Past, Present and Prospective Energy Transitions."

5. Zeng Bo et.al., "An analysis of previous blackouts."

6. Keith Bradsher, "China's Utilities Cut Energy Production Defying Beijing," *New York Times*, May 24, 2011, accessed May 18, 2020, https://www.nytimes.com/2011/05/25/business/energy-environment/25coal.html.

7. "Guojia tongji ju: Guanyu 1956 nian quanguo gongye shengchan qingkuang" [National Statistics Bureau: Industrial output in 1956], January 29, 1957, in *Zhonghua renmin gonghe guo jingji dang'an,* ed. Zhongguo shehui kexueyuan, 983–988.

8. *Huadong dianli gongye*, 15.

9. Guo-liang Luo, Yan-ling Li, Wen-jun Tang, Xiao Wei, "Wind Curtailment."

10. Zhongying Wang, Haiyan Qin, and Joanna Lewis, "China's Wind Power Industry."

11. Dahai Zhang et al., "Present Situation and Future Prospect of Renewable Energy in China."

12. Guo-liang Luo, Yan-ling Li, Wen-jun Tang, Xiao Wei, "Wind Curtailment," 1192.

13. See National Development and Reform Commission, "China's Policies and Actions on Climate Change," China Climate Change Info-Net, November 2014, http://en.ccchina.org.cn/archiver/ccchinaen/UpFile/Files/Default/20141126133727751798.pdf.

14. Guo-liang Luo, Yan-ling Li, Wen-jun Tang, Xiao Wei, "Wind Curtailment," 1194.

15. Moore, "A Friend Remembers Y. S. Sun," 111.

16. *Huadong dianye zhi*, 13.

17. Liu, Zhou Yang, and Xinzhou Qian, "China's Risky Gamble."

18. Liu, Zhou Yang, and Xinzhou Qian, "China's Risky Gamble."

19. Beeson, "The Coming of Environmental Authoritarianism," 283.

20. Minxin Pei, *China's Trapped Transition*, 9.

21. Chatterjee, "The Asian Anthropocene."

22. Oreskes and Conway, *The Collapse of Western Civilization*, 51–53.

23. Guo-liang Luo, Yan-ling Li, Wen-jun Tang, Xiao Wei, "Wind Curtailment," 1197.

24. Guo-liang Luo, Yan-ling Li, Wen-jun Tang, Xiao Wei, 1196–1198.

25. Dahai Zhang et al., "Present situation and future prospect of renewable energy in China."

26. Han Lin, *Energy Policies and Climate Change in China*, 186.

27. Li Mengyin et al., "Nengyuan zhuanxing jincheng tubiao" [Progress charts on energy transition], Do-Energy, August 20, 2019, https://doenergytw.blogspot.com/2019/06/blog-post.html?fbclid=IwAR0MkYonixEwZ5Q1auMQ5-l4LZrAc-j31aCg3tzRJJsvKFdhsqX99h3D-tZs.

28. Wang Zihao, "Han Guoyu ti nengyuan zhengce zhichi hedian" [Han Kuo-yu raises energy policy and supports nuclear power], *Events in Focus*, August 22, 2019, accessed on October 27, 2019, https://www.eventsinfocus.org/news/3372.

29. "Qianzhan jichu jianshe jihua—luneng jianshe" [Forward-Looking Infrastructure Plan—Green Energy], Executive Yuan, last modified March 23, 2017, accessed on October 27, 2019, https://www.ey.gov.tw/File/694A3325D84DAB64.

Archives

Academia Historica, Taipei, Taiwan
Academia Sinica, Archives of the Institute of Modern History, Taipei, Taiwan
Academia Sinica, Lin Pin-yen Personal Collection, Taipei, Taiwan
American Heritage Center, Laramie, Wyoming, USA
Beijing Municipal Archives, Beijing
John Heinz History Center, Pittsburgh, Pennsylvania, USA
Seeley G. Mudd Manuscript Library, Princeton, New Jersey, USA
National Archives Administration, Taipei, Taiwan
National Archives and Records Administration, College Park, Maryland, USA
National Archives and Records Administration (Denver), Broomfield, Colorado, USA
National Diet Library, Tokyo, Japan
Shanghai Municipal Archives, Shanghai, China
Sun Yun-suan Memorial Museum, Taipei, Taiwan
University of Tokyo Economics Library, Tokyo, Japan

Press Sources

Chinese

Fangzhi shibao [Textile Times]
Renmin ribao [People's Daily]
Shen Bao [Shanghai News]
Xinshengbao [Taiwan Shin Sheng Daily News]

Books, Articles, and Unpublished Manuscripts

Andreas, Joel. *Rise of the Red Engineers: The Cultural Revolution and the Origins of China's New Class.* Stanford, CA: Stanford University Press, 2009.
Ang, Yuen Yuen. *How China Escaped the Poverty Trap.* Ithaca, NY: Cornell University Press, 2016.

Argersinger, R. E. "Voltage Standardization from a Consulting Engineer's Point of View." *Transactions of the American Institute of Electrical Engineers* 46 (1927): 172–174.

Bao, Guobao. "Shoudu dianchang zhi zhengli ji kuochong" [The retrofitting and expansion of the Capital Power Station]. *Gongcheng* 4:2 (1929): 269–271.

Beckert, Sven. *Empires of Cotton: A Global History*. New York: Knopf Doubleday, 2015.

Beeson, Mark. "The Coming of Environmental Authoritarianism." *Environmental Politics* 19:2 (2010): 276–294.

Beijing gongdianju dangshiju zhi bangongshi [Beijing power administration party history office], eds. "Jiefang qianxi Beiping dianli gongsi de huchang douzheng" [The factory protection campaigns of the Beiping electric company at the eve of liberation]. *Beijing dangshi yanjiu* 6 (1989): 39–42.

Beijing shi dang'an guan [Beijing Municipal Archives]. *Beiping heping jiefang qianhou* [Archival documents on the events before and after the peaceful liberation of Beiping]. Beijing: Beijing shi dang'an guan, 1988.

Bell, Lynda S. "From Comprador to County Magistrate: Bourgeois Practice in the Wuxi County Silk Industry." In *Chinese Local Elites and Patterns of Dominance*, edited by Joseph W. Esherick and Mary Backus Rankin, 113–139. Berkeley: University of California Press, 1990.

Beltran, Alain. "Introduction: Energy in History, the History of Energy." *Journal of Energy History/Revue d'Histoire de l"Energie* 1 (December 2018). https://energyhistory.eu/en/node/84.

Bian, Morris. *The Making of the State Enterprise System in Modern China*. Cambridge, MA: Harvard University Press, 2005.

Broggi, Carles Brasó. *Trade and Technology Networks in the Chinese Textile Industry: Opening Up Before Reform*. New York: Palgrave Macmillan, 2016.

Byrnes, Corey. *Fixing Landscape: A Techno-Poetic History of China's Three Gorges*. New York: Columbia University Press, 2018.

Carin, Robert. *Power Industry in Communist China*. Hong Kong: Union Research Institute, 1969.

Centeno, Miguel Angel. "The New Leviathan: The Dynamics and Limits of Technocracy." *Theory and Society* 22:3 (1993): 307–335. https://doi.org/10.1007/bf00993531.

"Changzhou shachang jinxing jinxun" [Recent developments in Changzhou Cotton Mill]. *Wujin yuebao* 4:2 (1921): 9–11.

Chatterjee, Elizabeth. "The Asian Anthropocene: Electricity and Fossil Developmentalism." *The Journal of Asian Studies* 79:1 (February 2020): 1–22. https://doi.10.1017/S0021911819000573.

Chen, Deyu, and Chen Dingkun. "Jingjintang dianli wang zhi jingji diaodu" [Economic dispatching in the Beijing-Tianjin-Tangshan power grid]. *Renmin dianye* 8 (February 1951): 93–96.

Chen, Weiguo, et al., eds. *Zhongguo jindai mingren gupiao jiancang lu* [Record of stock certificates collections by famous people in Modern China]. Shanghai: Shanghai daxue chubanshe, 2012.

Chen, Yun. "Gongsi heying zhong ying zhuyi de wenti" [On a few problems related to public-private joint management]. In *Chen Yun wenxuan di er ji* [Selected works of Chen Yun, vol. 2], 294–297. Beijing: Renmin chubanshe, 2015.

Chen, Yun. "Jieshou Shenyang zhi jingyan" [The experience of taking over Shenyang]. In *Chen Yun wenxuan di yi ji* [Selected writings of Chen Yun, vol. 1], 374–379. Beijing: Renmin chubanshe, 2015.

Chen, Yunlin, ed. *Guancang minguo Taiwan dang'an huibian*, vol. 40 [Collection of Republican Taiwan documents by the Number Two Historical Archives]. Beijing: Jiuzhou chubanshe, 2007.

Chen, Zhigang, ed. *Xiangtan dianji chang zhi* [The annals of the Xiangtan Electrical Manufacturing Works]. Xiangtan: Xiangtan dianjichang, 1992.

Chen, Zhongxi. "Dianqi shiye yu dianqi shiye jianshe" [Electrical power industry and its establishment]. *Ziyuan weiyuanhui jikan* 4:3 (September 1944): 1–16.

Cheng, Linsun. "Lun kangri zhanzheng qian Ziyuan weiyuanhui de zhong gongye jihua" [On the NRC's heavy industrialization plan during the War of Anti-Japanese Resistance]. *Jindai shi yanjiu* 2 (1986): 38–51.

——. "Ziyuan weiyuanhui yu Zhongguo jihua jingji de qiyuan" [The NRC and the origins of China's planned economy]. *Ershiyi shiji* 4 (2004): 88–100.

Cheng, Yu-feng, and Cheng Yu-huang, eds. *Ziyuan weiyuanhui dang'an shiliao chubian* [Preliminary selection of archival materials of National Resources Commission]. Taipei: Academia Historica, 1984.

——. *Ziyuan weiyuanhui jishu renyuan fu mei shixi shiliao* [Archives on the National Resources Commission Technicians' training in the United States]. Taipei: Academia Historica, 1988. [Cited in notes as FMSXSL.]

Cheng, Yu-kwei (Zheng Youkui). *Foreign Trade and Industrial Development of China: An Historical and Integrated Analysis through 1948*. Washington, DC: The University Press of Washington DC, 1956.

Chūgoku tsūshinsha chōsa bu. *Shanhai denryoku kōshi no soshiki to jigyō* [Organization and Operations of the Shanghai Power Station]. Shanghai, 1938.

Clarke, William. "China's Electric Power Industry." In *Chinese Economy Post-Mao: A Compendium of Papers Submitted to the Joint Economic Committee*, 95th Congress, 2nd session, 1978.

Collins, H. M. *Changing Order: Replication and Induction in Scientific Practice*. Chicago: University of Chicago Press, 1992.

Coopersmith, Jonathan. *The Electrification of Russia, 1880–1926*. Ithaca, NY: Cornell University Press, 1992.

Corfitzen, William. "It Works Both Ways: Chinese Engineers in America." *Reclamation Era* 32 (November 1946): 244–245.

Cotton, John S. "Yangzi jiang sanxia shuilifadian jihua (futu)" [Plans for hydroelectric power on the Three Gorges of the Yangtze River]. Translated by Zhang Guangdou and Zhu Shulin. *Taiwan gongchengjie* 1:1 (1947): 14–15.

Denki Shinpōsha. *Hoku chūshi denki jigyō benran* [Handbook on the electrical industries in North and Central China]. Tokyo: Denki Shinpōsha, 1939.

"Dianqi gongye biaozhun" [Standardization in the electrical industries]. *Zhongguo diangong* 1:1 (July 1943): 2.

"Dianqi shiye qudi guize cao'an" [Draft regulations for the electrical industries]. *Dianye jikan* 4 (1931): 32–61.

Dikötter, Frank. *Exotic Commodities: Modern Objects and Everyday Life in Modern China*. New York: Columbia University Press, 2006.

Ding, Xiangli. "Transforming Waters: Hydroelectricity, State Making, and Social Changes in Twentieth-Century China." PhD diss., SUNY Buffalo, 2018.

Dodgen, Randall A. *Controlling the Dragon: Confucian Engineers and the Yellow River in Late Imperial China*. Honolulu: University of Hawaii Press, 2001.

"Doitsu denki jigyō no kigyō keitai" [Corporate trends in Germany's electrical power sector]. *Chō denryoku* 180 (December 1936): 1–3.

Dong, Shu, trans. "Minying dianye shi jiao guanying dianye wei jia lun" [Privately run power plants are superior to government-run power plants]. *Dianye jikan* 4:2 (1933): 1–2.

"Duiyu Shijingshan fadian chang yici fasheng zhongyao shigu de jiancha zongjie" [Conclusion of inspections following multiple serious accidents at Shijingshan Power Station]. *Renmin dianye* (May 1951): 9–16.

Duus, Peter. "Zaikabo: Japanese Cotton Mills in China, 1895–1937." In *The Japanese Informal Empire in China, 1895–1937*, 65–100. Princeton, NJ: Princeton University Press, 2014.

Edgerton, David. *The Shock of the Old: Technology and Global History since 1900*. London: Profile Books, 2006.

Ekbladh, David. *The Great American Mission: Modernization and the Construction of an American World Order*. Princeton, NJ: Princeton University Press, 2012.

Elvin, Mark. "Three Thousand Years of Unsustainable Growth: China's Environment from the Archaic Times to the Present." *East Asian History* 6 (1993): 7–46.

Esherick, Joseph W., and Mary Backus Rankin, eds. *Chinese Local Elites and Patterns of Dominance*. Berkeley: University of California Press, 1990.

Faure, David. "The Control of Equity in Chinese Firms within the Modern Sector from the Late Qing to the Early Republic." In *Chinese Business Enterprise in Asia*, edited by Rajeswary Ampalavanar Brown, 50–69. London: Routledge, 1995.

Feng, Zizai. "Zhejiang Wuxing sichouye gaikuang" [An overview of the silk industry in Wuxing Zhejiang]. *Shiye tongji* 1:3–4 (June 1933): 95–102.

Fleming, Roscoe. "Damming the Yangtze Gorge." *Popular Mechanics* 85 (March 1946): 100–102.

Fravel, M. Taylor. *Active Defense: China's Military Strategy Since 1949*. Princeton, NJ: Princeton University Press, 2019.

Freeburg, Ernest. *The Age of Edison*. New York: Penguin, 2013.

Gao, Ming. "Yijiusiwu zhi yijiuliuwu Shanghai dianli gongye yanjiu, 1945–1965" [The research on electrical industry in Shanghai, 1945–1965]. PhD diss., Shanghai Jiaotong University, 2014.

Ge, Zuhui. "Ziyuan weiyuanhui zhongyang diangong qicaichang gaikuang" [Overview of the Central Electrical Manufacturing Works of the National Resources Commission]. *Gongye qingnian* 1 (1941): 51–52.

Ghosh, Amitav. *The Great Derangement*. Chicago: University of Chicago Press, 2016.

"Gongye jishu gaijin jianxun" [Bulletins on the improvement of industrial technology]. *Ziyuan weiyuanhui gongbao* 5:6 (December 1943): 48–49.

Goto-Shibata, Harumi. *Japan and Britain in Shanghai, 1925–1931*. New York: St. Martin's Press, 1995.

Graham, Loren. *The Ghost of the Executed Engineer: Technology and the Fall of the Soviet Union*. Cambridge, MA: Harvard University Press, 1993.

Gruhl, Werner. *Imperial Japan's World War Two: 1931–1945*. Piscataway, NJ: Transaction Publishers, 2011.

Gui, Naihuang, Liu Jinyu, and Yang Guohua. "Tanglangchuan jihua ji Dianbei shuili fadian zhi zhanwang" [Praying Mantis River plan and the hydropower outlook for Northern Yunnan]. *Ziyuan weiyuanhui jikan* 4:3 (September 1944): 80–83.

Guo, Dewen, and Sun Keming. "Kangzhan banian lai zhi dianqi gongye" [The electrical equipment industry during eight years of War of Resistance]. *Ziyuan weiyuanhui jikan* [National Resources Commission Quarterly] 6: 1–2 (1946): 124–131.

Guo, Zhicheng. "Fakanci" [Foreword]. *Dianye jikan* 1 (1930): 1.

Hamilton, Clive. "The Anthropocene as Rupture." *The Anthropocene Review* 3:2 (2016): 93–106. https://doi.org/10.1177/2053019616634741.

He, Da. "Miansha fangzhi yu xiaofei dianli liang" [Power consumption of cotton milling]. *Huashang shachang lianhehui jikan* 9:1 (1931): 30–40.

Hein, Laura Elizabeth. *Fueling Growth: The Energy Revolution and Economic Policy in Postwar Japan*. Cambridge, MA: Harvard University Asia Center, 1990.

Ho, Chiah-sing. "Dongli jiqi jinkou yu jindai Zhongguo gongye hua (1910–1937)" [The import of prime movers and the industrialization of modern China (1910–1937)]. *Guoshi guan guankan* 39 (March 2014): 1–48.

Hou, Defeng, et al. "Yangzijiang fadian gongcheng dizhi wenti zhi jiantao" [Considerations on geological problems at the Yangtze Gorges site]. *Dizhi pinglun* 13:1/2 (1948): 159–160.

Howell, Sabrina. "Jiayou (Add Oil!): Chinese Energy Security Strategy." In *Energy Security Challenges for the 21st Century: A Reference Handbook*, edited by Gal Luft and Anne Koria, 191–218. Santa Barbara, CA: ABC Clio, 2009.

Hu, Ping. *Haijiao qiying: Taiwan wushi niandai hongse geming yu baise kongbu* [Flag shadow on the cape: Red revolution and white terror in 1950s Taiwan]. Nanchang: Ershiyi shiji chubanshe, 2013.

Huadong dianli gongye zhi bianzuan weiyuanhui. *Huadong dianli gongye zhi* [Gazetteer of the East China Electrical Power Industries]. Beijing: Zhongguo shuili gongye chubanshe, 1996.

Huang, Hui. "Zhongguo zhi shuili ziyuan ji shuili fadian zhi zhanwang" [The hydro-electric potential of China and its outlook]. *Ziyuan weiyuanhui jikan* 4:3 (September 1944): 18.

Huang, Xi. *Zhongguo jinxiandai dianli jishu fazhan shi* [A history of electric power technology in modern China]. Jinan: Shandong jiaoyu chubanshe, 2006.

Huang, Xiaoming. *The Rise and Fall of the East Asian Growth System, 1951–2000: Institutional Competitiveness and Rapid Economic Growth.* London: Routledge Curzon, 2004.

Huang, Yuxian. "Peitong Safanqi fukan Sanxia shuili fadian jihua baogao" [Report on accompanying Savage on a repeat visit to the Sanxia hydroelectric project]. *Hubei wenshi ziliao* 1 (1997): 18–23.

Huashang shachang lianhe hui [Association of Chinese Textile Mills]. *Zhongguo shachang yilanbiao* [Table of cotton mills in China]. Shanghai, 1933.

Hudson, Mark J. "Placing Asia in the Anthropocene: Histories, Vulnerabilities, Responses." *Journal of Asian Studies* 73:4 (November 2014): 941–962. www.jstor.com/stable/43553461.

Hughes, Thomas. "The Evolution of Large Technological Systems." In *The Science Studies Reader,* edited by Mario Biagioli, 202–224. London: Routledge, 1999.

——. *Networks of Power: Electrification in Western Society.* Baltimore, MD: Johns Hopkins University Press, 1983.

Huzhou difangzhi bianzuan weiyuanhui, ed. *Huzhou shizhi* [Huzhou Gazetteer]. Beijing: Kunlun chubanshe, 1999.

Ide Taijiro. "Shina senryō chi no denki jigyō gaikyō" [Overview of the power sector in occupied territories of North China]. *Chō denryoku* 201 (September 1938): 1–5.

Isenstadt, Sandy, Margaret Maile Petty, and Dietrich Neumann. *Cities of Light: Two Centuries of Urban Illumination.* London: Routledge, 2014.

Ishikawa, Yoshijiro. "Naka shina wa dō naruka" [What should be done with Central China]. Osaka: Daitō denki shinkō kai, 1938.

Jackson, Isabella. *Shaping Modern Shanghai: Colonialism in China's Global City.* Cambridge: Cambridge University Press, 2017.

Jarman, Robert L., ed. *Shanghai: Political and Economic Reports, 1842–1943. British Government Records from the International City,* vol. 14. London: Archive Editions, 2008.

Jiang, Yan. "Beiping gongren yingjie jiefang de douzheng" [The struggle for liberation by the workers of Beiping]. *Gonghui bolan* 1 (2002): 58–59.

Jianshe weiyuanhui [National Construction Commission]. "Jianshe: Ling fadian zhoulü biaozhun guize" [Orders on the guidelines for electrical power standards]. *Jiangsu sheng zhengfu gongbao* 573 (1930): 15.

——. "Jianweihui niding jiandu minjian dianye yuanze" [The NCC stipulates regulations to regulate private power plant operators]. *Gongshang banyuekan* 18 (1929): 17–20.

——. *Quanguo dianchang tongji* [Statistical investigation on the nation's electric industry]. Nanjing, 1933.

———. *Quanguo fadianchang diaocha biao.* Nanjing, 1929.

Jiao tong da xue xiao shi bian xian zu. *Jiaotong Daxue xiaoshi, 1896–1949 nian* [A history of Jiaotong University, 1896–1949]. Shanghai: Shanghai jiaoyu chubanshe, 1986.

"Jibei dianli gongsi Pingjintangcha gedi xiaodian liang" [North Hebei power company power sales in Beiping, Tianjin, Tangshan, Chahaer]. *Tianjin tongji jingji yuebao* 28 (1948): 36.

Joint Economic Committee. *Chinese Economy Post-Mao: A Compendium of Papers Submitted to the Joint Economic Committee*, 95th Congress, 2nd session, 1978.

Kajima, Jun. "Shanhai denryoku sangyō no tōgō to kōiki nettowāku" [The consolidation of Shanghai's electrical industries and its broader regional networks]. In *Gendai chūgoku no denryoku sangyō "fusoku no keizai to sangyō soshiki* [The electrical power industry of modern China: The economy of insufficiency and industrial organization], ed. Tajima Toshio, 91–114. Tokyo: Shōwadō, 2008.

Kale, Sunila S. *Electrifying India: Regional Political Economies of Development.* Stanford, CA: Stanford University Press, 2014.

Kanemaru, Yūichi. "Cong pohuai dao fuxing—cong jingji shi lai kan tongwang Nanjing zhi lu" [From destruction to restoration—Evaluating the road to Nanjing from economic history]. *Ritsumeikan keizaigaku* 46:4 (February 1998): 854–867.

———. "Shina jihen chokugo Nihon ni yoru kachū denryoku sangyō no chōsa to fukkyū keikaku" [The investigation and restoration of the electrical industries in Central China by Japan immediately after the Second Sino-Japanese War]. *Ritsumeikan keizaigaku* 53:5–6 (2010): 148–170.

Kaple, Deborah A. "Soviet Advisors in China in the 1950s." In *Brothers in Arms: The Rise and Fall of the Sino-Soviet Alliance, 1949–1963*, edited by Odd Arne Westad, 117–140. Stanford, CA: Stanford University Press, 1998.

Kasza, Gregory. *The Conscription Society: Administered Mass Organizations.* New Haven, CT: Yale University Press, 1995.

Kerkvliek, Benedict. "Everyday Politics in Peasant Society (And Ours)." *Journal of Peasant Studies* 36:1 (2009): 227–243.

Kinzley, Judd. *Natural Resources and the New Frontier: Constructing Modern China's Borderlands.* Chicago: University of Chicago Press, 2018.

Kirby, William C. "The Chinese War Economy." In *China's Bitter Victory: The War with Japan*, edited by James C. Hsiung and Steven I. Levine, 185–213. Armonk, NY: M.E. Sharpe, 1992.

———. "Continuation and Change in Modern China: Economic Planning on the Mainland and Taiwan, 1943–58." *Australian Journal of Chinese Studies* 24 (1990): 121–141.

———. "Engineering China: Birth of the Developmental State, 1928–1937." In *Becoming Chinese: Passages to Modernity and Beyond*, edited by Wen-hsin Yeh, 137–161. Berkeley: University of California Press, 2000.

———. "Technocratic Organization and Technological Development in China: The Nationalist Experience and Legacy, 1928–1953." In *Science and Technology in Post-Mao*

China, edited by Denis Fred Simon and Merle Goldman, 23–44. Cambridge, MA: Harvard University Press, 1989.

Kline, Ronald. *Steinmetz: Engineer and Socialist*. Baltimore, MD: Johns Hopkins University Press, 1992.

Köll, Elisabeth. *From Cotton Mill to Business Empire: The Emergence of Regional Enterprises in Modern China*. Cambridge, MA: Harvard University Asia Center, 2003.

——. *Railroads and the Transformation of China*. Cambridge, MA: Harvard University Press, 2019.

Lapwood, Ralph, and Nancy Lapwood. *Through the Chinese Revolution*. Letchworth, Hertfordshire: Garden City Press, 1954.

Larkin, Brian. "The Politics and Poetics of Infrastructure." *Annual Review of Anthropology* 42 (2013): 327–343. https://doi.org/10.1146/annurev-anthro-092412-155522.

Lassman, Thomas C. "Industrial Research Transformed: Edward Condon at the Westinghouse Electric and Manufacturing Company, 1935–1942." *Technology & Culture* 44:2 (April 2003): 306–339.

Latour, Bruno. *Science in Action: How to Follow Scientists and Engineers through Society*. Cambridge, MA: Harvard University Press, 1987.

Lee, Leo Ou-fan. *Shanghai Modern*. Cambridge, MA: Harvard University Press, 1999.

Lenin, Vladimir. "Report on the Work of the Council of People's Commissars." December 22, 1920. Seventeen Moments in Soviet History: An On-line Archive of Primary Sources. Accessed September 12, 2018. http://soviethistory.msu.edu/1921-2/electrification-campaign/communism-is-soviet-power-electrification-of-the-whole-country/.

Leshan shi dang'an guan [Leshan Municipal Archives]. "Sun Yunxuan zuan 'riwei kaifa dongbei dianli zhi gaishu" [Sun Yunxuan's report on the development of electrical power in the Northeast]. *Minguo dang'an* 35 (1994): 23–29.

Li, Cheng, and Lynn White. "Elite Transformation and Modern Change in Modern China and Taiwan: Empirical Data and the Theory of Technocracy." *China Quarterly* 121 (March 1990): 1–35.

Li, Daigeng. *Xin Zhongguo dianli gongye fazhan shilue* [A survey history on the development of the power-generating sector in New China]. Beijing: Qiye guanli chubanshe, 1984.

——. *Zhongguo dianli gongye fazhan shiliao: Jiefang qian de qishi nian* [Sources on the history of the development of China's electrical industries: The seventy years before liberation]. Beijing: Shuili dianli chubanshe, 1983.

Li, Fengwu, and Li Bin. "Liyong luo gangxian daiti luo tongxian" [Replace bare copper wire with bare steel wire]. *Dianshijie* 12 (1958): 156.

Li, Nanyang, ed. *Fumu zuori shu* 1 [Old writing of my father and mother: Letters and diaries of Li Rui and Fan Yuanzhen, 1938–1960]. Hong Kong: Time International Publishing, 2005.

Li, Shizhong. "Fangzhi yuanliao" [Textile materials]. *Huashang shachang lianhehui jikan* 10:1 (1932): 1–16.

Liang Mu [pseud.]. "Riben zhi guoying dianli" [The nationalization of electricity in Japan]. *Hanxue yuekan* 6 [Sweat blood monthly] (June 1936): 108–109.

Lieu, D. K. *Preliminary Investigation on Industrialization.* Shanghai: The China Institute of Economic and Statistical Research, 1933.

——. *The Silk Reeling Industry in Shanghai.* Shanghai: The China Institute of Economic and Statistical Research, 1933.

Lin, Han. *Energy Policies and Climate Change in China: Actors, Implementation, and Future Prospects.* London: Routledge, 2019.

Lin, Lanfang. "Zhanhou chuqi ziyuan weiyuanhui dui taidian zhi jieshou (1945–1953)—yi jishu yu rencai wei zhongxin" [The National Resources Commission's takeover of Taiwan Power Company in the early postwar period (1945–1952): Technology and engineers]. *Bulletin of the Institute of Modern History Academia Sinica* 79 (March 2013): 87–135.

Lin, Pin-yen. *Taiwan jingyan de kaiduan: Taiwan dianli zhushi huishe fazhan shi* [The beginning of the Taiwan experience: the history of Taiwan denryoku kabushiki kaisha]. Taipei: self-published, 1997.

Liu, Jinyu. "Kunhu dianchang choubei jingguo" [Planning process of Kunming Lakeside Electrical Works]. *Ziyuan weiyuanhui yuekan* 1:5 (1940): 313–319.

——. "Kunhu dianchang Penshuidong fadiansuo jianshe jingguo" [The construction process of the Penshuidong power station of Kunming Lakeside Electric Works]. *Ziyuan weiyuanhui jikan* 4:3 (1944): 131–149.

Liu, Lanbo. "Yikao qunzhong, tigao jishu, gaijin lingdao: wei shixian diyige wunian jihua er nuli" [Depend on the masses, advance technology, improve the leadership: Strive to achieve the first five-year plan]. *Renmin dianye* 20 (October 1955): 1–12.

Liu, Richard, Zhou Yang, and Xinzhou Qian. "China's Risky Gamble on Coal Conversion." New Security Beat. January 9, 2020. Accessed January 15, 2020. www.newsecuritybeat.org/2020/01/chinas-risky-gamble-coal-conversion/.

Liu, Taotian. "Fangzhiye gaikuang diaocha" [A survey of the textile industry]. *Jiaoyu yu zhiye* 179 (September 1936): 741–756.

Liu, Yan, et al., eds. *Huzhou shi dianli gongye zhi* [Gazetteer of the electrical industries in Huzhou]. Beijing: Zhongguo dianli chubanshe, 2004.

Liu, Yenbin, Zhou Wen, and Cha Renbo. "Woguo jiechu de dianli xitong zhuanjia—Cai Changnian xiansheng" [An outstanding electrical power systems expert of our nation—Mr. Cai Changnian]. *Dongli yu dianqi gongchengshi* [Power and Electrical Engineers] 11 (September 2009): 53–61.

Liu, Yingyuan, and Wang Wenbing. "Shijingshan fadian chang ruhe Guanche minzhu guanli chao e wancheng renwu" [How Shijingshan Power Station implemented democratic management to exceed production targets]. *Renmin dianye* 6 (December 1950): 18–21.

Lu, Hanchao. *Beyond the Neon Lights: Everyday Shanghai in the Early Twentieth Century.* Berkeley: University of California Press, 1999.

Lu, Yuezhang. "Zhongguo zhi ranliao ziyuan ji huoli fadian zhi zhanwang" [China's fuel resources and outlook of fossil-fuel electrical power]. *Ziyuan weiyuanhui jikan* 4:3 (September 1944): 27–37.

Luo, Guo-liang, Yan-ling Li, Wen-jun Tang, and Xiao Wei. "Wind Curtailment of China's Wind Power Operations: Evolution, Causes, and Solutions." *Renewable and Sustainable Energy Reviews* 53 (2016): 1190–1201. https://doi.org/10.1016/j.rser.2015.09.075.

Mao, Zedong. "Ba jundui bian wei gongzuo duiwu" [Transform the army into a workforce]. In *Mao Zedong xuanji di si juan* [Selected works of Mao Zedong, vol. 4]. Beijing: Renmin chubanshe, 1994.

Meiton, Fredrik. *Electrical Palestine: Capital and Technology from Empire to Nation.* Berkeley: University of California Press, 2019.

Min, Jiangyue. "Safanqi kaocha changjiang sanxia qianhou" [The events before and after John Lucian Savage's survey]. *Hubei wenshi ziliao* 1 (1997): 5–17.

Mitchell, Timothy. *Carbon Democracy: Politics in the Age of Oil.* New York: Verso Press, 2011.

——. *Rule of Experts: Egypt, Techno-Politics, Modernity.* Berkeley: University of California Press, 2002.

Mitter, Rana. *Forgotten Ally: China's World War II, 1937–1945.* Boston: Mariner Books, 2014.

Moore, Aaron Stephen. *Constructing East Asia: Technology, Ideology, and Empire in Japan's Wartime Era, 1931–1945.* Stanford, CA: Stanford University Press, 2013.

Moore, Joe. "A Friend Remembers Y. S. Sun." In *Wo suo renshi de Sun Yunxuan: Sun Yunxuan bashi dashou jinian teji*, 110–116. Taipei: Sun Luxi, 1993.

Muscolino, Micah. *The Ecology of War in China: Henan Province, the Yellow River, and Beyond, 1938–1950.* Cambridge: Cambridge University Press, 2015.

Musgrove, Charles. "Building a Dream: Constructing a National Capital in Nanjing, 1927–1937." In *Remaking the Chinese City: Modernity and National Identity, 1900 to 1950*, edited by Joseph Esherick, 139–158. Honolulu: University of Hawai'i Press, 2002.

Nakashi kensetsu seibi iinkai [Committee for the Compilation of Information for Economic Construction in Central China], ed. *Denki yōgu kōgyō hōkokusho* [Report on electrical appliances industry]. Shanghai: Kōain, 1940.

Nanjing difangzhi bianzuan weiyuanhui. *Nanjing dianli gongye zhi* [Gazetteer of Nanjing's electrical industries]. Nanjing: Jiangsu guji chubanshe, 1997.

Nathan, Andrew J. "China's Changing of the Guard: Authoritarian Resilience." *Journal of Democracy* 14:1 (January 2003): 6–17.

Nye, David E. *Electrifying America: Social Meanings of a New Technology, 1880–1940.* Cambridge, MA: MIT Press, 1991.

Oreskes, Naomi, and Erik Conway. *The Collapse of Western Civilization: A View from the Future.* New York: Columbia University Press, 2014.

Pauley, Edwin Wendell. *Report on Japanese Assets in Manchuria to the President of the United States.* Washington, DC: U. S. G. P. O, 1946.

Pearson, Peter J. G. "Past, Present and Prospective Energy Transitions: An Invitation to Historians." *Journal of Energy History/Revue d'Histoire de l'"Energie* 1 (December 2018). https://energyhistory.eu/en/node/57.

Pei, Minxin. *China's Trapped Transition: The Limits of Developmental Autocracy.* Cambridge, MA: Harvard University Press, 2006.

Peng, Zhen. "Zuohao chengshi gongzuo, yingjie jiefang gaochao" [Do our work in the cities well, welcome the climax of Liberation]. In *Beiping heping jiefang qianhou* [Archival documents on the events before and after the peaceful liberation of Beiping], 13–14. Beijing: Beijing shi dang'an guan, 1988.

Peng Zhen Zhuan bianxie zu [Peng Zhen biography editorial team], ed. *Peng Zhen nianpu* [The annals of Peng Zhen]. Beijing: Zhongyang wenxian chubanshe, 2002.

Pepper, Suzanne. *Civil War in China: The Political Struggle.* Berkeley: University of California Press, 1978.

Perry, Elizabeth. *Patrolling the Revolution.* Lanham, MD: Rowman & Littlefield, 2006.

——. *Shanghai on Strike: The Politics of Chinese Labor.* Stanford, CA: Stanford University Press, 1993.

Phalkey, Jahnavi, and Tong Lam. "Science of Giants: China and India in the Twentieth Century." *BJHS: Themes* 1 (2016): 1–11.

Pomeranz, Kenneth. *The Great Divergence: China, Europe, and the Making of the Modern World Economy.* Princeton: Princeton University Press, 2000.

Qiu, Xiuzhi, ed. *Wo suo renshi de Sun Yunxuan: Sun Yunxuan bashi dashou jinian zhuanji* [The Sun Yun-suan I know: Commemorative volume for Sun Yunxuan's eightieth birthday]. Taipei: Sun Luxi, 1993.

"Quanguo dianchang tongji zongbiao" [A statistical table of power plants throughout the nation]. In *Shenbao nianjian* [Shenbao yearbook]. Shanghai, 1934.

"Quanguo minying dianye qingyuan shimo ji" [A record of the beginning and end of the petitioning efforts by the private electric corporations across the nation]. *Dianye jikan* 1:1 (1930): 1–29.

"Ranliao gongyebu: diyi ci quanguo dianye huiyi jueyi" [Ministry of Fuel Resources: the motion of the first national electrical industries conference, March 2, 1950]. In *Zhonghua renmin gonghe guo jingji dang'an ziliao xuanbian* [Selections from the economic archives of the People's Republic of China], 209. Beijing: Zhongguo wuzi chubanshe, 1996.

"Riben quanguo dianli tongji" [Statistics on the national electric industry in Japan]. *Zhongyang yinhang banyuekan* 2:8 (1933): 1507.

"San nian li Sulian zhuanjia gei women de bangzhu he jinhou women ruhe geng hao de xiang Sulian xuexi" [Assistance rendered by the Soviet Union in the past three years and how we should learn from the Soviets from now on]. *Renmin dianye* 22 (December 1952): 2–3.

Savage, John Lucian. "Yangzijiang sanxia jihua chubu baogao (shang)" [Preliminary report on the plans for the Three Gorges of the Yangtze River (first half)]. *Minguo dang'an* 4 (1990): 19–27.

——. "Yangzijiang sanxia jihua chubu baogao (xia)" [Preliminary report on the plans for the Three Gorges of the Yangtze River (second half)]. *Minguo dang'an* 1 (1991): 41–50.

Schatz, Ronald. *The Electrical Workers: A History of Labor at General Electric and Westinghouse, 1923–60.* Urbana-Champaign: University of Illinois Press, 1987.

Schivelbusch, Wolfgang. *Disenchanted Night: The Industrialization of Light in the Nineteenth Century* [Lichtblicke: Zur Geschicte der kunstlichen]. Translated by Angela Davies. Berkeley: University of California Press, 1988.

Schwarcz, Vera. *The Chinese Enlightenment: Intellectuals and the Legacy of the May Fourth Movement of 1919.* Berkeley: University of California Press, 1986.

"Sekitan haikyū tōsei shikō rei" [Order to implement coal-rationing system]. *Chō denryoku* 222 (June 1940): 87.

Seow, Victor. "Carbon Technocracy: East Asian Energy Regimes and the Industrial Modern." PhD diss., Harvard University, 2014.

Shanghai dianye guanli ju [Shanghai Power Administration]. "Shanghai diqu dianli xitong tiaozheng fuhe de jingyan" [The experience of load adjustment in the power system of the Shanghai area]. *Renmin dianye* 22 (November 1955): 12–17.

Shanghai Municipal Council. *The Minutes of the Shanghai Municipal Council*, vol. 28. Shanghai: Shanghai Classics Publishing House, 2001.

——. *Report for the Year 1907.*

Shanghai shi dang'an guan [Shanghai Municipal Archives], ed. *Riwei Shanghai shi zhengfu* [Shanghai Municipal Government of the bogus Japanese regime]. Shanghai: Shanghai Municipal Archives, 1984.

Shanghai shi dianli gongyeju shi zhi bianzuan weiyuanhui [Shanghai electrical industries gazetteer committee], ed. *Shanghai shi dianli gongye zhi* [Gazetteer of the electrical industries in Shanghai]. Shanghai: Shuili chubanshe, 1993.

Shao, Qin. *Culturing Modernity: The Nantong Model.* Stanford, CA: Stanford University Press, 2003.

Shapiro, Judith. *Mao's War against Nature: Politics and the Environment in Revolutionary China.* Cambridge: Cambridge University Press, 2001.

"Shelun: Diangong yu jindai zhanzheng" [Editorial: The electrician and modern warfare]. *Zhongguo diangong* 1:4 (1943): 1–2.

Shen, Grace. *Unearthing the Nation: Modern Geology and Nationalism in Republican China.* Chicago: University of Chicago Press, 2014.

Shen, Sifang. "Dianli guantian" [Electrical irrigation]. *Nongye gongbao* 3 (1929): 14–15.

——. "Tongye cihou zhi ershi nian" [The next twenty years of our industry]. *Dianye jikan* 1 (1930): 69–70.

——. "Zhengli shoudu dianchang gongzuo zhi yi duan" [Part of the retrofitting work on the Capital Power Station]. *Gongcheng* 4: 2 (1929): 266–268.

Shen, Sifang, and Li Yanshi. "Benhui chuxi Deguo erci shijie dianli dahui daibiao Li Yanshi Shen Sifang er jun baogao" [Report by Li Yanshi and Shen Sifang following their participation as official representatives of our association at the Second World Power Conference in Germany]. *Dianye jikan* 3 (1930): 3–12.

———. "Riben chansi ye zhi dianqi hua" [The electrification of silkworm and silk industry in Japan]. *Fangzhi zhoukan* 1:5 (1931): 132–138.

Shiroyama, Tomoko. *China during the Great Depression: Market, State, and the World Economy, 1929–1937*. Cambridge, MA: Harvard University Press, 2008.

"Si tai yin mai zi" [Steinmetz]. *Mingdeng* 160–161 (1930): 418–419.

Smil, Vaclav. *China's Past, China's Future: Energy, Food, Environment*. New York: Routledge Curzon, 2004.

Sneddon, Christopher. *Concrete Revolution: Large Dams, Cold War Geopolitics, and the US Bureau of Reclamation*. Chicago: University of Chicago Press, 2015.

Sogō, Shinji. *Shigen kaihatsu hokushi tokuhon* [Resource exploitation reader for North China]. Tokyo: Daiyamondosha, 1937.

Steffen, Will, Paul J. Crutzen, and John R. McNeill. "The Anthropocene: Are Humans Now Overwhelming the Great Forces of Nature?" *AMBIO: A Journal of the Human Environment* 36:8 (2007): 614–621. https://doi.org/10.1579/0044-7447(2007)36[614:TAAHNO]2.0.CO;2.

Steinmetz, Charles Proteus. *Dianli shiye gaishu* [Overview of electrical industries]. Translated by Chen Zhang. Shanghai: Shangwu yinshuguan, 1931.

Suehiro, Akira. *Catch-up Industrialization: The Trajectory and Prospects of East Asian Economies*. Translated by Tom Gill. Singapore: NUS Press, 2008.

Sulian dianye zhuanjia baogao, vol. 2 [Reports of Soviet electrical experts]. Beijing: Ranliao gongyebu chubanshe, 1951.

Sun, Danchen. "Safanqi yu Li Bing fuzi" [Savage and Li Bing father and son]. *Zhongyang zhoukan* 8:20 (1946): 9.

Sun, Yat-sen. "How China's Industry Should Be Developed, 1919." In *Prescriptions for Saving China: Selected Writings of Sun Yat-sen*, ed. Julie Lee Wei, Ramon H. Myers, and Donald G. Gillin, 237–240. Stanford, CA: Hoover Institution Press, 1994.

———. "A Plea to Li Hung-chang (June 1894)." In *Prescriptions for Saving China: Selected Writings of Sun Yat-sen*, ed. Julie Lee Wei, Ramon H. Myers, and Donald G. Gillin, 3–18. Stanford, CA: Hoover Institution Press, 1994.

Suzuki, Jun. *Nihon no kindai: Shin gijyutsu no shakaishi* [Modern Japan: Social chronicles of new technology]. Tokyo: Chūō Kōron sha, 1999.

Taiwan Power Company. "Kan Fengshan weiwen ben gongsi ruying yuangong" [A visit to Fengshan to call on our employees who are participating in the training camp]. *Taidian lijin yuekan* 39 (April 1950): 62.

Tajima, Toshio, ed., *Gendai chūgoku no denryoku sangyō fusoku no keizai to sangyō soshiki* [The electrical power industry of modern China: The economy of insufficiency and industrial organization]. Tokyo: Shōwadō, 2008.

Tajima, Toshio, "Kahoku ni okeru kōiki denryoku nettowāku no keisei" [The formation of wide area electrical network in North China]. In *Gendai chūgoku no denryoku sangyō "fusoku no keizai to sangyō soshiki* [The electrical power industry of modern China: The economy of insufficiency and industrial organization], ed. Tajima Toshio, 115–150. Tokyo: Shōwadō, 2008.

Tan, Ying Jia. "Repairing China's Power Grid Amidst Perpetual Warfare." In *The Persistence of Technology*, ed. Stefan Krebs and Heike Weber, 53–70. Bielefeld: Transcript Verlag, 2021.

Tang, Mengxiong. "Yingguo Lankaixia mianye gongsi shiyan zidong buji zhi zhengshi baogao" [Report on automated looms by British Lanchashire Cotton Industries]. *Huashang shachang lianhehui jikan* 10:1 (1932): 190–203.

Tao, Lizhong, trans. "Dongli gongye zhi fangkong" [Air defense for power industries]. *Ziyuan weiyuanhui yuekan* 3:4–6 (1941): 229–233.

Tao, Lizhong, and Liu Jinyu. "Kunhu dianchang Penshuidong fadiansuo jianshe jingguo" [The construction process of the Penshuidong power station of Kunming Lakeside Electric Works]. *Ziyuan weiyuanhui jikan* 4:3 (1944): 131–149.

Tawney, R. H. *Land and Labor in China*. London: George Allen & Unwin Ltd., 1932.

Thomson, Elspeth. *The Chinese Coal Industry: An Economic History*. London: Routledge Curzon, 2003.

"Tongji: Ziyuan weiyuanhui fushu changkuang zhanhou neiqian qicai shuliangbiao" [Statistics: Equipment transported by the affiliated mines and industries of the National Resources Commission after the war]. *Ziyuan weiyuanhui gongbao* 1:1 (July 1941): 76.

Volti, Rudi. "Worker Innovation: Did Maoist Promotion Contribute to China's Present Technological and Economic Success?" In *Mr. Science and Chairman Mao's Cultural Revolution: Science and Technology in Modern China*, edited by Chunjuan Nancy Wei and Darryl E. Brock, 333–343. Lanham, MD: Rowman and Littlefield, 2013.

Wang, Der-wei David. *Fin-de-siécle Splendor: Repressed Modernities of Late Qing Fiction, 1849–1911*. Stanford, CA: Stanford University Press, 1999.

Wang, Hongchao, and Li Zhongzhi, eds. *Beijing shi dianli gongye zhi* [Gazetteer of Beijing electrical industries]. Beijing: Dangdai zhongguo chubanshe, 1995.

Wang, Shoutai. "Guanyu zhiding woguo dianqi biaozhun guifan zhi guan jian" [Opinions on the establishment of standards for electrical appliances of our nation]. *Zhongguo diangong* 1:1 (July 1943): 3–4.

Wang, Xiang. "Xinhai geming qijian de Jiangzhe sizhiye zhuanxing" [The transformation of the silk industry in Jiangsu and Zhejiang during the Xinhai Revolution period]. *Lishi yanjiu* 6 (2011): 21–36.

Wang, Zhongying, Haiyan Qin, and Joanna Lewis. "China's Wind Power Industry: Policy Support, Technological Achievements, and Emerging Challenges." *Energy Policy* 51 (December 2012): 80–88. https://doi.org/10.1016/j.enpol.2012.06.067.

Wei, Julie Lee, Ramon H. Myers, and Donald G. Gillin, eds. *Prescriptions for Saving China: Selected Writings of Sun Yat-sen*. Stanford, CA: Hoover Institution Press, 1994.

Wong, R. Bin. *China Transformed: Historical Change and the Limits of European Experience*. Ithaca, NY: Cornell University Press, 1997.

World Power Conference. *Transactions Third World Power Conference*. 7 volumes. Washington, DC: Government Printing Office, 1936.

Wright, Tim. *Coal Mining in China's Economy and Society, 1895–1937.* Cambridge: Cambridge University Press, 1984.

Wrigley, E. A. *Energy and the English Industrial Revolution.* Cambridge: Cambridge University Press, 2010.

Wu, Shellen X. *Empires of Coal: Fueling China's Entry into the Modern World Order, 1860–1920.* Stanford, CA: Stanford University Press, 2015.

Wugong, "Longmao shachang tingye pingyi," *Fangzhi zhoukan* 2:46 (October 1932): 1.

Xiangtan dianji changzhi bianzuan weiyuanhui. *Xiangtan dianji chang zhi* [The annals of the Xiangtan Electrical Manufacturing Works], 1–5. Xiangtan: Xiangtan dianjichang, 1992.

Xieersufu. *Liening Sidalin de Sulian dianqi hua* [Leninist-Stalinist electrification of USSR]. Beijing: Ranliao gongye chubanshe, 1954.

Xu, Junrong, and Huang Zhiming, eds. *Baise kongbu mimi dang'an* [Secret archives of the white terror]. Taipei: Dujia wenhua, 1995.

Xu, Yi-chong. *Electricity Reform in China, India and Russia.* Cheltenham, UK: Edward Elgar, 2004.

Xu, Ying. *Beiping eryue weicheng ji* [Two month siege on Beiping]. Beijing: Beijing chubanshe, 1993.

———. *Dangdai Zhongguo shiye renwu zhi* [Collective biography of industrialists in modern China]. Shanghai: Wenhai chubanshe, 1948.

Xu, Yingqi. "Zhongyang diangong qicai chang di si chang gaikuang" [Overview of the Number 4 Factory of the Central Electrical Manufacturing Works]. *Ziyuan weiyuanhui yuekan* [National Resources Commission Monthly] 2:4–5 (1940): 22–25.

Xu, Yongfeng. "Yijiu wuling Shanghai erliu hongzha ji yingdui" [The February 6, 1950, bombardment and countermeasures]. *Lishi yanjiu* 4 (2014): 101–115.

Xue, Yi. "Minguo shiqi shouci kexue kance Changjiang sanxia luelun" [A brief discussion about the first scientific survey of the Three Gorges]. *Wuhan daxue xuebao* 59:4 (July 2006): 463–469.

Xue, Yueshun, ed. *Ziyuan weiyuanhui dang'an shiliao huibian—dianli bufen* [Collection of archival materials on the National Resources Commission—Electrical power]. Taipei: Academia Historica, 1992.

Yang, Jisheng. *Tombstone: The Great Chinese Famine, 1958–1962.* New York: Farrar, Straus and Giroux, 2012.

Yang, Yan. "Gongbuju zhudao xia jindai shanghai dianli zhaoming chanye de fazhan, 1882–1893" [The development of the electric lighting industry in Shanghai, 1882–1893]. *Bulletin of the Institute of Modern History Academia Sinica* 81 (September 2013): 53–98.

Yao, Yuming. "Lue lun jindai Zhejiang sizhi ye shengchan de yanbian ji qi tedian" [A brief discussion on the evolution and special characteristics of Zhejiang silk industry]. *Zhongguo shehui jingji shi yanjiu* 4 (1987): 54–59.

Yeh, Emily, and Joanna Lewis. "State Power and the Logic of Reform in China's Electricity Sector." *Pacific Affairs* 77:3 (Fall 2004): 437–465.

Yeh, K. C. *Electric Power Development in Mainland China: Prewar and Postwar*. Santa Monica, CA: The RAND Corporation, 1956.

Yu, Depei, ed. *Taiwan dianyuan kaifashi: Koushu shi* [The history of electrical power development in Taiwan: An oral history]. Taipei: Taiwan Research Institute, 1997.

Yu, Maochun. *The Dragon's War: Allied Operations and the Fate of China, 1937–1947*. Annapolis, MD: Naval Institute Press, 2006.

Yun, Zhen. "Diangong qicaichang zhi choubei jingguo ji xianzhuang" [The planning and current state of the Central Electrical Manufacturing Works]. *Ziyuan weiyuanhui yuekan* [National Resources Commission Monthly] 1:1 (1939): 24–25.

——. "Dianli diangong zhuanjia Yun Zhen zishu" [An autobiographical account of electrical and electrical engineering expert Yun Zhen, part one]. *Zhongguo Keji Shiliao* [China Historical Material on Science and Technology] 21:3 (2000): 189–206.

——. "Dianli diangong zhuanjia Yun Zhen zishu er" [An autobiographical account of electrical and electrical engineering expert Yun Zhen, part two]. *Zhongguo Keji Shiliao* 21:4 (2000): 352–367.

——. "Dianli diangong zhuanjia Yun Zhen zishu san" [An autobiographical account of electrical and electrical engineering expert Yun Zhen, part three]. *Zhongguo Keji Shiliao* 22:2 (2001): 168–184.

——. "Dianli jianshe zhi fangzhen" [Guidelines on the development of electrical power]. *Jingji jianshe jikan* 1 (1942): 65–68.

——. "Dianqi gongye jianshe zhi zhanwang" [Outlook of the electrical equipment industry]. *Xin Zhonghua* 2:8 (1944): 40–45.

——. "Huadong gongyebu dianqi gongyechu chuzhang Yun Zhen yijiuwuyi nian jiuyue nianqiri zai jieyue yongtong yundong dahui shang yanjiang" [Speech by director of electrical equipment industries of the Ministry for Industry in Eastern China Yun Zhen on the copper conservancy campaign on 27 September 1951]. *Shanghai gongshang* [Shanghai Industry and Commerce] 34 (1951).

——. "Sanshinian lai Zhongguo zhi dianji gongye" [Electrical machinery industries in China during the past thirty years]. In *Minguo shiqi ji dian jishu* [Mechanical engineering and electrical engineering in the Republic of China], ed. Zhang Baichun, 255–264. Changsha: Hunan jiaoyu chubanshe, 2009.

——. "Sun Yunxuan xiansheng zai dalu qijian" [Mr. Sun Yun-suan during his time on the mainland]. In *Wo suo renshi de Sun Yunxuan: Sun Yunxuan bashi dashou jinian zhuanji* [The Sun Yun-suan I know: Commemorative volume for Sun Yunxuan's eightieth birthday], ed. Qiu Xiuzhi, 54–58. Taipei: Sun Luxi, 1993.

——. "Xuesheng yundong de genben yanjiu" [Fundamental research on student movements]. *Shaonian Zhongguo* 1:11 (May 1920): 14–24.

——. "Zhongyang diangong qicai chang ershiba niandu shiye zong baogao" [General report on the Central Electrical Equipment Plant for 1939]. *Ziyuan weiyuanhui yuekan* [National Resources Commission Monthly] 2:4–5 (1940): 10–22.

——. "Ziyuan weiyuanhui yu meiguo kenwuju dingyue sheji Sanxia shuidian gongcheng" [Agreement between Resources Commission and Bureau of Reclamation on the Design of the Three Gorges Dam]. *Hubei wenshi ziliao* 1 (1997): 28–30.

Yunnan sheng difangzhi bianzuan weiyuanhui, ed. *Yunnan shengzhi juan 37, dianli gongye zhi* [Yunnan provincial gazetteer, vol 37: Electrical industries]. Kunming: Yunnan renmin chubanshe, 1994.

"YVA jihua de shijian" [Implementing the YVA Plan]. *Yijiusiqi huabao* 4:11 (1946): 10.

"Zai Yin qicai neiyun shoufei banfa" [Regulations on the fees for the transportation of equipment in India], *Ziyuan weiyuanhui gongbao* 4:2 (February 1943): 22–23.

Zeng, Bo, et.al. "An analysis of previous blackouts in the world: Lessons for China's power industry." *Renewable and Sustainable Energy Reviews* 42 (2015): 1151–1163. https://doi.org/10.1016/j.rser.2014.10.069.

Zeng, Zhaojian. "Ji Sanfanqi xiansheng" [Remember Mr. Savage]. *Xinshijie* 2 (1947): 40–42.

Zhang, Baichun, ed. "Kangzhan qian de diangong jishu" [Electrical technology before the war of resistance]. In *Minguo shiqi ji dian jishu* [Mechanical engineering and electrical engineering in the Republic of China], 156–226. Changsha: Hunan jiaoyu chubanshe, 2009.

——. "Yun Zhen xiansheng tan zhizao dianji" [Yun Zhen speaks about the manufacturing of electrical machinery]. In *Minguo shiqi ji dian jishu* [Mechanical engineering and electrical engineering in the Republic of China], 76–81. Changsha: Hunan jiaoyu chubanshe, 2009.

Zhang, Chenghu. "Guoying dianli jiqi shiye zhi chengzhang yu zhanwang" [The growth and outlook for state-owned electrical machine industry]. *Ziyuan weiyuanhui jikan* 5:2 (1945): 210–216.

——. "Zhongguo dianxian gongye" [Wire industry of China]. *Ziyuan weiyuanhui jikan (diangong zhuankan)* [National Resources Commission quarterly (Special issue on electrical industries)] 5:2 (May 1945): 195–200.

Zhang, Dahai, et al. "Present Situation and Future Prospect of Renewable Energy in China." *Renewable and Sustainable Energy Reviews* 76 (2017): 865–871. https://doi.org/10.1016/j.rser.2017.03.023.

Zhang, Qian, ed. *The Minutes of Shanghai Municipal Council*, vol. 23. Shanghai: Shanghai Municipal Archives, 2001.

Zhang, Renjie. "Benhui wei jiejue Beiping diandeng dianche liang gongsi xuanan" [Construction Commission's resolution of the dispute between the Beiping Light and Power Company and Tramway Company]. *Jianshe weiyuanhui gongbao* 38 (1934): 45.

Zhang, Tianzhu, and Ren Jingye. "Tianjin dianye ju diyi fadian chang zenyang ba fadian shebei chuli cong 73.3 baxian tigao dao 100 baxian" [How did Tianjin Power Bureau increase instantaneous output from 73.3 percent of generating capacity to 100 percent?]. *Renmin dianye* 6 (December 1950): 7–17.

Zhang, Wangliang. "Shachang you zhengqi dongli gaiyong dianqi dongli zhi guan jian" [The key to transitioning from steam power to electric power in cotton mills]. *Huashang shachang lianhehui jikan* 8:4 (1930): 1–3.

Zhangjiakou diangongju [Zhangjiakou Power Bureau], ed. *Zhangjiakou dianye zhi* [Gazetteer of electrical industries at Zhangjiakou]. Zhangjiakou: Zhangjiakou Power Bureau, 1994.

Zhejiang sheng dianli gongyezhi bianzuan weiyuanhui, eds. *Zhejiang sheng dianli gongye zhi* [Zhejiang gazetteer on electrical power]. Beijing: Shuili chubanshe, 1995.

Zheng, Youkui, Cheng Linsun, and Zhang Chuanhong. *Jiu Zhongguo ziyuan weiyuanhui—shishi yu pingjia* [The National Resources Commission of old China—historical facts and appraisals]. Shanghai: Shanghai shehui kexueyuan chubanshe, 1984.

"Zhi Henan sheng minzheng ting Zhang Boying tingzhang jielue wei kenqing fahuan Kaifeng dianchang" [Request for the restitution of Kaifeng Power Station addressed to Director Zhang Boying of the Henan Civil Affairs Department]. *Dianye jikan* 1:1 (1930): 77–78.

Zhonggong zhongyang dangxiao dangshi jiaoyuanshi, ed. *Zhonggong dangshi cankao ziliao*, vol. 7 [Reference materials on the history of the Communist Party of China]. Beijing: Renmin chubanshe, 1980.

"Zhonggong zhongyang Huabeiju dui Pingjin dixiadang zai jieguan chengshi zhong ying zuo de gongzuo zhishi" [Orders from the North China Bureau to underground party in Beiping and Tianjin regarding the takeover of cities]. In *Beiping heping jiefang qianhou* [Archival documents on the events before and after the peaceful liberation of Beiping], 19–25. Beijing: Beijing shi dang'an guan, 1988.

Zhongguo dianli renwu zhi bianshen weiyuanhui [Committee on the who's who guide to China's electrical industries], comp. *Zhongguo dianli renwuzhi* [Who's who guide to China's electrical industries]. Beijing: Shuili dianli chubanshe, 1992.

Zhongguo dianqi gongye fazhanshi bianzuan weiyuanhui. *Zhongguo dianqi gongye fazhanshi* [History of development of electric industry of China (Comprehensive history)]. Beijing: Jixie gongye chubanshe, 1989.

Zhongguo dier lishi dang'an guan [Number Two Archives], comp. *Zhonghua minguo dangan ziliao huibian* [Selections from the archival materials of the Republic of China]. Nanjing: Jiangsu renmin chubanshe, 1981.

Zhongguo renmin xieshanghuiyi xinan diqu wenshi ziliao xiezuohuiyi [Conference of the collaborative committee for archival sources in Southwest China of the Chinese People's Political Consultative Conference], ed. *Kangzhan shiqi neiqian xinan de gongshang qiye* [Industries that relocated to the southwest during the war of resistance]. Kunming: Yunnan renmin chubanshe, 1988.

Zhongguo shehui kexueyuan [Chinese Academy of Social Sciences], ed. *Zhonghua renmin gonghe guo jingji dang'an ziliao xuanbian* [Selections from the economic archives of the People's Republic of China]. Beijing: Zhongguo wuzi chubanshe, 1996.

Zhongguo shehui kexueyuan, Zhongyang dang'anguan, ed., *Zhonghua renmin gonghe guo jingji dangan ziliao xuanbian: Gongyejuan* (1953–1957) [A collection of archival materials on the PRC economy: Industry, 1953–1957]. Beijing: Zhongguo chengshi jingji shehui chubanshe, 1998.

Zhongyang ranliao gongyebu bianyi shi dianye zu. "Xiang Sulian zhuanjia men xuexi" [Learning from the Soviet experts]. *Renmin dianye* 22 (1952): 20–21.

Zhongyang renmin zhengfu ranliao gongyebu jihua si, ed. *Dianye jihua jiangxiban jiangyi* [Notes from the lectures on the electrification plan]. Beijing: Ranliao gongyebu, 1952.

Zhu, Xinyu. *Zhejiang sichou shi* [The history of silk in Zhejiang]. Hangzhou: Zhejiang renmin chubanshe, 1985.

Zhu, Yingui, and Daqing Yang, eds. *Shijie nengyuan shi zhong de Zhongguo: Dansheng yanbian liyong ji qi yingxiang* [China in global energy history: Emergence, transformation, use, and its influence]. Shanghai: Fudan University Press, 2020.

Zhu, Yongpeng. "New Energy Sources: Direction of China's Energy Industry." *Quishi (English edition)* 2:2 (April 2010). Last modified September 20. 2011. Accessed October 23, 2012. english.qstheory.cn/magazine/201002/201109/t20110920_111406.htm.

Zhu, Youzhi, and Guo Qin. Hunan jindai shiye renwu zhuanlue [Collected biographies of modern industrialists in Hunan]. Changsha: Zhongnan daxue chubanshe, 2011.

INDEX

Page numbers in *italics* refer to figures or maps. Page numbers in **bold** refer to tables.

Accra, Ghana, 112, 117
Adams, Henry, 60–61
advanced organic economy, 5
aerial bombardment, 33, 83–84, 97, 166, 183
Agnew, P. G., 106
AIEE (American Institute of Electrical Engineers), 90, 106
air pollution, 184, 188, 193
air raids, 62, 83, 97, 163, 188
Alabama, TVA projects sites in, 114
alcohol, 74, 100
Aldridge, T. H. U., 22–24, 29
All-China Labor Congress, 140, 151
Allgemeine Elektricitäts-Gesellschaft (AEG), 35
Allies, 97, 99
alternating current (AC), 7, 22
Aluminum Company of America (Alcoa), 122
aluminum, 168
American Institute of Electrical Engineers (AIEE), 90, 106
Anhui Province, 24, 48, 71, 177
Anshan Steel Works, 181
Anthropocene, 4, 17, 21, 37, 110–111, 136, 184, 189
Arcturus Radio Tube Company, 92
Argersinger, R. E., 90
American Standards Association (ASA), 106–107
Association of Cocoon Producers, 41

Association of Private Chinese Electric Corporations, 49
Aviation Affairs Commission, 129

Bao Guobao, 144, 153–154, 158, 182
Bao Steel, 175
Baoding Military Academy, 155
Barton, Sidney, 30
Bayernwerk, 90
Beatty, W. C., 131
Beckert, Sven, 20–21
Beihai Park, 154
Beijing, 2, 11, 15, 30, 74, 149, 152, 152–156, 177, 181–182, 187; and August 15 power blockade, 56–57; Beiping-Tianjin-Tangshan network, 139, 144, 180; *Kahoku Dengyō* in, 144; managing power companies in, 70; "peaceful liberation" of, 149; power blockade in, 149; power consumption, 180; power lines in, 70; rail convoy of coal, 145; renaming, 67; Second National Electrical Industries Conference in, 179–180; and water works, 145
Beiping Chinese Merchants' Electric Light Company, 66–68
Bell, Lynda S., 39
Berlin University, 45
bianxiang lueduo, 49
"Big Four" banks, 96
Birmingham, 37
blank-slate mentality, 114

boilers, 46, 58, 77, 138, 169
boiler-turbine systems, production of, 177
Bonneville Power Authority, 141–142
Boshan Power Plant, 74
Britain, 27, 32, 37, 52, 87; capitalists from, 7; power generators from, 90
British Insulated Cables Company (BICC), 91–92
British Overseas Airways Corporation, 117
Broggi, Carles, 33
Bureau of Hydropower Construction, 138
Bureau of Reclamation, 120
Burkill, A. W., 32

Cabet, Etienne, 44
Cable and Construction Company, 92
Cai Changnian, 152
Cai Xiaoqian, 159
Calcutta, 24, 27, 29, 76, 98, 108, 117
Canton Electric Light Company, 24
Capital Power Works, 58
carbon: carbon economy, 5, 10, 60, 184; carbon-intensive developmental model, 84; economy, 5, 10, 60, 184; limits of carbon-fueled growth, 192; lock-in, 6, 17, 189–192; resource extraction, 5; technocracy, 5, 205n
Carin, Robert, 11
Centeno, Miguel, 163
Central China Waterworks and Electricity Company, 64, 71, 73
Central Electrical Manufacturing Works, 86–87, 93, 108, 152; Chinese peasant analogy, 95–96; as *de facto* national standards institute, 103, 105, 109–110, 115–116; electrical equipment manufacturing by, 87; filing for compensation, 97; financial relief for, 97–99; low-interest loans, 96; and military communications equipment, 93; national standards institute, 103; overcoming scarcity, 99–103; products of, *88*; remuneration revenue, 92;

shutdown of, 94; subunits of, 94; Xiangtan construction, 94
Central Machine Works, 93, 146
Central Radio Manufacturing Works, 100
Central Steel Works, 93
Chalmers, Allis, 108
Champe Lu, 132
Chang, John Key, 36
Changchun, 138
Changsha, 93
Changshou, 101, 126–127, 131
changsi (factory-made silk), 40
Changxing Coal, 53
Changzhou School, 43
Chatterjee, Elizabeth, 192
chemicals, research staff for, 99–100
Chen Liangfu, 98, 106, 118–120, 124
Chen Shenyan, 156
Chen Yi, 157, 166
Chen Yun, 140, 151, 181
Chen Zhongxi, 65, 142, 159, 161
Cheng Linsun, 11
Cheng Wanli, 167, 169–170
Chiang Kai-shek (Jiang Jieshi), 8–9, 12, 44–45, 49, 138, 143, 148, 159, 161, 166
China Cotton Journal, 32
China Purchasing Agency, 98
Chinese Civil War, 17, 84, 114, 138, 165; phases of, 139–140; shaping outcome of, 161–162
Chinese Communist Party, 1, 3, 8, 178
Chinese Merchants Tramways Company, 21
Chinese Merchants' Electric Company, 21, 66, 74, 163, 176
Chinese National Currency. *See* CNC dollars
Chinese New Year, 180
Chinese Society for Electrical Engineers, 152
Chongqing, 62, 94, 96, 101–102, 107, 115–116, 124, 126–134

Chu Yinghuang, 105, 115
civilian life, militarizing, 164
Clarke, William, 12
clean energy, 1, 190, 195
Climate and Sustainability Policy
Research Group, 193
CNC dollars, 82, 89, 91, 96–97, 130, 135,
145–147
coal: coal-fired power, 64, 80, 84, 116,
141, 193, 195; consumption, 66, 78,
168, 191, 215n49; conversion, 191–192;
deposits, 9, 15, 63–64, 67, 81–82, 85;
mines, 13, 64, 70, 82–83, 141, 153, 188;
producer, 82; supply, 29, 64, 75–76
Cold War, 10, 190
collaborator regimes, 84
Columbia River Basin, 131
Communist Party Committee, 169–170,
172, 232n76
Communists, 9, 11, 17, 85, 137–140, 142–
144, 147–165, 169–170, 181–183
Compagnie Francaise de Tramways et
d'Eclairage Electrique de Shanghai, 21
concrete gravity dam, 126
conductivity, 110, 168
ConEdison, 122
Confederation of Machine-Woven Silk
Producers, 54
Construction Commission, 49, 56
Conway, Erik, 192
cooperation, technical, 4, 109
Coopersmith, Jonathan, 13
copper: conservation campaign, 164, 168,
181–182; deposits, 86, 95, 168; mines,
86, 95, 97; refinery, 95–97, 100; rods,
92, 97; wires, 86, 89, 95–98, 168
cotton mills, 20–21, 33, 37, 209n41;
British, 18, 36; Chinese, 20, 27,
30, 34; decreasing number of, 31;
electrification of, 35–37; Japanese,
19–20, 25–27, 34; power grid hookup,
27–28; self-supplied power, 33–34; of
Shanghai, 31, 33, 36–38, 177

Cotton, John S., 128
cotton: fiber lengths of, 28; harvest
flooding, 177; yarn, 25, 28, 209n41
Coulee Dam, 126
cow-hide glue, 100
creative destruction, 190
Crutzen, Paul, 4, 111, 184

Dachang, 42, 54–57
Dachen Islands, 171, 177
Dadu River (Dadu he), 127
Dagongbao, 144
Dahuizhan (grand battle), 188. *See also*
Wangting Power Station
Dai Li, 158–159, 185
Dai Ligang, 185
Dai Shixi, 56
Daido Electric, 66
Dajia River Integrative Development, 134
dams, 5, 81, 119, 121–122, 124, 128, 158
Dasheng (Cotton Mill), 34–36, 53
Dasheng Number 1 Cotton Mill, 25, 35
Dayouli, confiscating, 48
Dazhao Electric Company, 51
de Rossi, 30
decentralization, franchising mode of, 59
demand curve, flattening, 171
demand-management operations, 178
Democratic Progressive Party (DPP), 186
Democratic Unions Planning
Committee, 148
development, accelerated, 183–184
"development-at-all-cost" mindset, 188
diandeng chang (illumination
companies), 24
dianqi gongsi (power companies), 24
diesel generators, 48
Ding Chenwei, 100
direct current (DC), 21
Dodgen, Randall, 6
Dong Shu, 52
Dongfang Steel, 175
Donghua, 19, 32

Du Dianying, 146
Du Yuesheng, 74, 176
Dwight P. Robinson Company, 45

Earth system, 4, 21, 136, 196
East Asia Co-Prosperity Sphere, 62
East Asia Development Board, 76, 89
East Asian Electrical Industries
    Corporation, 68-69
East China Power Bureau, 164
East China Revolutionary University,
    170
East China Textile Administration, 174
East China Textile Management Board,
    173
East Hebei Electric Shareholding
    Company, 69
Eastern Textile District, 18
economic dispatching, 180
economic sovereignty, 4, 20, 36, 85, 87;
    defense of, 103–111
economies of speed, 36
Edgerton, David, 13
Edison, Thomas, 21
Ekbladh, David, 113, 131
Electric Bond & Shares Company, 32
electric company, 29, 33, 39, 46, 55–57,
    145, 154, 213n62
"Electric Power Development in
    mainland China: Prewar and Postwar"
    (K. C. Yeh), 11
Electric Service Code, 50
Electric Utility Regulation Board, 48, 53
Electrical Works, Number 4 factory of,
    96, 101. *See also* Central Electrical
    Manufacturing Works
electrical engineers, 6, 12, 44, 94, 144
electrical equipment industry, 86–87, 91,
    93, 95, 105, 110–111, 115
electrical industry, 3, 6, 8, 10, 12–16, 20,
    39, 48, 52, 58, 64, 69–70, 78, 95, 149,
    171, 178, 184
electrical machines, 94, 105, 174

electrical power, 9, 18, 24, 30, 32–35, 42,
    66–67, 70, 74, 78, 85, 112, 142, 153, 164,
    171, 184, 183, 185
electrical resources, battle for, 153–158
Electricity Law, 194
electricity: as bargaining chip, 29–37;
    networked, 7; nationalization of, 46;
    purchasing, 38; theft of, 46, 53; shaping
    of, 1–2; as tool of strength, 3
Elvin, Mark, 6
energy, increasing availability of, 102–102
energy policy trilemma, 187
energy security, 195
energy transition(s), 1, 4, 10, 17, 20, 26, 32,
    60, 84, 184–188, 191–192
engineer-bureaucrats, 61–62, 86, 145,
    156, 196, 205n15; coal-dependent
    energy infrastructure, 80, 82–83; and
    Communists, 140, 158, 178; defining,
    5–7; developing battle damage repair
    protocols, 101; of Guomindang, 9, 43,
    62, 65, 149, 152, 183; and hydropower
    turbine retrofitting; as industry
    regulators, 16; of NRC, 85, 91, 98,
    104; reporting progress, 58; retreating
    to Southwest China, 63, 78; and
    self-reliance, 87; sidelining of, 178–179;
    technology transfer agreements, 91; at
    TVA, 112–114
Esherick, Joseph, 14
Ewo Filature, 41
Executive Yuan, 49

Factory Protection Committees, 152, 156
factory-made silk (*changi*), 40
Falinsky, S. V., 106
Fan Yuanzhen, 143
Fang Dongmei, 44
Fang Gang, 81
Far East Intelligence Bureau of the
    Comintern, 156
Faure, David, 26
Fearon Road Power Station, 23, 26

February 28 incident, 143
"fertilizer for dam" financing scheme, 126
Fessenden, Stirling, 32
"Final Days of the Anti-Japanese
    Resistance, The" (map), 62, *63*
First World Power Conference, 54
Five-Anti Campaign, 176
five-year plan, 191–192
flour mill, 24, 130, 169, 173–175
Fontana Dam, 124
food processing, 169
fossil fuels, 1, 111, 190
France: capitalists from, 7;
    delegates from, 51
Free China, 87
Freeberg, Ernest, 22
French Concession, 48
French Indo-China, 83, 95
French Power Company, 163
Fu Zuoyi, 149, 154–156
Fushun Power Plant: after Soviet
    takeover, *139*

Gansu Province, 116
Gansu, wind power bases in, 189
Gao Ming, 216n66, 235n34
Ge Helin, 157
Ge Zuhui, 101
General Cable Corporation, 132
geological studies, 133
Georgia, TVA projects sites in, 114
Germany, power generators from, 90
Ghosh, Amitav, 21
Glasgow, 37
glassworkers, 92
Go Green with Nuclear, 194
GOELRO, 84, 178, 181, 183
Gold Dust River, 95
Gong Debo, 159
Gong Zhaohuan, 25
Government Institute of Technology. *See*
    Shanghai Jiaotong University
Grand Canal, 5–6

Grand Coulee Unit, 132
Great Acceleration, 4–5, 21, 37, 85, 110, 184
Great Britain: delegates from, 51;
    standardization problem, 218n11;
    strike leverage, 30–31
Great Collapse of 2093, 192
Great Depression, 35, 54
Great Leap Forward, 5, 184, 188, 191
green energy, transition toward, 191, 195
Green Gang, 74, 176
gross domestic product (GDP), 5
Guangxi Province, 82, 94
Gu Zhengwen, 159–161
Gu Zhun, 170
Guan County (Guanxian), 127
Guan Zhiqing, 41–42, 55
Guangqin, 174
Guangyi, 42
Guangzhou Electric Company, 24
Guangzhou Power Bureau, 153
Guilin, 94
Guiyang, 94
Guo Keti, 144
Guo Zhicheng, 51
Guodian Corporation, 1
guomin jingji huifu qi ("recovery period
    for the national economy"), 9
Guomindang Party School, 107–108
Guomindang, 156, 161, 171, 183,
    188; collapse of, 78, 140–141;
    and Communists, 140, 149;
    defeat in Chinese Civil War, 114;
    engineer-bureaucrats of, 9, 43, 62, 65,
    149, 152, 183; engineering elite, 17, 152,
    157; government of, 34, 74, 94, 176;
    National Hydroelectric Bureau of, 136;
    regime of, 8–9, 11, 16–17, 40, 46–49,
    58–61; retreating to Southwest China,
    86, 94; turning to coal, 84–85

Hai Phong–Kunming Railroad, 82
Halske, Siemens, 93
Hamilton, Clive, 4

Han Kuo-yu (Han Guoyu), 195
Hanawa Yutaro, 75
handspun silk (*tusi*), 40
Hangzhou, 144
Hangzhou-Xiaoshan, 56
Harada Kumakichi, 73
Harbin Engineering University, 116
Harumi Goto-Shibata, 29
Hassōden, 69
Hebei, 68; wind power bases in, 189
Hein, Laura Elizabeth, 66
Henan, 68, 150
Hengda Cotton Mill, 32
Hengfeng Cotton Mill, 26–27, 38
hiki-age (repatriation), 141
Himalayan Hump, 95
Ho, Chiah-Sing, 89
Hong Kong, 195
Hou Defeng, 133
Hsu Yuan, 30
Hu Shi, 8, 115
Huadong dianye guanli ju. See East China
    Power Bureau
Huang Bingchang, 24, 176
Huang Hui, 34–35, 80, 115, 143, 159
Huang Shih-hsiu (Huang Shixiu), 185, 194
Huang Xiaoming, 3
Huang Yuxian, 124, 130, 135, 157
Huang Zuoqing, 41
Huasheng Company, 25
huasige (Chinese silk poplin), 41
Huaxin Spinning Mill, 25–26
Hughes, Thomas, 12–13, 64, 90
Hunan Province, 81
Hunt, William, 107
Huzhou, 37–38: competition, 42, 51;
    dispute with Woo-shing Power, 54;
    first power station in, 42; local power
    relations, 39–40; organic economy of,
    60; replicating Japanese production,
    41; saving silk from, 40–42; silk
    industry electrification, 16, 40,
    58–60; silk merchants, 39, 41; tariffs,

35, 54–55. *See also* Changxing Coal;
    Dachang; Wuxing County
hydraulic lock-in, 6
hydro first, thermal second, 67, 190
hydropower: dams, 84, 158; development,
    14, 81, 84, 113, 123–24, 130, 134, 136;
    potential, 11, 17, 18; projects, 114, 118,
    126, 129, 136, 143, 191; stations, 81,
    112, 116, 127–128, 134–136, 143, 184;
    turbine, 101, 111

Ichigo Offensive, 102
Ide Taijiro, 66
incandescent lamps, 86
Industrial Bureau, 170
Industrial Technology Research Institute,
    194
Inner Mongolia, wind power bases in, 189
innovation theory, 187
Institute of Pacific Relations, 33
International Library of Technology, 28
International Radio Station, 91
International Settlement, 18
International Telephone & Telegraph
    Company (IT&T), 92
Ishikawa Yoshijiro, 72

J. G. White & Company, 142
Jackson, Isabella, 20
Japan Electric Association, 66
Japan: big four cotton mills, 34; capitalist
    from, 7; coal supply leverage, 75;
    cotton mills, 19–20, 25–27, 34;
    competition with Chinese silk,
    38–41; electric company executives in
    North China, 68; Chinese electrical
    industries controlled by, **79**; electrical
    industry of, 8–9; power generators
    from, 90; recovery options from,
    73–74; restoring public services, 72;
    seizing electrical utilizes, 68; Shanghai
    electrical crisis, 71–78
Japanese Imperial Army, 62

Japanese Mill Owners' Association, 31

Jardine, Matheson and Company, 40–41

Jiajia gua hongdeng, yingjie Mao Zedong, 154

Jialing River (*Jialing jiang*), 124, 133

Jiang Guiyuan, 118

Jiangsu Province, 43, 48, 71, 73, 170; Changzhou, 33, 43–45, 48, 58; wind power bases in, 189

Jiaxing Electrical Power Company, 51

Jiaxing, 144

Jin Lisheng, 41

Jinan, 70

joint venture, 26, 40, 69, 109

Jones, Fred O., 131

Journal of Asian Studies, 4

Kahoku Dengyō, 144

Kaifeng, 70

Kanemaru Yūichi, 64, 73

Kanto Earthquake, 72

Karachi, 99

Kasza, Gregory, 3

Kentucky Dam, 124

Keswick, W. J., 75

Kinzley, Judd, 14

Kirby, William, 12

Kline, Ronald, 44

Kōain (East Asia Development Board), 76, 89

Kōchū Kōshi, 66, 68

Köll, Elisabeth 34

Kong Xiangxi (H. H. Kung), 107–108

Korean War, 164, 167

Kovalyov, Ivan, 179

Kunming Lakeside Electrical Works, 81–84

Kunming: aerial bombardment of, 82–84, 97, 183; center for applied research, 93; copper mining, 95; copper refinery, 96–97, 100; hydropower, 80–81, 84; site for Central Electrical Manufacturing Works, 86

labor force, 164–165, 171; coordinating, 172

laissez-faire, 54, 105, 109

Lake Tai, 38

Lane, E. W., 133

laogui (old devil), 26, 38, 208n27

Lapwood, Ralph, 155

Larkin, Brian, 13

Lassman, Thomas, 99

Latour, Bruno, 5

Legislative Yuan, 49

Lend-Lease Act, 97

Lend-Lease Administration, 98, 120

Leninist-Stalinist Electrification of USSR (Shershov), 183

Li Bing, 131

Li Daigeng, 11, 170–172, 174, 176–177

Li Guoqin, 98

Li Hongzhang, 8, 25

Li Peng, 12

Li Pingshu, 21

Li Rui, 143

Li Yanshi, 40, 45, 51–53, 59

Liang Mu, 67

Lianyungang Railroad Bureau Power Station, 62

Liaodong Military Zone, 151

Liebknecht, Wilhelm, 183

light bulbs, 89, 101–102

Lilienthal, 115

Lin Biao, 149

Lin, Han 193

Lin Jin, 105, 115

Lin Lanfang, 141

Linfen, 150

Lion Rapids Dam, 158

Lion Rapids Reservoir, 129

Lisheng, 41–42

Little Fengman, 137, 149

Little, Robert, 21

Liu Dajun (D. K. Lieu), 38

Liu Dengfeng, 159

Liu Jinyu: as director of Kunming Lakeside Electrical Works, 81–83, 122,

Liu Jinyu (*continued*)
140–141; prosecution under spying charges, 158–161, 182, 185; takeover of Taiwan Power Company, 140–141, 158, 188; visit to the TVA, 122
Liu Lanbo, 177
Liverpool, 37
localism, 37
London, dependency on, 97–98
Longmao Cotton Mill, 32–33, 36
Longxihe, 101, 111, 116
Lou Xichen, 166
Lower Yangtze, 11, 15–16, 34–35, 37, 45, 48, 58–59, 63–64, 71–73, 122, 143, 148, 165
Lu Yuezhang, 82
Lu Zuofu, 109
Luo Ronghuan, 149

Mabian River, 127
Machangding, 161
machinery, obsolete, 20, 25, 27
Majiezi, 81, 96
Manchester, 37
Manchuko, 68
Manchukuo Ministry of Industries, 67
Manchuria Electric Company, 138
Mao Dun, 18–19
Mao Renfeng, 159
Mao Zedong, 138, 148, 154, 164–165
"March of the Volunteers," 44
Marx, Karl, 183
mass mobilization, 17, 150, 164, 182, 189
May Fourth Movement, 8, 40, 42
May Thirtieth Movement, 20, 29
McManigal, R. D., 108, 109
McNeill, John, 111, 184
Mechanized Textile Bureau, 25
Meiton, Fredrik, 14
Mengjiang Electrical Corporation, 147
Mentougou coal mine, 145
metals, 74, 76, 110, 116, 168
military metabolism, 14
military-industrial complex, 110–111

Min River, 124
Ming Dynasty, 6
Mingdeng, 43
Mingliang coal mines, 82–83
Ministry of Ecology and Environment, 193–194
Ministry of Economic Affairs, 91
Ministry of Electric Industries, 181
Ministry of Environmental Protection, 192
Ministry of Fuel Resources, 5, 10, 168–169, 172, 177–178, 183
Minoshima Cotton Mill, 28
minying, term, 49
Miscellaneous Grain Association, 77
Mitchell, Timothy, 13–14
Mitsui Bussan, 29
Mitsui Company, 76
Monteith, A. C., 106
Moore, Joe, 191
Morrison-Knudsen, 131
mu, 45, 93
Mudanjiang, 150
Mumford, Lewis, 60–61
Municipal Committee, 165
Muscolino, Micah, 5, 14

Naigai, 19, 34
Naitō Kumaki, 68–71
Nakashima Kesago, 73
Nanjing, 81, 86, 89, 109, 134, 144, 156, 160, 167, 170, 173; electric company nationalization, 46; generating capacity, 58; government, 46–52, 55, 58–59, 65, 67, 90, 141; National Defense Planning Commission, 65; National Voltage Standards, 90; in second Sino-Japanese War, 62–63, 71–73, 80; Tianjin-Nanjing railroad, 68. *See also* Guomingdang
Nanjing Power Works, 45–46, 48, 53
Nanjing Road (Shanghai), 22, 167
Nanshi Power Company, 176
Nantong, 33–36, 53, 55
Nathan, Andrew, 3

National Congress of Model Workers, 178

National Construction Commission (NCC), 27, 34, 46, 48, 50, 53, 56–57, 65–66, 90, 93.

National Defense Planning Commission, 7, 65

National Development and Reform Commission (NDRC), 189

National Grid (Britain), 90

National Hydroelectric Commission, 120–121

National Resources Commission (NRC), 62, 101, 142, 177, 206n19; figures from, **79**; founding of, 7, 64–65; technological improvements, 100–101

national salvation through industrialization, 47

National Statistical Investigation of Electrical Power Industries, 27

National Voltage Standards, 90

Nationalist Air Force, 165

nationalization, versus privatization, 42–43

natural gas, 195

Nehru, Jawaharlal 181

Nei Rongzhen, 149

National Beiping University, 147

Networks of Power (Hughes), 12–13

New Deal, 114

New Delhi, dependency, 97–98

New Fourth Army, 170

New Taipei City, 194

New York, dependency on, 97–98

News Tornado, 185

Nie Jigui, 26

Nie Rongzhen, 149

Nie Yuntai, 26, 38

Nippon Electric Company, 66, 68

Niu Jiechen, 42, 54–55

Niusi zhoukan, 160

"North China Economic Construction Plan," 67

North China Electric Company, 63–64, 70

North China Electric University, 193

North China Electrical Industries Corporation (*Hoku-shi denryoku kōgyō kabushiki kaisha*), 66

North China Industrial Development Plan, 67

North China, electrification of Japanese-controlled, 67–71

North Hebei Power Company, 144–145, 153, 154, 158

North Hebei Power Company, 149

Nuclear MythBusters, 185, 194–195

Okabe Eiichi, 71

Okamoto Issaku, 75

Okazaki Katsuo, 76

Opium War, 7, 40

"optimizing power output with existing equipment," 181

Ordnance Department, 97

Oreskes, Naomi, 192

organic economy, 4–5, 20, 60

Organization of the Petroleum Exporting Countries (OPEC), 11–12

Oshikawa Ichiro, 71

Pacific Ocean, 116

Pacific Power and Light, 142

Palairet, C. M., 30

Pan, C. Y., 132

Paschal, George Reed, 126

path dependence, 187

Pauley Commission, 138

Pauley Reparation Mission, 143

Pauley, Edwin, 138

Peach Blossom Creek, 130

peak-load management: boiler-turbine system production, 177; cotton harvest flooding, 177; implementation of, 164, 171–174, 191

Pearl Harbor, 76, 116

Pearl River Delta, 136

Pearl Street Power Station, 21

Pearson, J. G., 187
Pei, Minxin, 192
Pekin Syndicate, 98–99
Peng Zhen, 149, 152, 155
Pennsylvania–New Jersey
    Interconnection, 90
Penshuidong Power Plant, 83–84
People's Liberation Army (PLA), 9, 140,
    144, 149–151, 153, 155–157, 162, 163,
    164–165, 180, 185
People's Republic, 163
Pepper, Suzanne, 143
petrochemical products, 191
Philadelphia Sesquicentennial
    Exposition, 44–45
Pingshan, 124
power blockade, 29, 30, 56, 140, 149–150,
    154–155
Power Conservation Committee, 7
power conversation campaigns, 178
power stations (*fadian chang*):
    argument for government-run,
    52–53; lack of uniformity in, 48;
    privately owned, 47
power vacuum, 72
Preece & Cardew, 24
Preece, Arthur H., 24
pricing system, implementation of, 57
"Principle of the People's Livelihood," 42
private interests, saving, 47–53
private operators, mobilizing, 46
private power plant owners, 49
Privately Owned Public Utility
    Regulation Law, 50
privatization, versus nationalization, 42–43
Project RAND, 11
protection committee, 152
Provisional Senate, 77
public-private cooperation, 113
public-private partnerships, 52, 94, 164,
    175–177
Pudong Electric, 19, 32, 73, 77–78, 164,
    168, 176

Pulin, confiscating, 48
Punjab, 124
Puodong Electric, 176

Qian Changzhao, 218n23
Qilu Electric Shareholding, 69
Qing Dynasty, 6
Qing Empire, 7–8
Qingdao Elektrizitätswerk, 34
Qingfeng cotton mill, 34
Qingyuandong, 127, 129–130, 135
Qishuyan Power Station, 39, 45, 53, 58–59
Qishuyan Zhenhua Power, 45
Qiushi, 1

Radio Corporation of America (RCA), 92
Rangoon, 99
rank-and-file workers, 10, 149, 157, 166,
    169, 178
Rankin, Mary, 14
rapid response, perils of, 188–190
raw materials, 100
Reformed Government of Shanghai,
    74–75
Reformed Government of the Republic
    of China, 71
"Regulations Governing Privately
    Owned Industries," Article 12 of, 56
Ren Yiyu, 147, 155
Renewable Energy Law, 189, 193
Renmin dianye, 179
residential lighting, 23
Riverside Power Station, 18, 29
Roosevelt, Franklin Delano, 114–115
rubber, export of, 99
Ruhr Valley, 51
"Rules of Standardization of Frequency
    and Voltages," 53
"run-of-river" installation, 83

S. Sakuragi, 30
safety, reaffirming importance of, 180
Savage, John Lucian, 17, 112–113, 123–137

scarcity, applied research and, 99–103
Second National Electrical Industries
     Conference, 179–180
Second World Power Conference, 51, 54,
     60
self-reliance, path to, 89–99
self-supplied power, *19*, 20, 32–33, 36,
     208n30, 210n69
Seow, Victor, 14, 205n15
Shaanxi, 150
Shandong Province, 68, 74, 142, 173
shangban (merchant-run), 49
Shanghai, 18; Communist takeover of,
     166; clothing industry, 33; cotton
     mills in, 24, 31, 33, 36–38, 177;
     electrical crisis in, 71–78; generating
     capacity, **174**; hyperinflation,
     146, 148; main power stations/
     textile mills, *19*; March 1941
     restriction orders, 76–77; peak-load
     management, 64, 171–174, 177; power
     output figures, **174**; power shortage,
     75–78; Public Utilities Party
     Committee, 167; silk filatures, 38–42;
     utilization time, **174**
Shanghai Boiler Factory, 177
Shanghai Cotton Exchange, 26
Shanghai Industrial Production
     Committee, 175
Shanghai Jiaotong University; as
     Government Institute of Technology,
     6, 43, 91, 157; as Nanyang Public
     School, 46
Shanghai Municipal Assembly, 147
Shanghai Municipal Council (SMC),
     15, 18, 21, 36; acquisition made by, 22;
     cutting off electricity, 16; daylight
     savings time, 75; gross revenue,
     50; imposing; power blockade,
     29; promoting electrification, 26;
     selling Electricity Department, 32;
     transnational colonialism within, 20
Shanghai Power Administration, 168–177

Shanghai Power Company, **19**, 22, 27, 32,
     36, 50, 58, 72, 75–78, 139, 144, 146–
     148, 156, 164, 167, 170–171
Shanghai Public Works Department, 156
Shangwan Flour Mill, 175
Shanxi, 68, 150
Shanxi-Chahar-Hebei base area, 147, 165
Shao, Qin, 34
Sheffield, 37
Sheldon, S. R., 43
Shen Sifang, 40, 43, 51, 53, 55, 59
Shen, Grace, 87
Shenbao, 76
Sheng Xuanhuai, 25
shengguan faicai, 74
shenghuo yongdian, 180
Shenxin Number 8, 32
Shenyang, takeover of, 151
Shershov, S. F., 183
Shijingshan Power Station, 70, 74, 145,
     152–155, 179–181
Shilongba, 81, 84
Shimen, 70
Shimizu Dōzō, 89
Shipai, 124, 125f
Shizui Village (Shizuicun), 81
Shuo County (Shuoxian), 150
Sichuan, 94
Sichuan-Yunnan railroad, 97
Siemens AG, 89
silk: energy economics of, 38; factory,
     56; filatures, 38–42; firms, 41, 54–58;
     importing filatures, 38; industry,
     16, 38–42, 46, 54–60; international
     demand for, 40–41; manufacturers,
     55; merchants, 39; mills, 54–55;
     production, 16, 35, 37–39, 41, 46, 60
Sino-American cooperation, 123, 128
Sino-British Textile Bureau, 27
Sino-Japanese War (second), 2, 63–64,
     84, 93, 116, 155
Sino-Russian Industrial University, 62
Sixth All-China Labor Congress, 140

siying (privately run), 49
Skoda, 177
sluice conveying method, 83
Smil, Vaclav, 5, 11–12
Sneddon, Christopher, 113
Social Construction of Technology
    (SCOT), 13
Sogō Shinji, 69
Songming, 83
Southern Manchurian Railway, 65
Southwest China, 2, 9, 60, 62–64, 74, 78,
    80, 84–86, 94, 100–102, 141
Soviet Union, 84, 107, 114, 143, 175, 179,
    183
spindles, 25–27, 30, 34, 37
Standard Oil, 167
standardization, 103–111; lack of, 87
standards, uniform, imposing, 48
state control, 54–61
State Grid Corporation of China, 187
state ownership, 47
state power, projecting, 178–184
steam engines, 5, 15, 18, 20, 25–28, 33,
Steffen, Will, 111, 184
Steinmetz, Charles Proteus, 43
Strauss, Benjamin, 127
street lamps, 18, 23
strikebreakers, 29, 31
Su Yang, 171, 176
sub-Saharan Africa, 112
Sugimoto Aki, 28
Sun Chuanfang, 45
Sun Danchen, 131
Sun Yat-sen (Sun Zhongshan), 8, 42, 44,
    46–47, 49, 51
Sun Yun-suan (Sun Yunxuan), 12, 62, 112,
    116, 118, 121, 124, 136, 141, 160, 190
Sun Zhifei, 176
Sun Zi, 115
Sun-Moon Lake (*Riyuetan*) Power
    Station, 142
sustainability, 195
Sutlej River, 124

Suzhou Creek, 18
Suzhou, 144
Suzuki Teiichi, 66
Sweat Blood Monthly, 67
Swift, Carelton B., 138

T. K. Ho, 75
T. V. Soong (Song Ziwen), 51, 153, 158
Ta Kung Pao (Dagongbao), 87, 109
Taiping Rebellion, 6
Taipower, 142–143, 158–160, 190–191, 94
Taiwan Peace Preservation Corps, 160
Taiwan Power Company, 136, 141–142,
    185. *See also* Taipower
Taiwan Strait, 136, 195
Taiwan: electrical power system of, 158;
    and fossil fuels, 186; gas dependence,
    195; economic takeoff, 190; energy
    future discussions, 193; 2017 power
    outage, 194; premier, 12
Taiyuan Steel Works, 181
Tajima Toshio, 14, 64, 70
Tang Mingqi, 115
Tang Wenzhi, 6, 115
Tanglang River, 127
Tangshan, 70
Tanizaki Jun'ichirō, 18
Tanomogi Keikichi, 66
Tao Lizhong, 83
Ta-tan (Datan), 185, 194–195
Tawney, R. H., 95
technocracy, defining, 163–164
technocratic orientation, shared, 12
technology transfer agreements, 110
Telegraph Works Company, 92
telephones, field, 87, 91, 93, 95–98, 100,
    102, 106, 110, 218n23
Tennessee River System, 122
Tennessee River, profile of, *121*
Texaco, 167
textile production, energy revolution in,
    24–29
Third Front policy, 191

Third World Power Conference, 54, 57, 60

Three Gorges Dam, 112, 157

Three-anti Campaign, 170

Tian Han, 44

Tianjin, 34, 70: Tianjin Power Bureau, 179; Tianjin Power Company, 69; Tianjin Power Plant, 181

Tiansheng Port Power Station, 35, 53

tiger hunts, 170

Tōka (*Donghua*) Number 1 mill, 32

Tokyo Electric Light Company, 66

Tomoko Shiroyama, 25, 39

Tongchang Xieji, 25

Tongming Electric Company, 35

Tongzhou County, 147, 154

toufei, 231n61

transnational colonialism, 20

trapped transition, 192

Truman, Harry, 138

Tsuji Hideo, 71–73

tungsten, 92, 97–98, 100

Tuo River (Tuojiang), 124

turbines: boiler-turbines, 62, 82; gas, 191; steam, 48; wind, 189

tusi (handspun silk), 41

TVA, 13–14, 17, 113, 115, 118–122, 128, 136, 142; Chinese replication, 123–137; Muscle Shoal Plant, 126; project sites, 114; 2017 blackout, 186

Ujikawa Electric, 66

Union of Chinese Cotton Mills, 27

United Gas Company, 57

United States: capitalists from, 7; common voltage usage in, 90–91; cooperation with, 103–104; delegates from, 51; Import and Export bank, 108; as power consumer, 1; power generators from, 90; technological diplomacy, 114–123

Universal Trading Company, 97–98

University of Nanking, 157

University of Wisconsin, 44

US Aid program, 191

US Army Air Transport, 117

US Army Hospital, 112

vacuum tubes, 92–95, 101

vanguard industry, 12

voltage standards: benefits of, 106; lack of, 48; tradeoffs involving, 104–105

Wah Chang Trading Company, 98

Wang Daohan, 177

Wang Jingchun, 92

Wang Kemin, 70

Wang Pingyang, 155

Wang Shoujing, 104

Wang Shoutai, 104

Wang Xiaohe, 148, 156

Wang Yangming, 6

Wang Yanqiu, 159

Wang Yimei, 41

Wang Youliang, 160

Wang Zihui, 74

Wangting Power Station, 177, 188

War of Anti-Japanese Resistance, 4, 8–9, 86, 103, 105, 110–111, 140, 168, 170

war, mobilization for, 15, 64–67, 80, 87, 109, 111

warfare, modes of, 1–2

Weng River (Weng jiang), 135–136

Weng Wenhao, 107, 116, 144

Western Hunan, 94

Westinghouse, 22, 44, 87, 99, 103–110, 115, 118, 122; Associated Companies Department, 105–106; technology transfer agreement, 157

Whampoa Military Academy, 8

"What Is the Purpose of National Salvation through Industrialization?" (speech), 52

White Terror (campaign), 158

White, Lynn, 12

White, Theodore, 148

wild chicken poplin, 41
Wilson Dam, 118
wind: curtailment, 189, 193; farms, 189; power, 189; turbines, 150
wires, 85, 87, 89, 91–92, 94, 96, 98, 101–102, 110
Woodhead, H. G. W., 126
Woo-shing Electric, 39, 42–46, 51, 54–57, 60, 89
worker's innovation, narrative of, 149–150
World War I, 27, 52, 90
World War II, 112, 138
Wright, Tim, 14, 36
Wrigley, E. A., 4–5
Wu Daogen, 115–116
Wu Guozhen (K. C. Wu), 160
Wu River (Wujiang), 124, 133
Wu Shi, 159, 161
Wu Xingzhou, 24
Wu Zuguang, 152
Wu, Shellen, 14
Wuchang (cotton mill), 34
Wuhu, Anhui Province, 24
Wuxi, 33–34, 39, 41, 173. *See also* Huzhou
Wuxing County, 41

Xiao Chaogui, 149
Xiashesi, 81, 93–94
Xiefing Yiji, 32
Xihe Cotton Mill, 29
Xiling Gorge (Xiling xia), 124
Xinhua News Agency, 149–150
Xinshengbao, 160
Xinxiang, 68
Xiong Bin, 146
Xiyuan (West Garden), 8
Xu Ying, 109, 144, 153–154
Xue Shouxuan, 39
Xuzhou, 70

Yada Shichitaro, 30
Yalu River, 177
Yan Huixian, 159, 161

Yan Xishan, 150
Yang Yan, 21–22
Yangshupu Road, 18, *19*, 26, 37, 144, 165
Yangtze Delta, 65, 71
Yangtze Gorges, map of, *125*
Yangtze Valley Administration (YVA), 131
yangzhuangsu (plain western silk), 41
Yanji, 150
Yaolong Electric Company, 101
Ye Jianying, 155
Ye Kongjia (K. C. Yeh), 11
Yellow River, 5–6
Yenching University, 155
Yichang, 124, 126, 128, 131–135
Yiliang County, 83
Yin Zhongrong (K. Y. Yin), 123
Yumen Oil Fields, 117
Young China Association, 44
Yu Bin, 160
Yu En-ying, 224n32
Yuen Ang, 3, 59
Yun Daiying, 44
Yun Zhen: career overview, 6; defection to Communists, 156–158; director of NRC Electric Bureau, 80–82; education, 43; electrical equipment manufacturing 86–87, 91–96, 99–100; electrical equipment manufacturing in the PRC, 181; interactions with revolutionaries, 44; interactions with TVA, 115, 122; negotiation with Westinghouse, 104–106; political persecution, 182; promoter of copper conservation, 168; as utility regulator 35, 57
Yunnan, 94

Zeng Guofan, 6, 26
zengchan jieyue ("increase production, practice economy"), 174
Zhabei Electric, 19, 21, 72–173, 163–164, 168, 176
Zhabei Power Company, 72

Zhang Baichun, 182
Zhang Dinghai, 166
Zhang Guangdou, 112, 116, 118–119, 123–124, 128, 135, 137
Zhang Guobao, 156
Zhang Jian, 25, 34
Zhang Jiazhi, 144
Zhang Renjie, 46
Zhang Wangliang, 27
Zhangde, 68
Zhangjiakou, 147, 155
Zhejiang Industrial School, 44
Zhejiang Labor Union, 170
Zhejiang Province, 44, 48, 53, 60, 71
Zhejiang University, 193
Zhenjiang, 144
Zheng Youkui (Y. K. Cheng), 10–11, 37
Zhengzhou, 44

Zhenhua Electric (Power Company), 45–46, 48, 51, 53
Zhenxin, 34
Zhifu Electric Shareholding Company, 69
Zhifu, 70
Zhongnanhai, 8
Zhou Enlai, 167, 182
Zhou Zhilu, 130
Zhou Zhirou, 160–161
Zhu Dajing, 56
Zhu Feng, 161
Zhu Qiqing, 92
Zhu Shen, 70
Zhu Yongpeng, 1
Zhuge Ming, 160–161
Zi River (Zishui), 123
Ziliujing, 45, 62, 116
Zimmerman, John E., 57

Lightning Source UK Ltd.
Milton Keynes UK
UKHW010756140921
390217UK00010B/184

9 781501 758959